Join Us on the Internet

WWW: http://www.thomson.com
EMAIL: findit@kiosk.thomson.com

thomson.com is the on-line portal for the products, services and resources available from International Thomson Publishing (ITP).

This Internet kiosk gives users immediate access to more than 34 ITP publishers and over 20,000 products. Through *thomson.com* Internet users can search catalogs, examine subject-specific resource centers and subscribe to electronic discussion lists. You can purchase ITP products from your local bookseller, or directly through *thomson.com*.

Visit Chapman & Hall's Internet Resource Center for information on our new publications, links to useful sites on the World Wide Web and an opportunity to join our e-mail mailing list. Point your browser to: **http://www.chaphall.com** or **http://www.thomson.com/chaphall/electeng.html** for Electrical Engineering

A service of I(T)P®

Cover design: Curtis Tow Graphics

Printed in the United States of America

Chapman & Hall
115 Fifth Avenue
New York, NY 10003

Chapman & Hall
2-6 Boundary Row
London SE1 8HN
England

Thomas Nelson Australia
102 Dodds Street
South Melbourne, 3205
Victoria, Australia

Chapman & Hall GmbH
Postfach 100 263
D-69442 Weinheim
Germany

International Thomson Editores
Campos Eliseos 385, Piso 7
Col. Polanco
11560 Mexico D.F
Mexico

International Thomson Publishing–Japan
Hirakawacho-cho Kyowa Building, 3F
1-2-1 Hirakawacho-cho
Chiyoda-ku, 102 Tokyo
Japan

International Thomson Publishing Asia
221 Henderson Road #05-10
Henderson Building
Singapore 0315

1 2 3 4 5 6 7 8 9 10 XXX 01 00 99 98

Library of Congress Cataloging-in-Publication Data

Elgerd, Olle Ingemar, 1925-
Electric power engineering / Olle I. Elgerd, P.D. van der Puije. -
- 2nd ed.
p. cm.
Includes bibliographical references and index.
ISBN 0-412-10551-9 (hc : alk. paper)
1. Electric engineering. I. Van der Puije, Patrick D., 1937-
TK146.E44 1997
621.31--dc21 97-69
CIP

British Library Cataloguing in Publication Data available

To order this or any other Chapman & Hall book, please contact **International Thomson Publishing, 7625 Empire Drive, Florence, KY 41042.** Phone: (606) 525-6600 or 1-800-842-3636.
Fax: (606) 525-7778. e-mail: order@chaphall.com.

For a complete listing of Chapman & Hall titles, send your request to **Chapman & Hall, Dept. BC, 115 Fifth Avenue, New York, NY 10003.**

Contents

CHAPTER 6
The Electric Power Network 239

APPENDIX A
Phasor Analysis 405

APPENDIX B
Spectral Analysis 422

APPENDIX C
The SI Unit System 432

APPENDIX D
Units of Energy and Power Conversion 435

Preface

This book is about electric energy: its generation, its transmission from the point of generation to where it is required, and its transformation into required forms. To achieve this end, a number of devices are essential—such as generators, transmission lines, transformers, and electric motors. We discuss the design, construction, and operating characteristics of the electric devices used in the transformation to and from electric energy.

This text is designed to be used in a one-semester course in electric energy conversion at the second-year level of the Bachelor of Engineering course. It is assumed that the student is familiar with the laws of thermodynamics and has taken a course in basic circuit analysis, including the application of phasors.

We begin with a discussion of how humankind has successfully harnessed the energy of wind, water, the sun, biomass, animals, geothermal sources, fossils, and nuclear fission to make its life comfortable. Some of the consequences of this activity on the environment are examined.

In Chapter 2, we review the basic physics of energy and its conversion. This may be, to some extent, a repetition of knowledge gained in high-school and first-year university courses. However, we believe that such review is necessary to establish a suitable base from which to launch the subject of electric energy conversion.

Chapter 3 begins with a review of the basics of charge, capacitance, current, Ohm's law, and power, then covers the fundamentals of magnetic fields, inductance, inductive coupling, and ends with a discussion of ferromagnetism.

In presenting the material we find it natural to follow the flow of electric energy in a modern power system. We start with electric generators in a power-generating station. We follow the journey of the energy as it is transformed from various primary sources in the prime mover and generator through step-up transformers and into the coarse mesh of the high-voltage power transmission system. When it reaches its destination it is again transformed in step-down transformers and fed into the fine mesh of the distribution system from which it is finally portioned out to the multitude of users.

Following the above organization, we have devoted Chapter 4 to electric generators—more specifically, the three-phase alternator or synchronous machine. This is the workhorse of the electric power generating system, and it is used to produce all but an insignificant portion of the total electric power. Chapter 5 tells the story of the electric power transformer. In Chapter 6, we discuss the operating features of the power network or grid, which transports the energy from the generator to the consumer. This includes the problem of "loadflow," the problem of parallel operation of generators, and important aspects of "system control."

Chapters 7 and 8 are devoted to the most important electric motors—devices that convert electric energy into mechanical forms. The most significant design features and operating characteristics of various types of motor, both dc and ac, are discussed.

In the final chapter, we present a number of electric machines used in special applications, such as the linear induction motor, which has applications in traction, and the brushless dc motor, which is used as an actuator in robotics and control systems. The treatment of these machines is necessarily brief as their basic characteristics were discussed in preceding chapters.

1

Energy: The Basis of Civilization

This book is an introductory text on electric energy—its generation, transmission, and conversion to and from other forms of energy. Electric energy is unquestionably the most versatile and universally useful form of energy available. The demand for electric energy is growing at a faster rate than for any other form of energy. However, electricity comprises only a fraction of the total energy demand. Before we turn our attention to the main topic, it would be useful to view the role of electricity from both general and historical angles.

1.1 Historical Perspective

Archaeological evidence indicates that humans learned to master, with increasing skill, the various forms of energy that nature provided. According to the evolutionary model of human development, *Homo erectus* populated the temperate zones about a million years ago. Archaeological evidence supports the view that for most of that period our forebears knew how to control and use fire. Later, they would develop the technology for making fire. By converting the energy stored in a chemical form in wood into heat they were able to improve the comfort of their habitat and also the quality of their diet by cooking their food. This gave *Homo erectus* and the other human ancestors a distinct advantage over other species and contributed immeasurably to their survival and further development.

The first of the ancient civilizations began around ten thousand years ago—about 500 human generations back. In terms of one human lifetime, this is a long period, but in relation to the total life span of our species, the attributes that we describe as "civilization" developed at the very last "moment." Why did our civilization develop with such relative suddenness? Let us identify a few important factors.

The invention of the bow and arrow 15,000 to 20,000 years ago was a very important factor. We may classify this device as "weapon," "implement," or "tool," but "energy converter" is probably the most accurate technical description. It enabled the hunter to transform a small portion of his muscle energy into a highly controllable form of kinetic energy (see Chapter 2)—that of a deadly missile. To an enormous degree it simplified the food-gathering tasks of early humans and afforded "leisure time" that could be used for other activities, such as art and religion.

Later, humans would abandon the nomadic life of the hunter and settle into the ancient city-states. Agriculture and animal husbandry were the prime prerequisites for this revolutionary change of lifestyle. Plants and animals have always provided the life-sustaining links between the energy of the sun and the energy required by humans. Their domestication provided control and also the possibility to improve these essential energy transformation processes.

Pottery and the development of metal extraction increased the demand for the primary source of energy—wood. The use of wood as a source of energy had the side effect of clearing the land for agriculture. In many areas, this resulted in the first recorded "energy crisis."

Around 3000 B.C. people learned how to harness the energy of the wind so that the pre-Christian era saw fleets of Phoenician, Egyptian, Greek, and Roman sailing vessels plying the Mediterranean. The added mobility of wind-driven ships contributed immeasurably to trade and communications. Domestication of the horse and the invention of the saddle and related devices for controlling horses added "horsepower" to travel over land. The water wheel, which was invented at about the beginning of the Christian era took the drudgery out of grinding food and sharpening weapons, tools, and agricultural implements.

It is worth noting that, up to the time of the Renaissance, the most important source of energy—fossil fuels—had not been tapped. The most hectic period of human civilization, the industrial revolution, was powered by coal, followed by petroleum and natural gas. When we contemplate the extent to which science and technology have changed (and also prolonged) human life in the last ten generations, the term "revolution" is most appropriate.

The quality of life that our modern technological civilization offers can be sustained and further improved only if we are able to keep the wheels of industry turning. The oil embargo of the early 1970s demonstrated the vulnerability of modern technological society to the shortage of energy. One of the most important future tasks must be to develop new sources of energy to maintain the high standard of living in the industrialized countries and to help meet the aspirations of the developing nations of the world. The new sources of energy must be safe and reliable and characterized by an acceptable economic and environmental costs. Primary electric energy[1] consumption in North America is about 5% of the total energy, but the prediction is that this share will increase in the future as it is poten-

[1] Primary electrical energy is defined as energy generated from geothermal, water, nuclear, solar, tide, wind, and wave sources.

tially more "environmentally friendly" than other forms of energy. It is therefore important to study the nature, production, and consumption of both primary and secondary forms of electric energy.

1.2 Energy Flow in Industrialized Societies

Figure 1.1[2] shows a pie chart of the sources of energy used in North America, as well as in Canada, Mexico, and the United States in 1993. The figure illustrates the following interesting facts:

- Hydrocarbon fossils, that is, oil, natural gas, and coal, constitute 95% of the primary energy sources of energy in North America.
- Primary electric, the energy obtained from damming rivers, nuclear power stations, geothermal sources, solar, tide, wind, and wave amounts to an unimpressive 5%.
- It would appear that more energy could be obtained from primary electric sources but these sources usually have a very high environmental cost. Dams drown millions

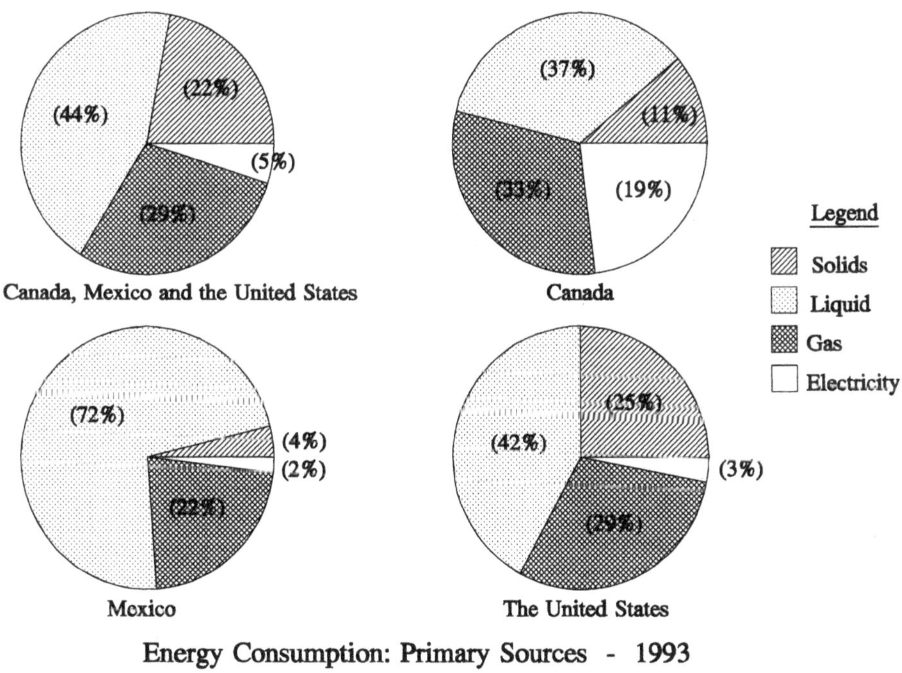

Figure 1.1

[2] Source: 1993 UN Energy Statistical Yearbook, ISBN 92-1-061161-6, United Nations, New York, 1995.

of hectares of land and cause the extinction of migrating fish by preventing them from reaching their spawning grounds upstream. The vegetation on the lake bed produces large quantities of methane as it rots. Methane is one of the so-called "greenhouse" gases. Nuclear power stations were always viewed with suspicion, but after accidents on Three-Mile Island and in Chernobyl, there was renewed nervousness over this source of energy. Geothermal, tide, solar, and wind sources have minimal environmental impact and also contribute a minuscule amount of energy.

- As the overall energy-conversion efficiency is only about 50%, many of the most important energy-conversion devices, for instance, the internal combustion engine, are characterized by a relatively low conversion efficiency. In driving an automobile the "useful work" would represent the energy needed to overcome the forces of friction, while "waste heat" is energy "lost" in the radiator and the exhaust pipe.
- In this connection it is worth pointing out that energy is never lost—it is transformed into new forms. When these new forms of energy appear as "low-grade" heat, we talk about "waste energy." The immutable laws of thermodynamics prevent us from "recycling" low-grade heat into a high-grade form.

Figure 1.2a[3] shows the total energy consumed in North America in 1993 as well as figures for Canada, Mexico, and the United States. It can be seen that 90.846×10^{18} joules were consumed.[4] This is such a large number that it is difficult to appreciate

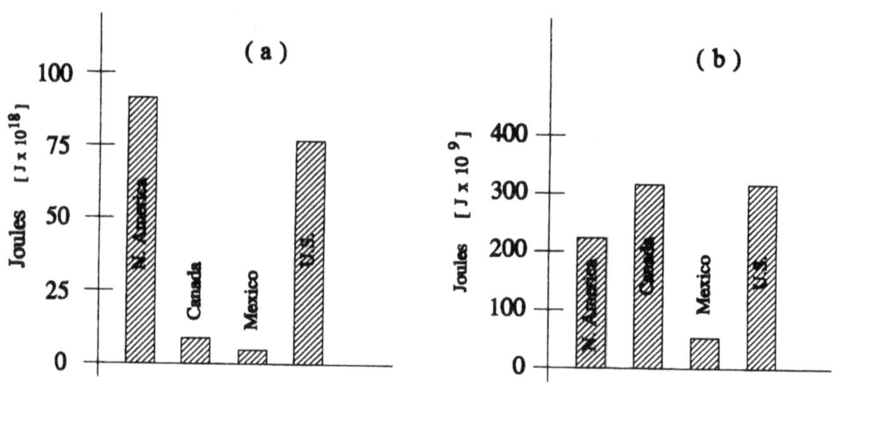

Total Energy Consumption - 1993 **Per-capita Energy Consumption - 1993**

Figure 1.2

[3] Source: 1993 UN Energy Statistical Yearbook, ISBN 92-1-061161-6, United Nations, New York, 1995.

[4] The joule [J] is the basic unit of energy, and J/s or watt [W] is the basic unit of power in the *metric* or SI system *(Système Internationale d'Unites)*. The system is sometimes referred to as MKSA—the letters standing for the basic units: *m*eter, *k*ilogram, *s*econd, and *a*mpere. Units of energy and power are discussed further in Chapter 2 and in Appendix D.

its meaning. Figure 1.2b shows the average rate of energy consumption per capita to be 220×10^9 J per year. Since there are 31.5×10^6, seconds in a year, we obtain a value of about 6984 J/s per person. But the rate of energy consumption of 1 J/s is equal to a power of 1 W. Hence each North American on average, consumes energy at nearly 7000 W every second of the day and night.

Probably the reader still has only a vague feel for the computed power. Is 7000 W a large or small value? If we convert this power into horsepower (hp), we obtain the figure, 9.4 hp. The average reader may have a better feel for this unit. In fact, a strong person working at a sweat-driving tempo can, with difficulty, sustain an output of about 0.1 hp.

Therefore in 1993, each person in North America was assisted by 94 "energy slaves." These slaves build our houses, grow our food, build our roads, power our factories, cars, ships, and airplanes, keep our homes cool or warm, and in general help us sustain our affluent life style.

Citizens of most countries have only a couple of energy slaves each. Each citizen of the United States and Canada has about 135 energy slaves, while Mexicans have about 21 each. In fact, North America, with only about 7% of the world's population, consumes about one-third of the energy produced.

1.3 The Growth of Energy Consumption

Figure 1.3[5] shows the rate of consumption of energy for the world since the year 1900 with a projection to the year 2020. The graph shows that energy is being consumed at an exponentially[6] rising rate. To determine the exponent, we plot the rate of consumption against time on a semilogarithmic paper as shown in Figure 1.4.

From the graph we determine the slope of the curve to be 0.0262. Thus the curve of energy consumption approximately fits the equation:

$$y = Ae^{\alpha t} \tag{1.1}$$

where $A = 32$ and $\alpha = 0.0262$.

[5] Sources: World Energy Council. *Energy for Tomorrow's World: The Realities, the Real Options and the Agenda for Achievement*, 15th WEC Congress, Madrid, Spain, September 1992; International Energy Agency. *Energy Statistics and Balances for non-OECD Countries*. Paris: OECD, 1996; Darmstadter, J. *Energy in the World Economy*. Baltimore, MD: The Johns Hopkins Press, 1971; Marecki, J. *Podstawy Przemian Energetycznych*, Warszawa: Wydawnictwa Naukowo-Techniczne, 1995.

[6] An exponential or geometric growth results when constant percentage increments are added periodically in a compounded fashion. For example, if you deposit \$100 in a bank that gives 7% interest compounded annually. After 1 year you have accumulated \$107, after 2 years \$114.49, after 3 years \$122.50, etc. In about 10 years you have doubled the initial investment, in 20 years you have quadrupled it, etc. A geometrically growing process is characterized by a doubling at regular intervals, referred to as the "doubling time" of the process. When plotted on a semilogarithmic graph paper (Figure 1.4) an exponential process gives a straight line with a slope.

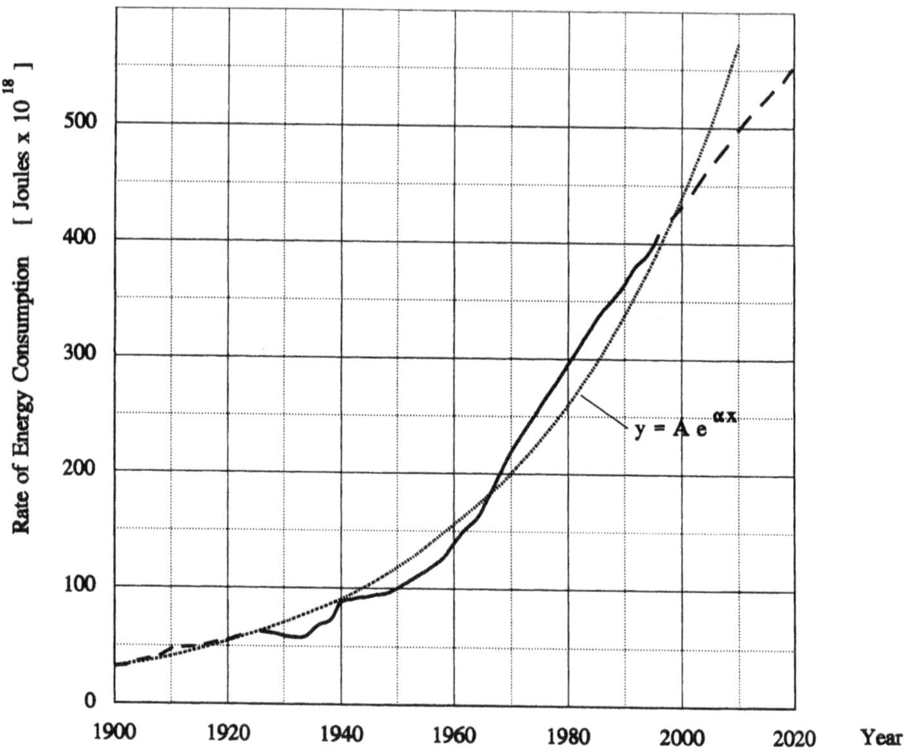

Figure 1.3

It can be seen from Figure 1.4, that the rate of consumption of energy doubled approximately every 26 years. Equation (1.1) is plotted on Figure 1.3 for comparison.

Figure 1.5[7] shows the growth of the human population since the turn of the century. It can be seen that the growth of the population—a natural process—is exponential. The approximate equation that fits the human growth curve since 1940 is

$$y = Be^{\beta x}, \tag{1.2}$$

where $B = 1.04$ and $\beta = 0.0180$.

If the present trend persists, the population will double in only 38 years. We make the following additional observations:

- It is clear that the demand for energy is growing faster than the population, which is itself increasing exponentially. This indicates a growing per capita energy consump-

[7] UN Demographic Yearbook. New York: United Nations; 1956 and 1994. The human population, *Scientific American*, Sept., 1974.

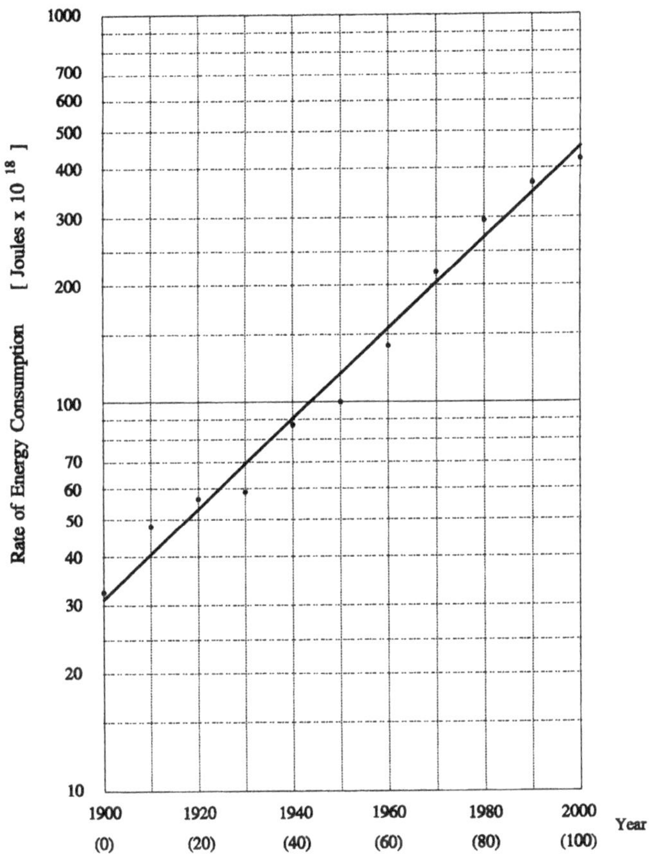

Figure 1.4

tion. However, the growth is not uniform—citizens of wealthy countries get to consume far more energy than those of poor countries.

- Processes that grow exponentially are treated with caution by most engineers and scientists, since they signify a situation that is uncontrolled, unstable, or a "runaway." Exponential growth processes usually lead to a catastrophic situation.[8] The exponentially growing energy demand on a worldwide basis caught up with the supply in the early 1970s. The tumultuous events that followed on the world energy market created shock waves that have not yet abated.

[8] Consider the following example: About 1% of the surface of a Florida lake is covered by water hyacinths. This pesky plant, like a cancer, grows at a geometric rate. Assuming that the doubling time is one year, the hyacinths would then grow to cover 2% of the surface in 1 year, 4% in 2 years, etc. The reader can easily confirm that it will take about 6 years to cover half the lake's surface but then, very suddenly, the total lake will be choked in only 1 additional year.

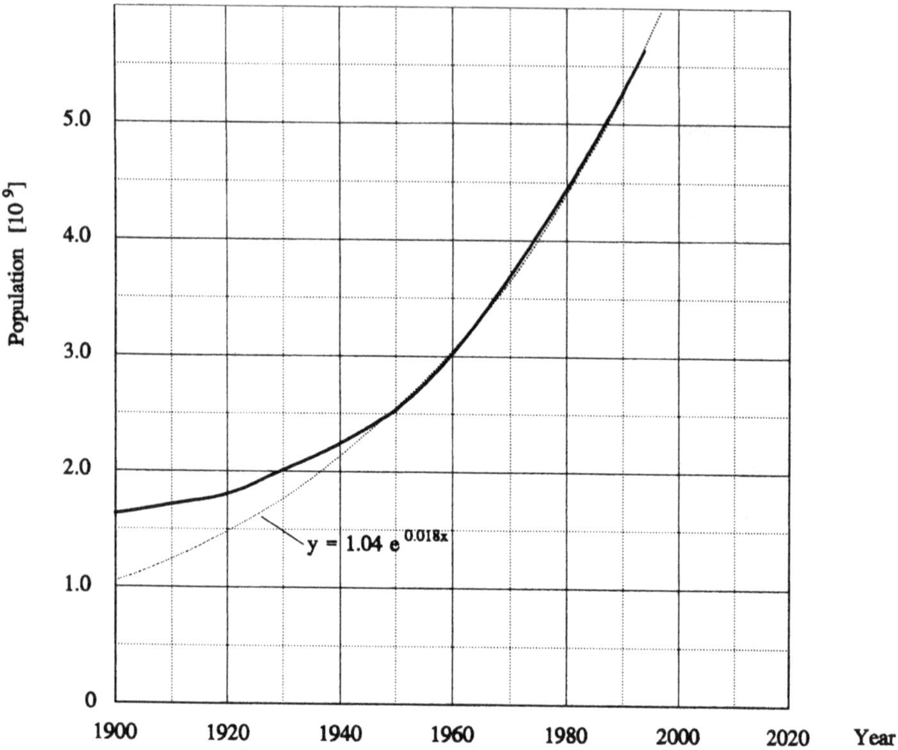

Figure 1.5

1.4 Electric Energy

Figure 1.6[9] shows that in terms of the "end-use" of the energy consumed in North America, only 11% of it is in the form of electricity. Comparable figures for Canada, Mexico, and the United States are also presented. Electric energy is important because in a multitude of ways it influences our lives daily. When on the rare occasion we experience an electric "blackout," we are reminded of our dependence on this form of energy.

The extreme versatility and usefulness of electricity stem from the following unique features:

- Instant availability
- Easy transmittability
- Easy controllability

[9] Source: 1993 UN Energy Statistical Yearbook, ISBN 92-1-061161-6, New York: United Nations, 1995.

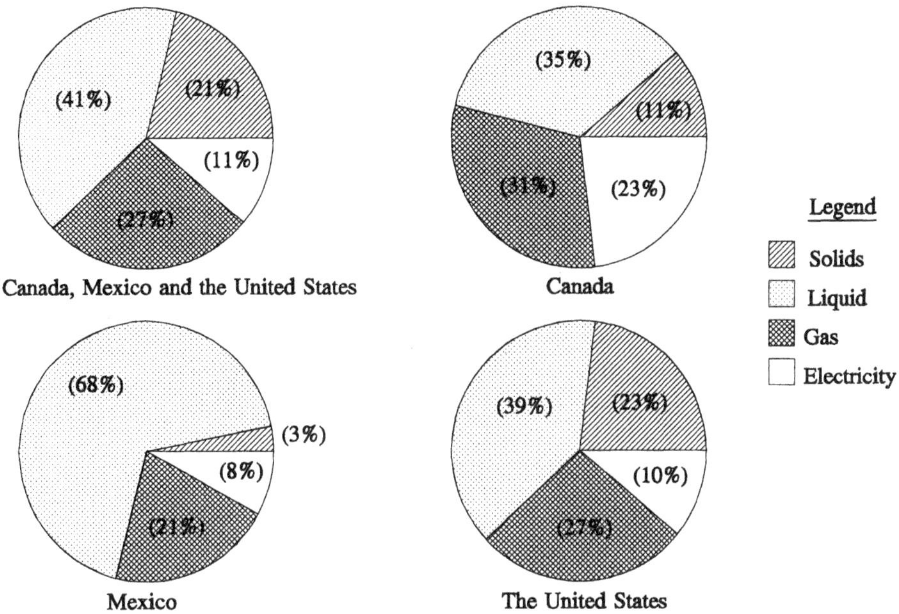

Energy Consumption: End Use - 1993

Figure 1.6

Electromagnetic waves carry intelligent commands to and experimental data from man-made robots exploring the outer reaches of our planetary system. The robots themselves can operate unaided for years *via* finely tuned electromechanical control systems powered by solar-electric cells. Reliable electric pacemakers give new lease of life to crippled heart patients. By "direct distance dialing" we can establish instantaneous voice communication with people on the other side of the world.

The student solving the end-of-chapter problems in this book is freed from the drudgery of numerical computation by an electronic calculator that fits in the palm of the hand. Modern electronic computers perform computations in a few seconds that would take the average human a lifetime to complete, and, unlike human beings, the computer will do the job with unerring accuracy.

Electric motors varying in size from fractions of a watt to tens of megawatts turn the vast majority of the wheels of industry. In all of these and a multitude of others, electric energy is involved.

Figure 1.1 shows the percentages of electrical and other forms of energy generated from primary sources in the three countries in North America. Figure 1.6

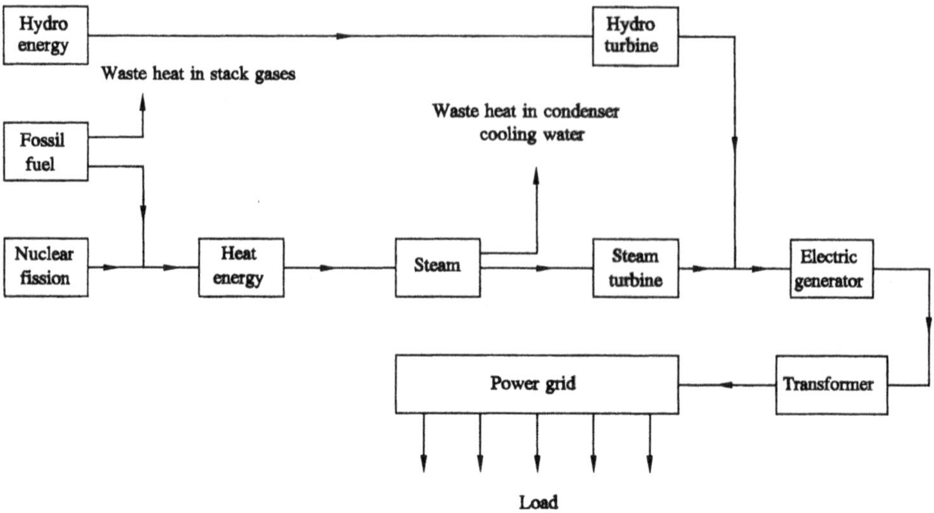

Figure 1.7

shows the percentages of electrical energy from primary as well as secondary sources, that is, it includes electrical energy generated by burning one of the other fuels. Figure 1.7 shows, in block diagram form, the steps required to convert the main sources of energy into electricity:

- Energy stored in water trapped behind a dam (hydroelectric)
- Energy stored in fossil fuels (conventional thermal electricity)
- Energy liberated from a nuclear reactor by fission

1.4.1 Hydroelectric Power

Figure 1.8a[10] shows the percentages of electrical energy derived from various primary sources in Canada in 1993. Figure 1.8b presents similar information for the United States in 1993. It can be seen that about 61% of the electricity generated in Canada is from hydraulic sources, while in the United States it is 9%. The potential energy of the water is converted in turbines that drive electric generators, the energy is fed into the power grid.

The simplicity of the hydroelectric process makes it by far the most reliable bulk electric generation process. Its efficiency is also very high—of the order of 90%. Because the initial cost can be very high, dam construction is often com-

[10] Sources: *Nuclear Facts: Seeking to Generate a Better Understanding*. The Canadian Nuclear Association. Toronto: Canada. *United States: Annual Energy Review 1994*. Washington, DC: Energy Information Administration.

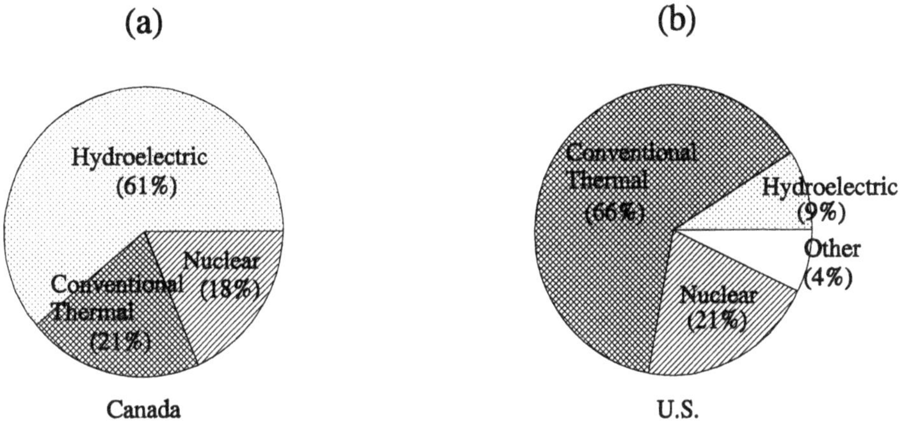

Electrical Energy Generated - 1993

Figure 1.8

bined with water irrigation projects. In the absence of any fuel costs, hydroelectric power is usually very attractive economically. Hydroelectricity is pollution-free *per se*, but dams have vast and often negative impact on the ecosystem.

An attractive feature of hydroelectric power is its high generation controllability, that is, the generation can be varied conveniently and quickly in response to changes in the load. In a hydroelectric plant, the electric power level is changed by simply opening or closing the water gates to the turbine. One basic feature of electric energy is that it cannot be stored in large quantities. It must, therefore, be generated the instant the customer demands it. As the power-generating authority has no control over the timing of the customer demand, it must monitor the system constantly, and instantaneously match its generation to the demand. The electric power demand varies widely throughout the day and week as shown in Figure 1.9. The ease with which a hydroelectric plant can be controlled makes it attractive as a "peaking unit." The hydroelectric generation process is also finding increased use in "pumped hydrostorage", an energy storage method described in Chapter 2.

1.4.2 Electricity From Fossil Fuel

It can be seen from Figure 1.8a that about 21% of the electric power generated in Canada in 1993 was obtained from burning fossil fuels. For the United States the percentage for the same year was 66%. The energy is released in the form of high-grade heat in the furnace. The heat is used to produce steam in the boiler at high temperature and pressure in a complex system of heat exchangers (see Figure

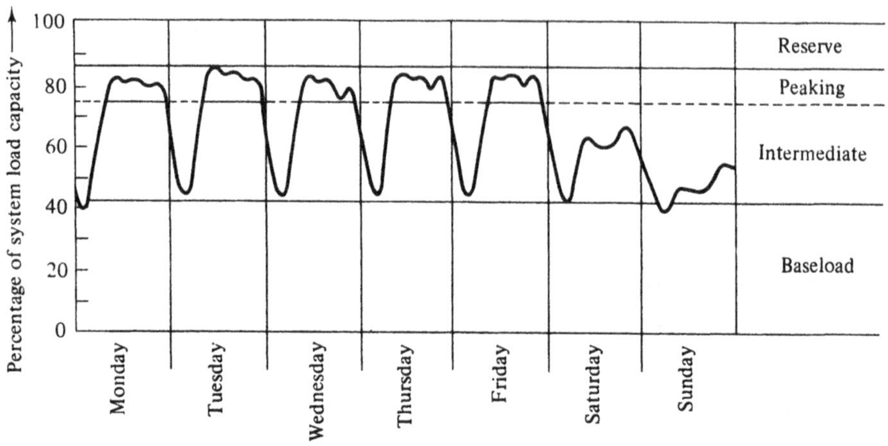

Figure 1.9 Weekly power demand variation of an electric utility.

2.13). The steam is led to the turbine through the steam drum (which serves as a low-capacity buffer steam storage device). In the turbine, part of the thermal energy is transformed into mechanical form. The steam turbine drives an electric generator from which electric energy is fed into the power grid.

The expanded steam is cooled in the condenser where it turns into water. The water is pumped back into the boiler, thus completing a closed "steam cycle." The steam-electric generation process is a very complex and roundabout way of obtaining electric energy—but it is the best one that technology offers when fossil fuels must be used as a primary energy source. Furthermore, the process has poor efficiency—about 35% to 40% at best. The efficiency can be increased by raising the pressure and temperature of the steam, but the strength of the metals used in the boiler, heat exchangers, pipes, and turbine set the upper limits to both temperature and pressure.

Most of the energy is lost as low-grade heat to the cooling water in the condenser. This water, when it exits from the condenser, has an elevated temperature in the approximate range of 10–20°C above ambient. Due to the large quantities needed it may have a negative environmental impact ("thermal pollution") when it is allowed to flow into relatively small bodies of water. For that reason, closed cooling ponds are often used, or the cooling may take place in cooling towers. Heat energy is also lost to the atmosphere *via* the stack gases.

The exhaust gases contain, in addition to waste energy, the chemical air pollutants, which constitute the greatest problem associated with the generation of power from fossil fuels, particularly when coal is the fuel. Due to its complexity, a steam plant has a much lower reliability than a hydroelectric plant, and its power level cannot be varied as conveniently nor as fast. Changes in its power output must be accomplished by changes in the combustion rate of the fuel. If those changes take place too rapidly, they result in unacceptable thermal stresses in the boilers and heat exchangers. As a consequence, once a steam-driven plant is "on-line" it is normal practice to try to keep its power level fixed by letting it carry the power system "baseload" (see Figure 1.9).

1.4.3 Electric Power Generation From Nuclear Reaction

A nuclear reactor is a thermal-type power plant. It differs from a fossil-fueled plant by the absence of a combustion chamber. The heat source is now a nuclear reaction. The nuclear reactor produces heat in a controlled fission process which transforms some of the mass of the nuclear fuel into energy according to Einstein's formula:

$$E = mc^2. \tag{1.3}$$

The energy appears in the form of high-grade heat which is then used to produce steam. The steam drives a turbine-generator in a conventional way. A nuclear power station requires cooling water for its condensers, like any thermal power plant, and will therefore have the potential for thermal pollution. It does, however, produce zero air pollution, a feature that greatly enhances its attraction in comparison to a conventional thermal unit. Due to the energy density of mass (see Section 2.12), a nuclear power station requires a minute amount of fuel.

Great care must be taken in the design of a nuclear power station to prevent radiation leaks. This adds to the initial cost and construction time required. The public reaction to a nuclear power station in the neighborhood is invariably negative, and this is based on the fear of radioactive leaks.

1.4.4 Electric Energy Storage

Figure 1.9 shows how the electric power demand undergoes hourly variations throughout a typical day and week. Ideally, electric energy should be generated at the constant average rate. This would require less installed generating capacity and would result in better economy and longer operating life of the equipment. This mode of electric power system operation would require electric energy storage facilities. A natural gas system operates approximately in this manner. In spite of a highly fluctuating demand the production of gas at the wellhead takes place at approximately a constant rate. The necessary storage takes place in storage containers (including caverns) and in the pipelines themselves.

An electric storage facility that could deliver, for example, 1000 MW during 2 hours must have a storage capacity of 2000 MWh, or $7.2 \cdot 10^{12}$ J. Such facilities do not presently exist. For example, the energy stored in an electrical "pipeline" in the form of electrostatic energy ($\frac{1}{2}Cv^2$, refer to Chapter 3) in the shunt capacitance represents a much smaller amount of energy.

Some energy is stored in kinetic form (see Example 6.1) in the rotating masses of the generators. However, if we drew any significant amount of energy from this storage, the frequency would drop at a rapid rate. This means that electric energy must be generated at the instant it is demanded. This also means that we must have enough generating capacity to be able to handle the peak load. This, of course, adds to the cost of the equipment.

Pumped-hydropower storage and compressed-gas storage, both described in Chapter 2, represent hybrid electro-mechanical storage facilities. They have become increasingly popular as supplies for "peaking power" (see Figure 1.9).

Energy storage in electric and magnetic fields, both discussed in Chapter 3, are the only "true" electrical storage methods known. The electric field storage, quantitatively described by equation (3.16):

$$w_e = \tfrac{1}{2}Cv^2 \tag{1.4}$$

requires either a very large capacitance, C, or a high voltage, v. The magnetic field storage, given by equation (3.76),

$$w_m = \tfrac{1}{2}Li^2, \tag{1.5}$$

calls for a large inductance and/or high current. This method of storage seems to offer the only possibility to handle amounts of energy in the 1000 MWh range. By supercooling the magnetic coil, one could conceivably obtain current values (and magnetic field densities) of the required magnitudes.

Large superconducting magnets have been built for laboratory experiments in particle physics. The largest magnet in existence has a storage capacity of 800 MJ. This represents only 0.22 MWh of electric energy. Within the limits of existing technology, this figure can be raised, and studies (Peterson, 1975) have mentioned energy storage figures of 10^7 MJ.

In the electric-generating machinery, as well as in the electric-transmission and storage—technology, considerable interest is presently focused on the use of superconductors. By cooling an electric conductor to temperatures close to absolute zero (-273°C) the conductor loses its resistance to electric current. Thus its ohmic losses drop drastically. Current densities of tens-of-millions amperes per square centimeters can be tolerated in such conductors.

1.5 Summary

Electricity, like no other form of energy, helps sustain our modern technological civilization. All signs indicate that it will assume greater importance in the future. In view of this, it seems reasonable to advocate the need for a better understanding by both electrical and nonelectrical engineers of the basic characteristics of electric energy technology. The objective of this book is to provide that knowledge.

EXERCISES

1.1 In 1976 the total U.S. "end use" of energy amounted to 40×10^{18} J. Assume that you could obtain all this energy by means of a 100% efficient nuclear reaction, according to Einstein's formula (1.3). How many kilograms of mass would be required?

1.2 Exponential growth in natural processes lies at the heart of many of the resource problems that we are now facing. To demonstrate the speed with which an exponential process "takes off" consider the following problem:

In 1626 the governor of the Dutch West India Co. was said to have bought from the Manhattan Indians the island that now carries their name for about $24. If this amount had been invested in a bank yielding 6% interest compounded annually, what would be the value of the investment in the year 2000?

References

Annual Energy Review 1994. Washington, DC: Energy Information Administration.

Coale, A.J. *The History of the Human Population* in "The Human Population", Scientific American Book, ISBN 0-7167-0515-X, 1974.

British Petroleum. *Statistical Review of World Energy. 1981.*

British Petroleum. *Statistical Review of World Energy. 1996.*

Darmstadter, J. *Energy in the World Economy.* Baltimore: The Johns Hopkins Press: 1971.

International Energy Agency. *Energy Statistics and Balances for non-OECD Countries.* Paris: OECD, 1996.

Marecki, J. *Podstawy Przemian Energetycznych.* Warszawa: Wydawnictwa Naukowo-Techniczne, 1995.

Peterson, H.A., et al. Superconductive energy storage: Inductor-convertor units for power systems. *IEEE Transactions*, 1975, 94: 1337–1348.

Rose, H., Pinkerton, A. *The Energy Crisis, Conservation and Solar.* Ann Arbor, MI: Ann Arbor Science, 1981.

U.N. Demographic Yearbook. New York: United Nations, 1956.

U.N. Energy Statistical Yearbook. New York: United Nations, 1993.

U.N. Demographic Yearbook. New York: United Nations, 1994.

World Energy Council Commission. *Energy for Tomorrow's World: The Realities, the Real Options and the Agenda for Achievement.* Draft Summary, Global Report, 15th WEC Congress, Madrid, Spain, September 1992.

2

Fundamentals of Energy

In this chapter we introduce the reader to some fundamental physical characteristics of energy. Many of the devices to be discussed in later chapters, such as motors, generators, and transducers, transform energy from electric to mechanical form or vice versa. In electric heaters, energy is being transformed from electric to caloric (thermal) form. In a storage battery, a transformation takes place between electrical and chemical forms of energy.

A proper understanding of electric energy technology is facilitated by a broad knowledge of energy in its many forms. The objective of this chapter is to tie together seemingly unrelated pieces of energy-related topics the reader has probably learned in several basic science and engineering courses—physics, statics, dynamics, and so on. We are concerned here mostly with nonelectrical forms of energy.

The SI unit system is used throughout this book. A brief description of this important unit system is given in Appendix C.

2.1 Energy and Gravitation

It is the fate of all inhabitants of the earth, including humans, to spend their lives under the constant influence of the earth's gravitational force. No other factor affects our lives so profoundly—even the type of creatures we are. Gravity is fundamental to all our sciences and technology, including the concept of energy. We find it natural, therefore, to choose gravity as the takeoff point in our story on energy.

Humankind learned the effects of gravity very early; for example, it is easier to walk downhill than to climb a mountain. Long before human hunters invented the bow and arrow, they had learned to chase herds of animals to the edge of precipices and then let the force of gravity relieve them of the dangerous job of killing the prey.

Quite probably—although we will never know for sure—the thunderous fall of a round boulder inspired one of the unknown early inventors to design the first wheel. Only the taming of the horse and the harnessing of the wind antedate mankind's use of falling water in the age-old and continuing process of accruing "energy slaves." The gravity-powered water wheel was the most important source of power for many centuries and in recent decades plays a dominant role in the process of generating electric energy.

Up to the present generation, the force of gravity confined humans to their earthly habitat. The historic trip to the moon marked the beginning of a new era when men and women learned to employ and control forces of such magnitude as to break free from the grip of gravity.

2.2 Gravitational Force Field

Our familiarity with the gravitational pull makes "force" one of the most easily accepted and best-understood concepts in physics. We, as engineers, make daily use of this concept in our designs, creations, and analyses; we rarely dwell on the fact that we know as little now as did our distant ancestors about what "really does the pulling." This is a philosophical question that will not be fully, or even adequately, answered in the foreseeable future.

What makes us different from preceding humanoids, such as the Cro-Magnon, is our ability to describe, measure, utilize, and control the forces of nature, even though our understanding of the precision and scope of these forces is restricted. Because gravitational forces serve as a logical and easy way of defining energy and power and because we can, by analogy, extend those concepts to electrical forms of energy and power, we find it useful to discuss briefly the physical character and mathematical modeling of the earth's gravitational field.

Careful measurements of the gravitational force f acting[1] on a mass m placed in the vicinity of the earth (see Figure 2.1) or some large body of mass M reveal that the force outside[2] the large body is everywhere radially directed. It is attractive and of magnitude

$$f = k\frac{Mm}{r^2} \qquad [\mathrm{N}]. \tag{2.1}$$

The symbols are defined as follows: f = magnitude of the force, in newtons [N]; r = distance between centers of the masses, in meters [m]; m, M = masses, in kilograms [kg]. Because the force is characterized by a magnitude and a direction, it is a *vector*. The *universal gravitational constant* k has the numerical value

$$k = 6.670 \times 10^{-11}\ [\mathrm{N \cdot m^2/kg^2}]. \tag{2.2}$$

It is useful to write equation (2.1) in the following alternate form:

[1] An equal but opposite force is felt by the earth.

[2] The gravity force inside the earth decreases approximately linearly with distance to the earth's center.

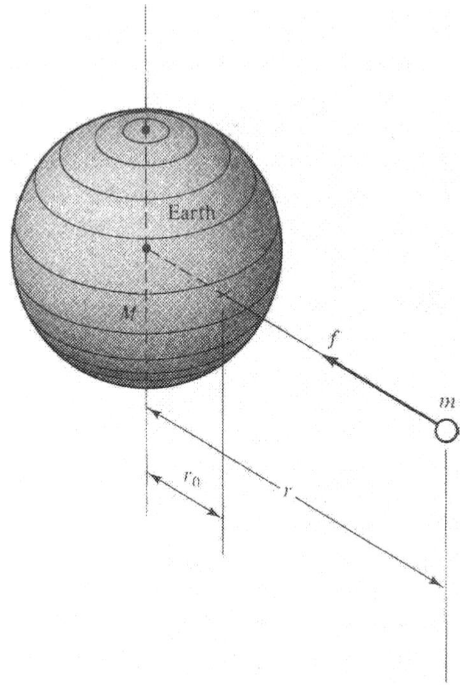

Figure 2.1

$$f = m\left(k\frac{M}{r^2}\right) \qquad [\text{N}]. \tag{2.3}$$

The expression inside the parentheses is a vector with physical dimension of *force per mass*, having the same direction as the force but a magnitude different from the force and independent of the mass m. We call it the *gravitation vector*, or *gravity* for short, and define it as

$$g \equiv k\frac{M}{r^2} \qquad [\text{N/kg}]. \tag{2.4}$$

In terms of this new vector we can write equation (2.1) in the following alternate form:

$$f = mg \qquad [\text{N}]. \tag{2.5}$$

The force has now been expressed as the product of mass m and a new entity, the gravitation vector g, which now embodies the gravitational character of the earth. The two formulas (2.1) and (2.5) are, of course, mathematically identical. But the latter permits the following physical interpretation:

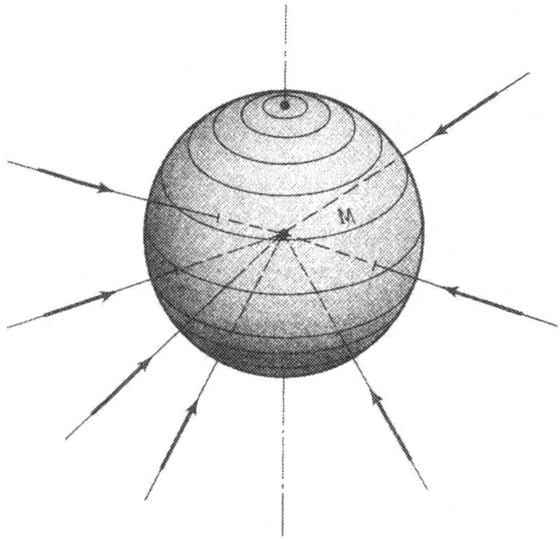

Figure 2.2 The earth's gravity field is everywhere radially directed.

The gravity g, being independent of m and solely depending on the presence and mass of the earth, is an "earth fixture" that exists around the earth independent of the presence of the mass, m. We refer to it as the *gravitational field.* Figure 2.2 is a "visual" representation of the gravitational field.

Most of our activities take place in a limited region close to the surface of the earth. In such a limited region, we can set $r \approx r_0$, where r_0 equals the earth's radius, the field lines are then essentially parallel to the gravitational field (see Figure 2.3) and thus the force is of constant magnitude:

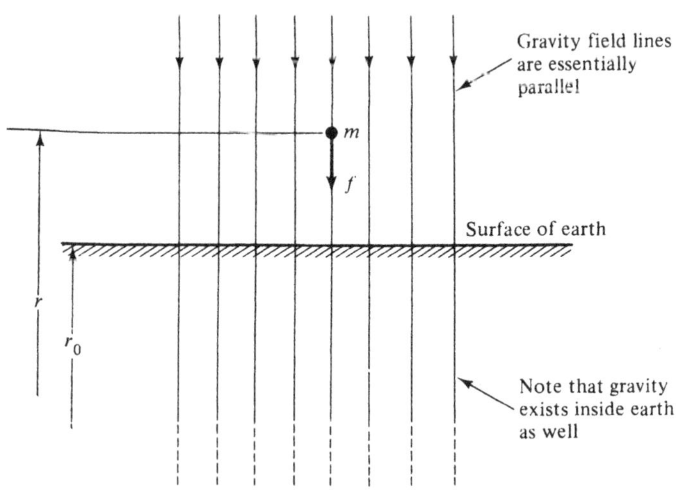

Figure 2.3

$$g \approx k\frac{M}{r_0^2} \approx g_0 = \text{constant}. \tag{2.6}$$

$$f \approx mg_0 = \text{constant}. \tag{2.7}$$

The *surface gravity* g_0 varies slightly around the earth, which is not perfectly spherical. (Furthermore, the centrifugal force caused by the rotation of the earth adds an indistinguishable component to the gravity force and varies with latitude.) By international agreement, the standard surface gravity is numerically defined as

$$g_0 = 9.80665 \text{ [N/kg]}, \tag{2.8}$$

or 9.81,[3] for short. Further, in terms of the surface gravity, the magnitude of the force can also be written in the alternate form:

$$f = k\frac{Mm}{r^2} = m\frac{kM}{r_0^2}\frac{r_0^2}{r^2} = mg_0\frac{r_0^2}{r^2} \qquad \text{[N]}. \tag{2.9}$$

2.3 Gravitational Energy Exchange: Definition of Energy

One dictionary defines energy (from the Greek *energos*, meaning active) as "capacity for doing work." This seems a reasonably clear definition until one finds that the same dictionary defines "work" as "transference of energy from one body to another." It is well known that the work or energy required to climb a hill increases in proportion to both the person's weight and the gain in elevation. Similarly, the energy that can be released from an impounded river increases with both the water flow and the height of the dam. Energy, certainly in a gravitational sense, is a quantity that depends on the product of force and distance.

In elevating a mass by moving it against[4] the force of gravity we need to expend energy. This energy is being stored. (Whether the storage takes place in the mass or in the field is a philosophical question.) When the mass is lowered the stored energy is released. For our purposes, we need to look at this well-known *energy exchange* process from a more quantitative point of view. Consider the energy expended in elevating the mass m in Figure 2.4 the vertical distance h. Guided by the "product rule" stated above, we define the energy or work incre-

[3] The physical unit for gravity is newtons per kilogram [N/kg]. If the mass m is released, the gravitational force will impart to it an acceleration a, which, according to Newton, follows the law, $f = ma$. By comparison with equation (2.5) we have $a = g$. This means that g must also have the physical unit for acceleration, meters per second squared [m/s^2].

[4] In the case of the water trapped behind a dam, its energy is supplied by the atmospheric force of evaporation, originally emanating from the sun.

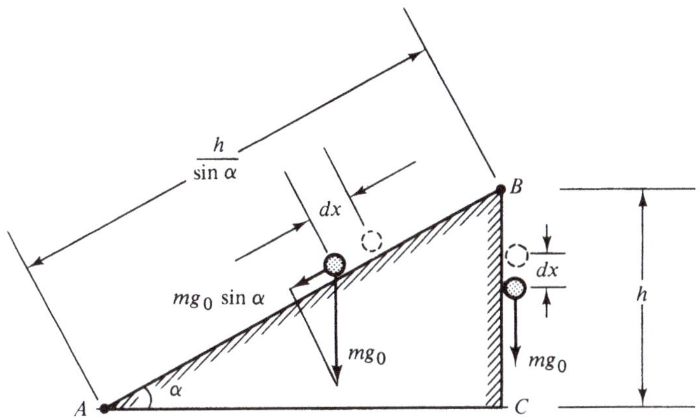

Figure 2.4

ment dw needed to move the mass the incremental distance dx as the product of force and distance according to

$$dw \equiv \text{force} \cdot dx. \tag{2.10}$$

By "force" we mean that component we encounter in reality. In moving vertically from $C \to B$ we must move against the full strength of the gravity force mg_0. In moving up the incline from $A \to B$ we encounter the smaller component, $mg_0 \sin\alpha$. We obtain the two alternate expressions for energy:

$$\begin{aligned} dw &= mg_0\, dx && (\text{along } C \to B); \\ dw &= mg_0 \sin\alpha\, dx && (\text{along } A \to B). \end{aligned} \tag{2.11}$$

The total energy expended in the two cases is obtained by integration:

$$\begin{aligned} w &= \int_0^h dw = \int_0^h mg_0\, dx = mg_0 h; \\ w &= \int_0^{h/\sin\alpha} dw = \int_0^{h/\sin\alpha} mg_0 \sin\alpha\, dx = mg_0 h. \end{aligned} \tag{2.12}$$

The energy is clearly independent of the path chosen and depends only upon the gain of elevation, h. (One can extend the proof to include any arbitrarily chosen path.)

The physical dimension for energy from the above definition is "force times distance." In the SI system, this unit of energy is given the special name *joule* [J]. The joule is a "derived" unit with the dimension [J] = [N · m] (newton-meter). It also has the dimension *watt-second* [W · s] (often used by electrical engineers).

Example 2.1

How much energy must be delivered by an elevator motor in lifting a load of 5 metric tons a vertical distance of 200 m (1 metric ton = 1000 kg)?

Solution: Formula (2.12) gives

$$w = 5000 \times 9.81 \times 200 = 9{,}810{,}000 \text{ [J]} = 9.81 \text{ [MJ]}. \qquad (2.1.1)$$

2.4 Gravitational Potential: Potential Energy

If we express equation (2.12) in terms of joules per kilogram (of the mass m) we obtain a measure v_g, defined as follows:

$$v_g \equiv \frac{w}{m} = \int g_0 \, dx = g_0 h \qquad \text{[J/kg]}. \qquad (2.13)$$

The new physical quantity is referred to as the gravitational potential. The increase in potential represents the added per unit increment in energy that we must impart to the mass to lift it from the lower to the higher level. A mass at a higher level is said to possess a higher potential gravitational energy than the same mass at a lower level. According to equation (2.13) each meter of altitude gain adds 9.81 [J/kg] to the potential of the mass. Expressed differently: For every 10.2 cm of altitude gain, its potential increases by 1 [J/kg]. Should we move the mass in a horizontal direction, perpendicularly to the gravity force, no change of energy takes place. That is, the potential energy of the mass will be constant for two points on the same horizontal level, and these points would then represent equipotential surfaces. In a building, the floors constitute equipotential surfaces, and the walls coincide with the gravitational field lines.

The integration of equation (2.12) was simple because we considered the gravity force to be constant. If we wish to compute the gravitational potential at greater distances from the earth's surface, then we must use the general force given by equation (2.9):

$$v_g = \frac{w}{m} = \int_{r_0}^{r} g \, dr = \int_{r_0}^{r} g_0 \frac{r_0^2}{r^2} \, dr = g_0 \frac{r_0}{r}(r - r_0) \qquad \text{[J/kg]}. \qquad (2.14)$$

[Note that if we set $r \approx r_0$ and $r - r_0 = h$, we get equation (2.13).] The equipotential surfaces will now consist of concentric spherical shells. (An earth satellite traveling a circular orbit retains a constant velocity because of this fact.)

Our definition of potential actually defines potential differences. The zero level can be chosen arbitrarily—just as we can choose arbitrarily the zero point of a temperature scale. In our definition we arbitrarily set the gravitational potential at the surface of the earth equal to zero. If we plot the potential (2.14) as a function of distance from the earth we obtain the graph shown in Figure 2.5. Note

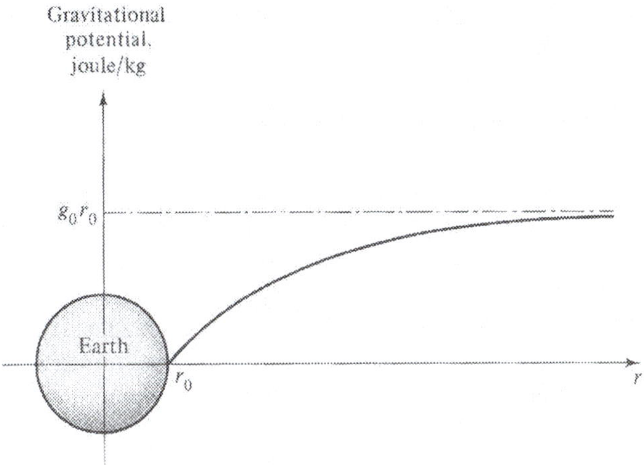

Figure 2.5

that the potential reaches an asymptotic value of $g_0 r_0$. This graph can be visualized as a "potential energy hill" to be climbed should one wish to leave the surface of the earth.

We make one final observation. The potential as we have defined it is equal to the line integral of the gravitation vector. It increases in value in the negative g direction. The path along which the integration is carried out has no significance—only the endpoints count. In most cases the integrations are easily performed if we choose the path (as we did) to coincide with the gravity vector.

Example 2.2

How much energy is needed to transport a 70-kg astronaut to an altitude of 150 km above the surface of the earth?

Solution: The radius of the earth is

$$r_0 = 6.38 \times 10^6 \text{ [m]}. \tag{2.2.1}$$

Formula (2.14) gives

$$v_g = 9.81 \frac{6.38 \times 10^6}{6.530 \times 10^6} 1.5 \times 10^5 = 1.438 \times 10^6 \qquad \text{[J/kg]}. \tag{2.2.2}$$

The approximate formula (2.13) gives

$$v_g \approx 9.81 \times 1.5 \times 10^5 = 1.472 \times 10^6 \qquad \text{[J/kg]}, \tag{2.2.3}$$

that is, an error of 2.5%. Thus, for each kilogram of "payload" we need to expend 1.437×10^6 J of energy. For our 70-kg astronaut we must "pay" an energy price of 100.6×10^6 J. (Note that this is the energy needed to get the astronaut to the required altitude. It is not the energy needed to keep him or her there. Compare this to Exercise 2.12.)

2.5 General Expressions for Energy

In our previous discussions, all forces involved were gravitational in origin. In practice, forces emanate from a number of sources, such as springs, pressurized gases, and friction. In the remaining sections of this book we are concerned solely with forces of electrical and magnetic origins.

In most technical applications, forces perform work in either a translational or a rotational sense. Rocket propulsion is an example of the former and a common electrical motor of the latter. We develop appropriate formulas for both cases, as shown in Figure 2.6. Guided by equation (2.10), we define the incremental energy or work performed by the force to be the product of the force and the incremental distances in each case:

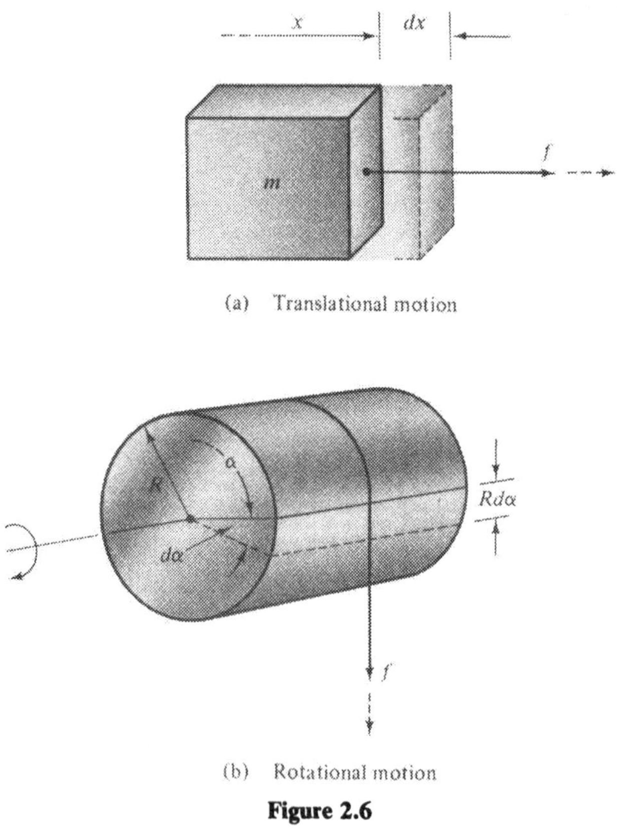

Figure 2.6

$$dw = f \cdot dx \qquad \text{(translational motion)}, \tag{2.15}$$

$$dw = fR \cdot d\alpha \quad \text{(rotational motion)}. \tag{2.16}$$

These formulas give the incremental work (or energy) performed by the forces. If the forces are permitted to act over finite distances, then we obtain the total energy by summation or integration:

$$w = \int f\, dx \qquad [\text{J}]. \tag{2.17}$$

$$w = \int fR\, d\alpha = R \int f\, d\alpha \qquad [\text{J}]. \tag{2.18}$$

When the forces are constant, we get

$$w = f \int dx = fx \qquad [\text{W}], \tag{2.19}$$

$$w = Rf \int d\alpha = Rf\alpha \qquad [\text{J}], \tag{2.20}$$

where x and α represent the total translational and rotational movements, measured in meters and radians, respectively.

2.6 Rate of Energy or Power

Assume that the forces f in Figure 2.6 perform the incremental work expressed in equations (2.15) and (2.16) in the incremental time dt. We now say that the forces are capable of doing work at a given rate or developing power p, expressed by

$$p \equiv \frac{dw}{dt} = \frac{f\, dx}{dt} = f\frac{dx}{dt} \qquad [\text{W}]; \tag{2.21}$$

$$p \equiv \frac{dw}{dt} = \frac{fR\, d\alpha}{dt} = fR\frac{d\alpha}{dt} \qquad [\text{W}]. \tag{2.22}$$

We identify the ratios dx/dt and $d\alpha/dt$ as the translational and rotational velocities s [m/s] and ω [rad/s], respectively. We have

$$p = fs \qquad [\text{W}] \qquad \text{(translational case)}, \tag{2.23}$$

$$p = fR\omega = T\omega \qquad [\text{W}] \quad \text{(rotational case)}, \tag{2.24}$$

where $T = fR$ represents the moment or torque exerted by the force f.

The unit of power is newton-meter per second [N · m/s], or joule per second [J/s], or watt [W]. Larger power units often used by electric power engineers are kilowatt [kW] and megawatt [MW], defined as follows:

$$1\ [\text{MW}] = 1000\ [\text{kW}] = 1{,}000{,}000\ [\text{W}].$$

Another popular unit of power that still lingers among engineers is horsepower [hp], defined as 1 [hp] ≡ 0.746 [kW]. The metric horsepower (*cheval vapeur*) is equal to 0.736 kW.

Example 2.3

The power rating of an electric motor drive for a mine elevator must be determined. The motor must be capable of elevating a 5-ton load up the 200-m vertical mine shaft at a velocity of 5 [m/s]. The acceleration and deceleration periods are 5 seconds each.

Solution: From the given specifications we compute the acceleration a: For the acceleration phase,

$$a = \frac{5\ [\text{m/s}]}{5\ [\text{s}]} = 1\ [\text{m/s}^2]. \tag{2.3.1}$$

For the deceleration phase,

$$a = -\frac{5\ [\text{m/s}]}{5\ [\text{s}]} = -1\ [\text{m/s}^2]. \tag{2.3.2}$$

During the acceleration period, the distance covered will be $\frac{1}{2} \times 1 \times 5^2 = 12.5$ [m]. Deceleration will require an equal distance. The distance over which the elevator travels at a constant velocity (steady state) is 175 [m], which will be covered in 35 [s]. In Figure 2.7, we have plotted the velocity during the entire 45-[s] lift cycle. We compute the force f for the three phases of the operation.

During the steady-state phase, the force f must be equal to the force of gravity:

$$f = f_{ss} = mg_0 = 5000 \times 9.81 = 49050\ [\text{N}] = 49.05\ [\text{kN}]. \tag{2.3.3}$$

During the acceleration and deceleration phases, the total force acting on the load follows from the application of Newton's law of motion as

$$f_{\text{acc}} - mg_0 = m \cdot 1; \tag{2.3.4}$$

$$f_{\text{dec}} - mg_0 = m \cdot (-1). \tag{2.3.5}$$

Thus,

$$f_{\text{dec}} = m(g_0 - 1) = 5000 \times 8.81 = 44{,}050\ [\text{N}]; \tag{2.3.6}$$

$$f_{\text{acc}} = m(g_0 + 1) = 5000 \times 10.81 = 54{,}050\ [\text{N}]. \tag{2.3.7}$$

The force exerted during each phase of the lift cycle is plotted in Figure 2.7.

Finally, we obtain the power p produced by the motor from equation (2.23) as follows.

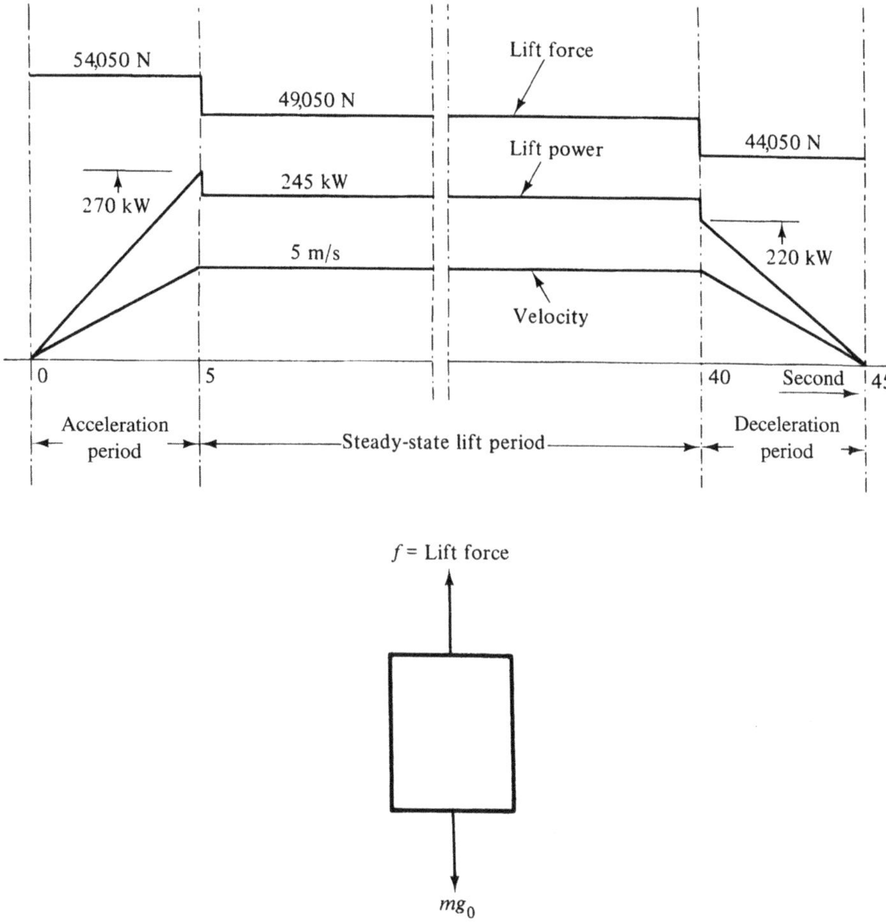

Figure 2.7

During the Steady-State Phase

$$p_{\text{dec max}} = 44.05 \times 5 = 220.3 \text{ [kW]}. \tag{2.3.8}$$

During the Acceleration Phase

The lift force is constant and the velocity increases linearly from zero to the maximum value of 5 [m/s]. The power required must therefore also increase linearly from zero to the maximum value:

$$p_{\text{acc max}} = 54.05 \times 5 = 270.3 \text{ [kW]}. \tag{2.3.9}$$

During the Deceleration Phase

A similar reasoning tells us that the power must decrease linearly from the maximum value:

$$p = p_{ss} = 49050 \times 5 = 245{,}250 \text{ [W]} = 245.3 \text{ [kW]}. \qquad (2.3.10)$$

The power produced by the elevator motor for the entire lift cycle is plotted in Figure 2.7. This example shows that the power produced by the motor varies considerably during the lift cycle. The motor control system must provide this varying power.

One final note: In our calculations, we did not consider friction, wind resistance, and so on. In a practical situation, these must be included, as they result in unavoidable energy and power losses. The size of the motor must be chosen to provide an adequate power margin over the basic requirements.

Example 2.4

Figure 2.8 shows a Pelton turbine used to transform the potential energy of water when a relatively large "head," h, is available. The control valve is used to control the flow rate of the water, i [kg/s]. The energy of the high-pressure high-velocity water is "caught" by the turbine blades. The blades must have such a velocity that when the water is leaving the trailing edges of the blades its velocity must be zero. Only then is all the potential energy of the water fully extracted. Derive an expression for the turbine power.

Solution: For every second the potential energy of i [kg] of water is released. Since the water falls through a potential difference given by equation (2.13), we have

$$v_g = g_0 h \qquad \text{[J/kg]}. \qquad (2.4.1)$$

A total energy released per second is

$$p = v_g i = g_0 hi \qquad \text{[J/s]}. \qquad (2.4.2)$$

This is power in watts. A numerical example is

$$h = 700 \text{ [m]}, \quad i = 10{,}000 \text{ [kg/s]}. \qquad (2.4.3)$$

Substituting, we get

$$p = 9.81 \times 700 \times 10{,}000 = 68{,}670{,}000 \text{ [W]} = 68.67 \text{ [MW]}. \qquad (2.4.4)$$

(Due to the unavoidable losses associated with the process, the turbine power will be less than this ideal figure.)

Where does this power (or energy) go? Were the turbine not constrained it would accelerate to a high speed, and the results would be destructive. In practice, the turbine drives an electric generator. Through a mechanism (discussed in Chap-

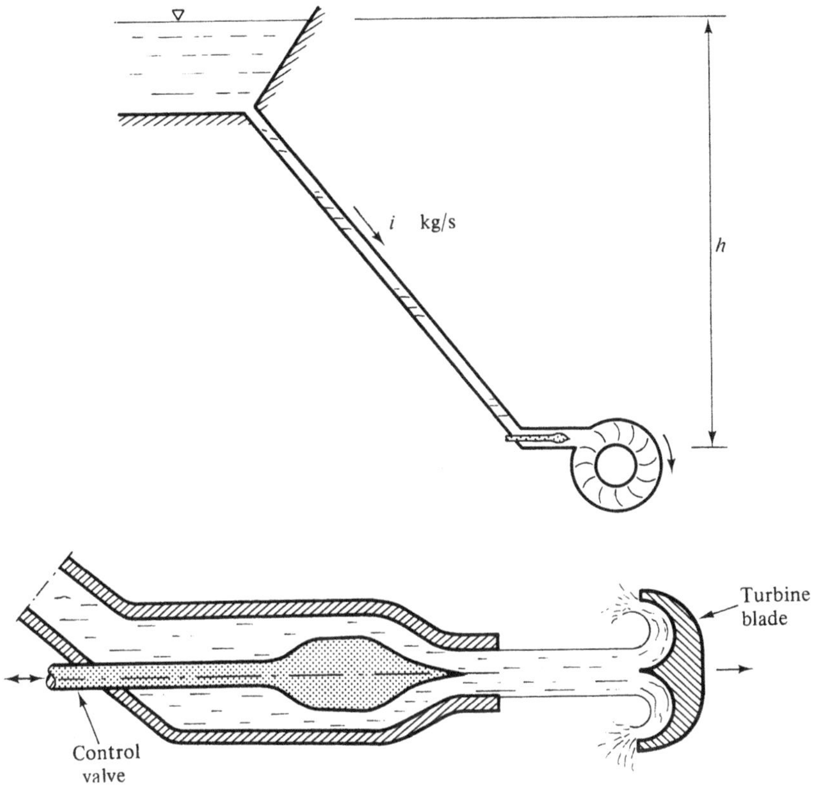

Figure 2.8

ter 4) the currents in the generator windings will create an electromechanical torque that counteracts the turbine torque exactly. The result is zero acceleration, that is, constant speed. At the same time the generator feeds 69 MW (minus losses) to the electric grid where it is distributed to millions of electric power consumers.

2.7 The Law of Conservation of Energy: First Law of Thermodynamics

We can compute the energy delivered by the elevator motor in Example 2.3 by two methods. The simpler method is to apply equation (2.19) to the three separate lift phases. We can also use (2.21), which gives the energy

$$w = \int p\, dt \qquad [\text{J}]. \tag{2.25}$$

(This integral clearly gives the area under the curve for power in Figure 2.7.) The reader can perform either of these calculations and obtain the energy delivered by the motor as

$$w = 9{,}810{,}000\ [\mathrm{J}] = 9.81\ [\mathrm{MJ}]. \tag{2.26}$$

This energy has not been lost! In fact, we never "lose" a single joule of energy anywhere. We simply transform energy from one form to another. This is the *First Law of Thermodynamics*, which states that energy can be converted from one form to another but never destroyed.

In the case above, the electric energy delivered by the motor is used to increase the gravitational potential energy of the elevator and its load. This potential energy is computed from equation (2.12) as

$$w = 200 \cdot 5000 \cdot 9.81 = 9{,}810{,}000 \qquad [\mathrm{J}],$$

which is exactly equal to the energy delivered by the motor.

Now we understand what really took place when we elevated the 5-ton load a distance of 200 m. The energy that was originally used to produce the electricity has been transformed into the potential energy of the elevator and its load.

For example, assume that the electricity was produced in a hydraulic plant where falling water is used as the primary source. If we neglect all energy losses associated with the generation and transmission plus the motor losses, we would need exactly 5 tons of water falling through 200 m (or 10 tons falling 100 m) to produce the energy calculated in the previous example.

If we lower the 5-ton load by a procedure called *regenerative braking*, we can recapture the potential energy (see Section 7.4.2). The electric motor will now be driven as a generator feeding energy back into the electric network.

If we lower the load by using friction brakes, the energy would be transformed into heat (see Section 2.10), but, again, it would not be "lost."

Finally, if we allow the load to fall freely, the energy would be transformed into kinetic energy (discussed in Section 2.9) as the elevator accelerated toward the bottom of the mine shaft. The kinetic energy will be transformed into a disaster at the base of the elevator shaft.

Potential energy provides a means for energy storage in electric power systems. The curve in Figure 2.9 (cf. Figure 1.9) shows how the electric energy demand in such a system varies throughout a typical day, with a peak usually in the early afternoon and a minimum just before dawn. As it is essentially impossible to store large amounts of electric energy (see Chapter 3), in an electric power system one must vary the generation so that it matches the demand at each instant. However, the varying generation requirement conflicts with the strong desire to run steam-driven generator stations at a constant energy output (see Section 1.4.2). It would be effective if the steam plant could be run at the constant average power output shown in Figure 2.9. This can be achieved by means of a pumped

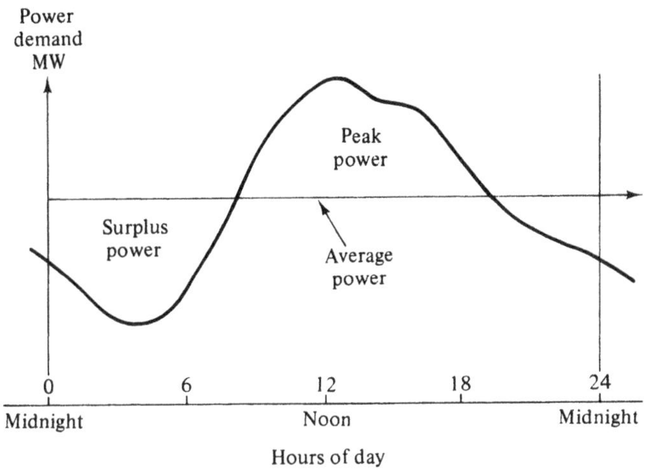

Figure 2.9

hydrostorage generation plant shown in Figure 2.10. An alternator is connected to a hydroturbine through which water can flow both ways.[5] If the water flows from the upper to the lower reservoir it will drive the alternator as a generator, feeding electric energy into the network. If electric energy is fed into the alternator from the network, then the alternator will run as a motor driving the turbine as a pump, pumping water back up to the upper reservoir.

The proper operating strategy would be to pump water during the night hours when surplus energy is available in the power system. This will increase the potential energy stored in the upper reservoir. In the afternoon hours, when the steam plant needs supporting power, we would reverse the flow and operate the facility as a *peaking generator*. As the peaking unit must deliver the difference between the varying demand and the constant output of the steam generator, its power output will not be constant. However (as was pointed out in Section 1.4.1) a hydraulic plant, unlike a steam plant, can be operated to provide varying output power.

Example 2.5

Assume that the average head (Figure 2.10), or level difference between the two reservoirs, is $h = 300$ [m]. How much water must the reservoirs store to produce the equivalent of 1200 megawatt hours [MWh] of energy? (Note that this amount of energy corresponds to a power output of 400 MW for a period of 3 hours or 200 MW for 6 hours.) In our analysis, let us assume the overall efficiency of a full "pump–turbine" cycle to be $\eta = 60\%$.

[5] Why is a Pelton wheel (Figure 2.8) not suitable for the job? How would you design a reversible turbine?

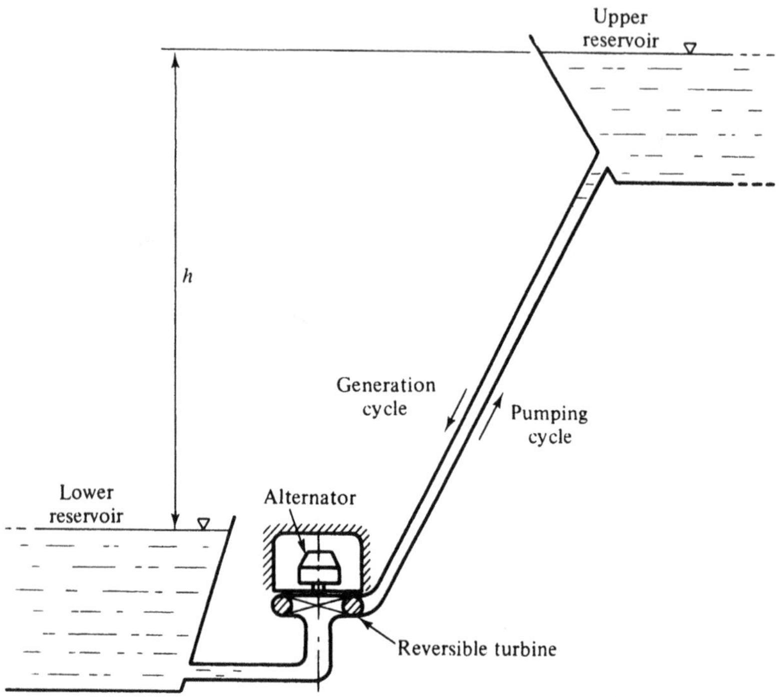

Figure 2.10

Solution: First we express the stored energy in joules:

$$\begin{aligned} 1200\ [\text{MWh}] &= 1200 \cdot 3600\ [\text{MWs}] = 1200 \cdot 3600 \cdot 10^6\ [\text{Ws}] \\ &= 4.32 \cdot 10^{12}\ [\text{J}] = 4.32\ [\text{TJ}]\ (\text{terajoules}). \end{aligned} \tag{2.5.1}$$

If the required quantity of water is m [kg], we then get from equation (2.12):

$$\eta \cdot m g_0 h = 4.32 \times 10^{12}; \tag{2.5.2}$$

$$m = \frac{4.32 \times 10^{12}}{0.6 \times 9.81 \times 300} = 2.45 \times 10^9 \text{ kg} = 2.45 \times 10^6 \text{ m}^3\ H_2O. \tag{2.5.3}$$

If each reservoir has a surface area of 1 km^2 the water level at each reservoir would change by 2.45 m during the total "pump–turbine" cycle.

2.8 Other Forms of Potential Energy

As we have seen, the earth's gravity field offers a means for potential energy storage. Other possibilities exist. One can, for example, store energy by compressing air and holding it in a suitable vessel. The compressed air can then be used to drive air turbines to recover the energy stored. Springs, torsion bars, and other elastic media can likewise be used to store energy but in smaller quantities. Let us demonstrate this with two examples.

Example 2.6

Consider the pressure vessel in Figure 2.11a. It has volume v, and is filled with air under pressure p. Compute the energy stored in the system.

Solution: Let the compressed air push an imaginary piston (Figure 2.11b) along a cylinder. The total energy stored in the cavity will be given up to the piston when the inside pressure p is equal to the outside pressure p_0, at which time the piston position $x = x_{max}$. To obtain the total stored energy we use equation (2.17) and integrate between $x = 0$ and $x = x_{max}$.

If the piston area is A [m^2] then the force on the piston is $(p - p_0)A$ [N]. From equation (2.17) we get

$$w_{stored} = \int_0^{x_{max}} (p - p_0)A \, dx \qquad [\mathrm{N \cdot m}]. \tag{2.6.1}$$

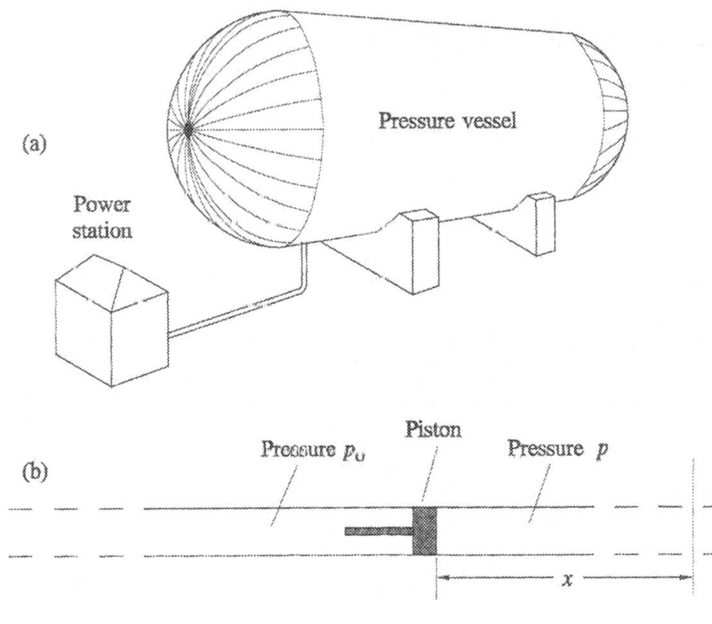

Figure 2.11

To perform the integration we must first find a relationship between p and x. For this purpose we use Boyle's law[6]:

$$pv = p_c v_c, \tag{2.6.2}$$

where the total gas volume v is

$$v = v_c + Ax \qquad [\mathrm{m}^3]. \tag{2.6.3}$$

Combination of equations (2.6.2) and (2.6.3) gives the required relationship between p and x as

$$p = \frac{p_c v_c}{v_c + Ax} \qquad [\mathrm{N/m^2}] \text{ or pascal } [\mathrm{Pa}]. \tag{2.6.4}$$

The maximum piston stroke $x = x_{max}$ is obtained by setting $p = p_0$ in equation (2.6.4). We get

$$x_{max} = \frac{v_c}{A}\left(\frac{p_c}{p_0} - 1\right) \qquad [\mathrm{m}]. \tag{2.6.5}$$

Substituting the expressions for p and x_{max} into the integral (2.6.1) puts it in a form that can be readily integrated. The integration gives the expression

$$w_{stored} = p_c v_c\left[\ln\left(\frac{p_c}{p_0}\right) - \left(1 - \frac{p_0}{p_c}\right)\right] \qquad [\mathrm{J}]. \tag{2.6.6}$$

Assuming the following numerical values:

$$v_c = 10^6\ [\mathrm{m}^3],$$

$$p_0 = 100\ [\mathrm{kPa}],$$

$$p_c = 1000\ [\mathrm{kPa}],$$

we get

$$w_{stored} = 1.40 \cdot 10^{12}\ [\mathrm{J}] = 1.40\ [\mathrm{TJ}]. \tag{2.6.7}$$

Compare this energy storage capacity[7] with that of the water storage facility in Example 2.5. (Note that the volume of the pressure vessel is about equal to the volume of water pumped in Example 2.5.)

[6] Boyle's law, stated in this form, applies strictly to gases in *static* equilibrium and at a constant temperature. We assume that the cavity is emptied at such a slow rate that we essentially have a "pseudo-static" situation.

[7] In a practical situation one would not, of course, reduce the air pressure down to 1 atmosphere ($\approx$100 [kPa]) during the generation cycle, for the same reason that one would not reduce the water-head h to zero in the pumped-water storage plant.

Example 2.7

In the introductory chapter, we suggested that the bow and arrow formed an *energy converter*. Compute how much potential energy can be stored in a drawn bow (Figure 2.12).

Solution: As the bowstring is pulled, a force f must be applied, which grows approximately linearly with distance x, that is,

$$f = kx \qquad [\text{N}], \tag{2.7.1}$$

where k is the *spring constant*, which determines the stiffness of the bow. It has the dimension newtons per meter [N/m].

From equation (2.17) we get

$$w_{\text{stored}} = \int_0^x kx\,dx = \frac{1}{2}kx^2 \qquad [\text{J}]. \tag{2.7.2}$$

The quadratic equation for energy stored is typical for all linear springs.

Numerical example: Compute the energy stored in a 180-N bow at its full 50-cm stroke.

The spring constant k is computed from the information that the force is 180 N for $x = x_{max} = 0.5$ m;

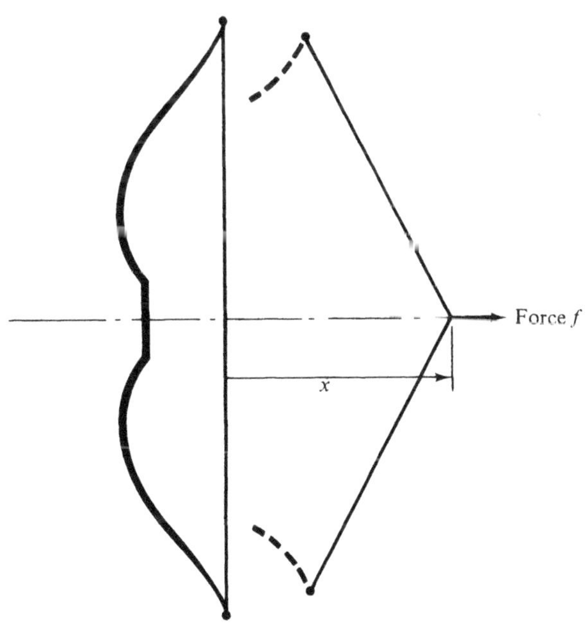

Figure 2.12

$$k = \frac{180}{0.5} = 360.0 \qquad [\text{N/m}]. \tag{2.7.3}$$

From equation (2.7.2) we get

$$w_{\text{stored}} = \frac{1}{2} \times 360.0 \times 0.5^2 = 22.5 \qquad [\text{J}] \tag{2.7.4}$$

2.9 Forms of Kinetic Energy

All the various forms of potential energy we have discussed belong to a class of static systems. In many important practical electric energy conversion problems, we encounter kinetic forms of energy—that is, energy associated with systems in motion.

Consider the mass m in Figure 2.6a. According to Newton's law, the mass will experience an acceleration d^2x/dt^2 in the x direction, when the force f is applied. Thus,

$$f = m\frac{d^2x}{dt^2}. \tag{2.27}$$

We introduce the velocity

$$s = \frac{dx}{dt} \tag{2.28}$$

as a new variable, and we can then write equation (2.27) in the form

$$f = m\frac{d^2x}{dt^2} = m\frac{ds}{dt} = m\frac{dx}{dt}\frac{ds}{dx} = ms\frac{ds}{dx} \tag{2.29}$$

or

$$f\,dx = ms\,ds. \tag{2.30}$$

Upon integration we obtain

$$\int_0^x f\,dx = \int_0^s ms\,ds, \tag{2.31}$$

that is,

$$\int_0^x f\,dx = \frac{1}{2}ms^2 \equiv W_{\text{kin}}. \tag{2.32}$$

This equation states that the work performed by the force f is absorbed by the mass in a new form of energy, called *kinetic energy*, which increases as the square of the velocity.

If we perform a similar analysis of the rotational mass in Figure 2.6b, we would obtain a symmetric expression for its kinetic energy as

$$w_{\text{kin}} = \frac{1}{2} I\omega^2 \qquad [\text{J}], \tag{2.33}$$

where I is the moment of inertia of the rotating mass and ω is its rotational speed expressed in per second [rad/s].

Example 2.8

The kinetic energy of the spinning rotors of synchronous generators is of very great importance in understanding the operation of the interconnected electric power system (see Chapter 6).

A turbogenerator rotor has the cylindrical form as shown in Figure 2.6b. It is spinning at 1800 rpm. The rotor dimensions are diameter = 1.0 [m], length = 3.0 [m]. The rotor is made from steel of density 7800 kg/m^3. How much kinetic energy does the spinning rotor have? (The rotor is, of course, coupled directly to a steam turbine, which will add significantly to the total kinetic energy.)

Solution: The moment of inertia of a solid cylinder with radius R is obtained from

$$I = \frac{1}{2} mR^2 \qquad [\text{kg} \cdot \text{m}^2], \tag{2.8.1}$$

where the mass, m, is computed as follows:

$$m = 7800 \cdot \pi \cdot 0.5^2 \cdot 3.0 = 18{,}380 \text{ [kg]}; \tag{2.8.2}$$

$$I = \tfrac{1}{2} \cdot 18{,}380 \cdot (0.5)^2 = 2298 \text{ [kg} \cdot \text{m}^2]. \tag{2.8.3}$$

The angular velocity is

$$\omega = \frac{1800}{60} 2\pi = 188.5 \qquad [\text{rad/s}] \tag{2.8.4}$$

We obtain the kinetic energy from equation (2.33):

$$w_{\text{kin}} = \tfrac{1}{2} \cdot 2298 \cdot 188.5^2 = 40.8 \cdot 10^6 \text{ [J]} = 40.8 \text{ [MJ]}. \tag{2.8.5}$$

Compared to the stored energy in Examples 2.5 and 2.6, this is not a very impressive figure. Nevertheless, energy storage in rotating masses ("superflywheels") has practical applications.

Example 2.9

An arrow weighing 20 g is shot from the bow described in Example 2.7. Compute its velocity when it leaves the bowstring.

Solution: If we disregard wind friction, we can assume that the total potential energy, 22.5 [J], is transformed into the kinetic energy of the arrow. We have

$$\tfrac{1}{2} \cdot 0.020 \cdot s^2 = 22.5. \tag{2.9.1}$$

Solving for the arrow velocity gives

$$s = 47.4 \text{ [m/s]}. \tag{2.9.2}$$

2.10 Caloric (Heat or Thermal) Energy

2.10.1 Ordered and Disordered Forms of Energy

We have referred to energy "losses" and "efficiency" of various energy transformation processes. We noted in Chapter 1 that the overall energy conversion efficiency of the most common energy conversion devices is only about 50%.

All practical energy transformation processes are inherently nonideal in the sense that there is always a certain portion of the energy that "disappears" in the form of caloric energy. The temperature of a solid, liquid, or gas is a measure of the intensity of the random motion of its elementary particles. The more heat a body absorbs, the more intense the particle motion and the higher its temperature.

Because caloric energy is associated with random particle motion it is referred to as a "disordered" form of energy. The level of "disorderliness" of energy is measured by its entropy. For example, electric energy (see Chapter 3) which is very highly ordered, has zero entropy. It is this "orderliness" of certain forms of energy that permits their controlled transformation into other forms and increases their "usefulness."

Nature has an inherent tendency to "go disorderly." It is said that the "entropy of the world tends toward a maximum." Nature itself therefore sets limits to the effectiveness of all energy transformation processes.

2.10.2 Reversible and Nonreversible Energy Transformations: Second Law of Thermodynamics

Consider the example of the bow and arrow discussed earlier. As the bowstring is released, the string and arrow will accelerate the air molecules in their path and increase the temperature of the surrounding air slightly. A small amount of the stored potential energy will be "lost" for this purpose, resulting in a somewhat lower arrow velocity than we computed earlier. Clearly, this results in an energy transformation efficiency less than 100%.

If this lost energy cannot be recaptured, the energy transformation process is called *nonreversible*. If there had been no heat loss the energy transformation process of the bow and arrow would be *reversible*. (We can conceive of the possibility of "catching" the flying arrow by means of an identical bow, stopping its flight and recapturing its kinetic energy.)

The usefulness of caloric energy increases generally with its temperature. Thus the heat in an acetylene flame is more valuable (high-grade heat) than the heat obtained from an automobile radiator (low-grade heat). We should not, however, understand this statement to mean that low-grade heat is useless. For example, a space-heating apparatus radiates heat that is at a lower temperature than is obtained from an automobile radiator, but nobody would suggest that this energy is "waste."

Furthermore, the fact that an energy process is nonreversible does not mean, necessarily, that the heat "lost" in the process cannot be used for other purposes. The transformation of low-grade heat into other useful forms usually results in low efficiency and may also involve long time lags between cause and effect. For example, a thermoelectric generator based on the Seebeck effect[8] will accept low-grade heat as input and deliver a low-voltage electric output. Or consider the low-grade heat supplied to a greenhouse. Biologically, this energy is transformed into the chemical energy of the plants, but the process takes time. The reader should contemplate how, over vast spans of time, fossil energy resources were formed.

An ordered form of energy can always be transformed with 100% efficiency into a disordered form (heat). For example, the flying arrow will eventually come to a stop, that is, all its kinetic energy will have been transformed to heat. All the potential energy of an elevator can be transformed into a smoking wreckage at the bottom of a mine shaft.

Conversely, a disordered form of energy (heat) cannot be transformed into an ordered form with a 100% conversion efficiency. This is the *Second Law of Thermodynamics*. This is the basic reason why all energy transformation processes that start with heat are always fairly inefficient. For a thermal power plant, where "disordered" heat energy is converted into a highly "ordered" electric energy, the efficiency seldom exceeds 35%.

2.10.3 The Caloric Energy Equivalent

If the 5-ton elevator in Example 2.3 were to drop 200 [m], its potential energy would first be transformed into kinetic energy during the descent. At the bottom of the mine shaft, the kinetic energy, in an instant, would be transformed into heat—as can be verified by the elevated temperature of the wreckage. How much heat would be generated by the 9810 [kJ] of stored potential energy? The answer

[8] Briefly, this effect is manifested as follows: When a closed loop of two different materials is formed, and if the two junctions are kept at different temperatures, a current will flow in the loop.

can be found from a knowledge of the equivalence between caloric energy and mechanical energy. Before the equivalence can be established, we need to define the unit of caloric energy:

The unit of caloric energy, kilocalorie (kcal), is defined as the amount of energy needed to raise the temperature of 1 kg of water by 1°K.

The conversion factor has been determined experimentally as

$$1\ [\text{kcal}] = 4.19\ [\text{kJ}]. \qquad \text{(Joule's constant).} \tag{2.34}$$

Example 2.10

How much energy, Q, is required to increase the temperature of 1 lb of water by 1°F?

Solution: We have

$$1\ [\text{lb}] = 0.4536\ [\text{kg}], \tag{2.10.1}$$

$$1\ [°\text{F}] = 0.5556\ [°\text{K}]. \tag{2.10.2}$$

Thus,

$$Q = 0.4536 \cdot 0.5556\ [\text{kcal}] = 0.252\ [\text{kcal}] \tag{2.10.3}$$

or

$$0.252 \cdot 4.19 = 1.06\ [\text{kJ}]. \tag{2.10.4}$$

This quantity of heat (252 cal) is referred to as a *British thermal unit* (BTU).

For some of the most important energy and power conversion factors, refer to Appendix D.

The caloric energy equivalent is very useful when determining the relative energy content of various fuels. We demonstrate this with the following example.

Example 2.11

As noted in Figure 1.6, fossil fuels account for the major portion of all electric energy generated in the United States.

How much electric energy can be derived from 1 kg of coal if it is known that this particular type of coal releases 13,100 BTU of caloric heat per pound when incinerated?

The heat is used to generate steam, which drives a steam turbine, which in turn propels the electric generator. Figure 2.13 depicts the process in somewhat more detail than Figure 1.7. As was mentioned in Chapter 1, the total efficiency of this process is quite low: most of the heat is lost through the stack gases and the condenser cooling water. For purposes of our analysis, we assume the overall energy transformation efficiency, η, to be 33% (including transmission losses).

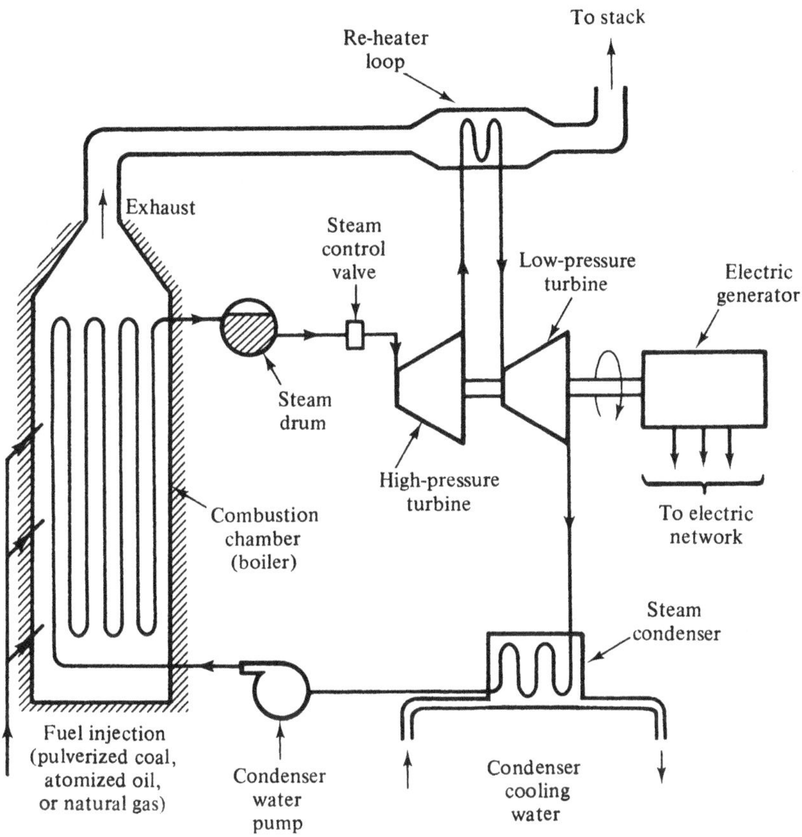

Figure 2.13

Solution: From Appendix D, we find that 1 lb of coal contains

$$1.055 \cdot 13{,}100 = 13{,}820 \text{ [kJ]} \tag{2.11.1}$$

of energy; 1 kg therefore releases

$$2.205 \cdot 13{,}820 = 30{,}470 \text{ [kJ]} \tag{2.11.2}$$

Taking the energy conversion efficiency into account we are left with

$$0.33 \cdot 30{,}470 = 10{,}060 \text{ [kJ]} \tag{2.11.3}$$

From Appendix D,

$$3.6 \times 10^3 \text{ [kJ]} \equiv 1 \text{ [kWh]}. \tag{2.11.4}$$

Therefore we get

$$2.79 \text{ [kWh] (of electric energy)}. \tag{2.11.5}$$

A medium-sized house requires a 5-kW compressor for running its central air conditioner. The energy derived from burning 1 kg of coal will therefore run this unit for slightly more than a half hour.

Example 2.12

A diet table states that a "normal" male, 190 cm tall and 90 kg in weight, requires a daily food intake of 2900 "calories." This energy is used to "fuel" the body for all of his various physical activities. To get a feel for the amount of energy involved, compute the height, h, of the flight of stairs our "normal" male must climb to "burn" this daily food intake.

Solution: When discussing diet, a "calorie" actually means a kilocalorie. From equation (2.34) we compute the joule equivalent:

$$w = 4.19 \cdot 2900 = 12{,}150 \text{ [kJ]}. \tag{2.12.1}$$

As the person's weight is 90 kg, we get from equation (2.12)

$$h = \frac{12.15 \times 10^6}{90 \times 9.81} = 13{,}760 \text{ [m]}. \tag{2.12.2}$$

Clearly, a normal sedentary male does not climb the equivalent of almost two Mount Everests daily. The explanation is, of course, that the body needs energy for all the vital internal processes plus keeping a normal body temperature. Also, the conversion efficiency is not 100%.

What is impressive in this example is the amount of energy actually contained in a normal diet.

2.11 Energy Dissipation

Consider a system consisting of a mass suspended by a spring. If excited, this system will perform oscillations. The energy of the system will oscillate between the spring and the mass. At the instant when it is moving at its maximum velocity, the total energy is stored in the form of kinetic energy in the mass. A quarter of a cycle later the velocity is zero, and the total system energy is now stored in the form of potential energy in the spring. If the mass oscillates at a frequency of f Hz the energy oscillates at $2f$ Hz.

In many such oscillatory systems it is desirable to stop or damp the oscillations. This can be accomplished if the energy is dissipated in the form of heat. A viscous damper (or shock absorber), can perform this function. It is shown in Figure 2.14.

As the damper is compressed, oil will flow as shown by the arrows. The force f increases with the rate of oil flow. Assuming laminar flow, the force is known to be approximately proportional to the velocity, that is,

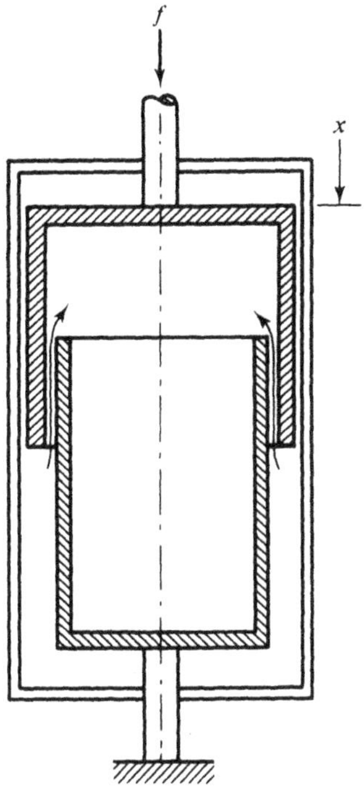

Figure 2.14

$$f = k_d \frac{dx}{dt} \qquad [\mathrm{N}], \tag{2.35}$$

where k_d is the damper coefficient. If the velocity is too high, the oil flow becomes turbulent, and the force changes its character radically.

From equation (2.17) we get for the energy absorbed by the oil in the form of heat as

$$w = \int f\,dx = \int k_d \frac{dx}{dt}\,dx = k_d \int \left(\frac{dx}{dt}\right)^2 dt \qquad [\mathrm{J}]. \tag{2.36}$$

The integrand is always positive (that is, the energy dissipated is independent of the direction of motion) and the rate of energy dissipation is

$$\frac{dw}{dt} = k_d \left(\frac{dx}{dt}\right)^2 \qquad [\mathrm{W}] \tag{2.37}$$

The electrical analog of a viscous damper is a resistor (see Section 3.14).

Example 2.13

The oscillating mass described above is mechanically coupled to the moving member of the shock absorber shown in Figure 2.14. Describe its motion with suitable mathematical expressions.

Solution: At any given moment, the total kinetic plus potential energy of the mass-spring system [according to equations (2.7.2) and (2.32)], is equal to

$$w_{\text{tot}} = \frac{1}{2}m\left(\frac{dx}{dt}\right)^2 + \frac{1}{2}kx^2 \qquad [\text{J}], \tag{2.13.1}$$

where x represents the spring elongation and k the spring constant. This energy will be dissipated in heat according to equation (2.37), and we can express this as

$$-\frac{d}{dt}\left[\frac{1}{2}m\left(\frac{dx}{dt}\right)^2 + \frac{1}{2}kx^2\right] = k_\text{d}\left(\frac{dx}{dt}\right)^2 \qquad [\text{W}]. \tag{2.13.2}$$

By performing the differentiation we obtain the following differential equation for the motion of the mass:

$$m\ddot{x} + k_\text{d}\dot{x} + kx = 0. \tag{2.13.3}$$

Integration of this equation produces the solutions shown in Figure 2.15. The reader is encouraged to confirm this.

2.12 Nuclear Energy

At the turn of the twentieth century Albert Einstein postulated the famous energy–mass relationship:

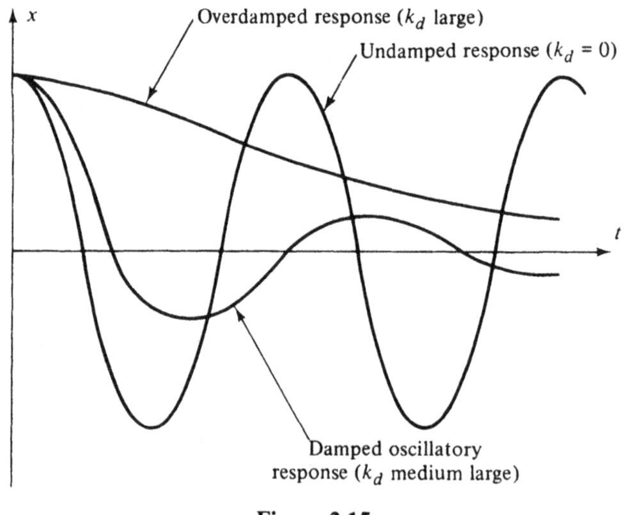

Figure 2.15

$$E = mc^2, \tag{2.38}$$

where E = energy in joules, m = mass in kilograms, and c = velocity of light in vacuum in meters per second. The formula indicates that mass—any mass—is equivalent to energy in enormous quantities. For example, 1 kg of mass, if completely converted into energy, according to Einstein's equation, would yield the following:

$$E = 1 \cdot (3 \cdot 10^8)^2 = 9 \cdot 10^{16} \text{ [J]}. \tag{2.39}$$

For comparison, from Figure 1.3 it can be seen that the total energy used in the world in 1994 was $390 \cdot 10^{18}$ J. That amount of energy could, in theory, be produced by the transformation of only 4333 kg of mass. Theory and practice are not always in agreement, however, and the fact is that today we do not know how to achieve a 100% efficient energy conversion of mass. However, we do know how to accomplish the feat partially.

Some materials, such as the uranium 235 isotope (235-U), are "fissionable." The nucleus of an atom of 235-U under bombardment by neutrons will absorb one neutron and form the isotope 236-U. Being highly unstable, this isotope fissions into two new atoms of xenon and strontium, plus additional neutrons. The total mass of the fission remnants is slightly less than the mass of the original atom, the difference in mass having been transformed into energy in quantities given by Einstein's equation. The energy appears in the form of heat, most of it absorbed by the remnants of the fission process.

The particular value of the uranium fission reaction is that it can be made both self-sustaining and controllable. By simply putting together a sufficient amount (core) of 235-U a critical mass[9] is reached, and the fission process starts. The rate of the reaction can be controlled by various means, such as the placement of control rods in the core.

The caloric energy released in the fission reaction will increase the core temperature, and the heat must be continuously removed. This is accomplished by pumping a coolant through the reactor core. The coolant may be water, gas, or a molten metal. The coolant serves the same purpose as the steam in the fossil-fueled process shown in Figure 2.13—it transports the energy out of the "combustion" area.

In a boiling-water reactor (BWR) the water leaves the reactor in the form of steam, which is led directly to the turbine. In a pressurized-water reactor (PWR) the water is prevented from boiling by the application of high pressure. The high-pressure, high-temperature water transfers its heat to a second body of water at a lower pressure in a heat exchanger or boiler, and the steam created is used to drive the electric turbogenerator. In a nuclear plant, the reactor essentially replaces the

[9] A simple analogy is offered by a camp fire. It is impossible to make a camp fire with a single log. A sufficient number must be put together before it burns properly. Similarly, the nuclear fuel will "burn" only if the reactor is sufficiently large.

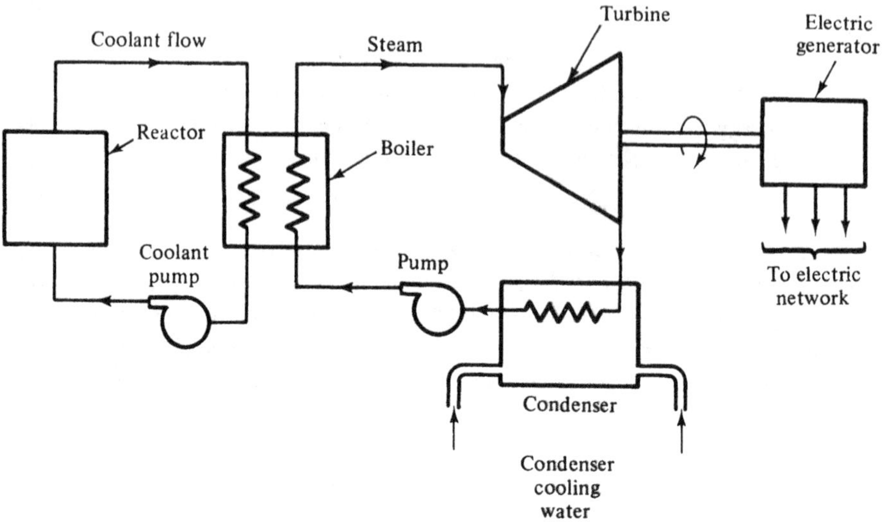

Figure 2.16

combustion parts (including stack, smoke, coal, sludge, and ash) of a fossil-fired unit (Figure 2.16).

Although the fission reaction converts approximately only 0.1% of the mass of the U-235 into energy, the energy "compactness" of nuclear fuel is still extremely impressive: 1 kg of 235-U produces the same amount of energy as approximately 3000 metric tons of coal! A 1000-MW coal-fired power plant requires a continuous feed of coal, usually transported by a fleet of railway cars, but the annual fuel charge of an equivalent nuclear station can be transported by a few trucks.

Controlled nuclear fission is now, after almost a third of a century of experimentation, well understood. We have not, as of this writing, succeeded, even in the laboratory, in controlling nuclear *fusion*. The process of fusion is responsible for the production of energy in the sun and when uncontrolled, is a hydrogen bomb. Light atoms are fused together into heavier ones with the simultaneous release of energy. Mankind will, it is hoped, eventually master this process. When this happens, the energy, like that in the fission process, will most likely be in the caloric form. To be useful it will have to be transformed into electricity; our electric transmission and distribution systems are not likely to become obsolete.

2.13 Solar Energy

The sun radiates to the earth amounts of energy that far exceed our present needs and even the energy needs of the foreseeable future. Solar energy fuels all our biologic processes and is the original source of water power, wind power, and, most importantly, our fossil fuels. Environmentally and economically, it is the most attractive source of energy. Why do we not use solar energy in the industrial, domestic, commercial, and transportation sectors? Three reasons account for this lack of use:

1. Solar energy arrives on earth at a very low intensity—of the order of only 1 kW/m^2 (measured in clear weather).
2. Solar energy does not arrive at all during night hours.
3. Solar energy is very difficult to transform into any other useful energy form, except low-grade heat.

Current technology provides only two ways to capture solar energy in large quantities:

1. By means of parabolic or spherical collectors solar energy can be concentrated into high temperature ($>$1000 K) caloric form. These high-intensity collectors work only in direct radiation—they are inoperative in cloudy weather.
2. By means of "greenhouse"-type collectors, the solar energy is transformed into low-temperature ($\approx$370 K) caloric form. These low-intensity collectors work, with reduced effectiveness, even on cloudy days.

These limitations put severe restrictions on the usefulness of solar energy.

Present technology permits us to utilize solar energy for domestic purposes, such as providing hot water for washing and space heating. A roof-mounted low-intensity solar collector measuring a few square meters, can satisfy this need for low-grade heat in every home. A recent study of a typical Florida home revealed that 60% to 70% of the total domestic energy need is for heating and cooling the live-in space and for heating water.

We see, therefore, a trend toward hybrid solar-electric energy service in the domestic market and maybe also in the commercial sector. In such a hybrid system, solar energy is used for those applications for which it is best suited—electricity is used for the rest. Solar energy can, conceptually at least, be used in conjunction with pumped hydraulic storage for peak power generation.

Example 2.14

The simplest type of low-intensity solar collector consists of a flat glass-covered box. The solar energy trapped in the box heats the water in a grid of copper tubes, which is part of a circulatory system. In order to serve the need during night hours, it is necessary to store enough hot water during the day. This is done in an insulated hot-water storage tank.

What should be the volume of this tank if the water temperature does not drop by more than 15 K during the night? It is assumed that the total heating load during the night period is 30 kWh.

Solution: If the tank capacity is x liters (1 liter H_2O weighs 1 kg) it can store x kcal/K; from Appendix D we see that 1 [kWh] is equivalent to 860 [kcal]. We have

$$x \cdot 15 = 30 \cdot 860 \tag{2.14.1}$$

$$x = 1720 \text{ [kg] } H_2O \ (\approx 454 \text{ gallons}) \tag{2.14.2}$$

2.14 Summary

In this chapter, we have reviewed the physical concepts of energy and power—all of the nonelectrical variety. Emphasis has been placed on the various forms in which energy appears and the possibilities for transformation from one form to another. In electric power engineering, the need for bulk energy storage is of great importance. We have discussed pumped hydro- and compressed gas storage possibilities.

Although energy is never lost, it will eventually after various transformations become degraded into low-grade waste heat that has little practical value. An energy engineer's skill should be directed toward the objective of "squeezing" the maximum use out of the energy resources as they are being transformed from the highest grade to the inevitable low-grade state.

EXERCISES

2.1 The moon has a diameter that is 27.2% of the earth's diameter. Its mass is 1.22% of that of the earth.

a) Find the surface gravity of the moon in percent of that of the earth.

b) On earth, each meter of altitude gain adds 9.81 J/kg to the gravitational potential of a mass. What would be the corresponding figure on the moon?

2.2 A rocket performs a vertical takeoff under the power of its thrust force. The latter equals 1.5×10^6 [N] and is assumed constant.

a) What power does the thrust engine deliver at takeoff?

b) What power does it deliver at the instant when the rocket velocity is 4800 km/h? Express all quantities in SI units and give your answer in both MW and hp.

2.3 Falling water exerts a torque of 9000 [N · m] on the turbine blades of a hydroturbine, which runs at a speed of 720 rpm. Compute the turbine power.

2.4 Consider the mine elevator discussed in Example 2.3. During the acceleration phase, the distance moved is 12.5 m. At the end of the acceleration phase we have increased the potential energy of the elevator load by

$$5000 \cdot 9.81 \cdot 12.5 = 613{,}100 \text{ [J]}.$$

The lift force during acceleration was computed to 54,050 [N]. According to equation (2.19), the lift motor delivers

$$54{,}050 \cdot 12.5 = 675{,}600 \text{ [J]}.$$

Why is there a discrepancy between the two values? Can you balance the energy equation?

2.5 A car has a "gas mileage" of 6 km/l at a speed of 90 km/h. Compute the gas mileage of the same automobile when traveling at 120 km/h.

Assumptions: (1) The windage and friction force acting upon the car increases as the square of the velocity; (2) The gas consumption of the car as measured in liters per hour is directly proportional to the power of the engine. (The efficiency of an automobile engine varies somewhat with engine speed, so this assumption is slightly erroneous. We disregard this fact in order to simplify the analysis.)

2.6 Figure 2.5 indicates that one needs to impart $g_0 r_0 = 9.81 \cdot 6.38 \cdot 10^6 = 6.26 \cdot 10^7$ [J/kg] to a mass to free it from the grip of the earth's gravity.

a) What vertical muzzle velocity ("escape velocity") must a bullet have never to return to the earth? For simplicity, neglect the air resistance in your analysis. (In reality the friction due to the air is very important and it will change your answer significantly.)

b) Use the data for the moon given (and computed) in Exercise 2.1 to find the escape velocity from the moon's surface!

2.7 Consider the pulley arrangement in Figure 2.17. When the system is released, it will accelerate in a clockwise direction. Find the vertical acceleration, d^2x/dt^2, of the masses.

Assumptions: (1) zero friction; (2) inertia of the pulley $= I$ kgm^2, (3) no slippage between pulley and string; (4) string is inelastic.

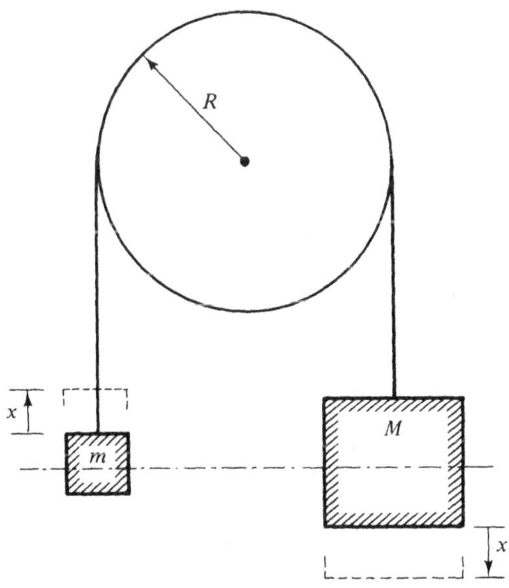

Figure 2.17

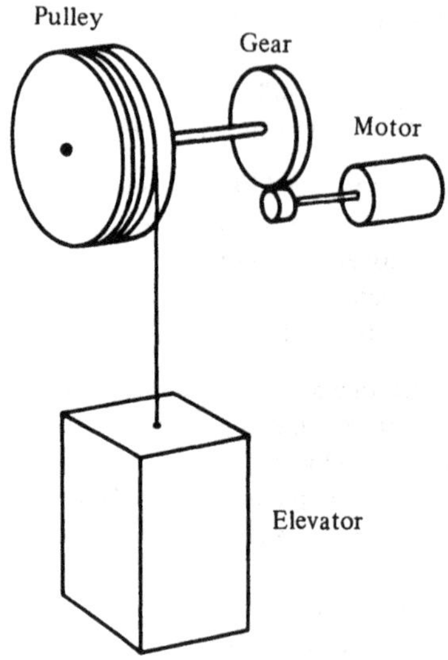

Figure 2.18

[HINT: Write an expression for the total energy of the system, w_{tot} as a function of x. As this energy must be constant you have $(d/dt)(w_{tot}) = 0$.]

2.8 Figure 2.18 depicts the drive system for an elevator. The elevator load is 10 metric tons (10,000 kg). The radius of the pulley is 1 m. The gear ratio is 25:1. The load is hoisted at the rate of 10 m/s.
Compute:
a) torque on the shaft of the pulley, in N · m,
b) motor torque, in N · m,
c) pulley speed, in rad/s,
d) motor speed, in rad/s and rpm,
e) motor power, in kW.
Neglect all losses.

2.9 In Example 2.8 the kinetic energy of a spinning rotor was computed. If the price of energy is five cents per kWh, how much would it cost to accelerate the rotor to its full speed?

2.10 If the 5-ton elevator in Example 2.3 were to drop 200 m and crash, what would be the temperature rise, ΔT, of the wreckage? The specific heat, c_T, of the elevator plus load is assumed to be 0.3 (meaning that it takes 0.3 kcal to raise the temperature by 1 K of 1 kg).

2.11 The total drop of a river from the mountain glaciers to the ocean is 2000 m. As the water tumbles down, its potential energy is transformed into heat. What is the change in temperature? Assume that there is zero heat loss to the environment during the descent.

2.12 In Example 2.2 we computed the potential energy needed to propel a payload to an altitude of 150 km. Assume now that we wish to keep it in a circular orbit, 150 km above the earth.

a) How much additional (kinetic) energy must be imparted to the payload to make this possible?

b) What respective shares do the potential and kinetic energies have of the total?

[HINT: By giving the payload an orbital velocity s of such a magnitude as to make the centrifugal force ms^2/r exactly equal to the force of gravity mg, the payload will circle the earth without "falling down."]

References

Anderson, E.E. *Thermodynamics*. Boston: PWS Publishing, 1994.

Karlekar, B.V. *Thermodynamics for Engineers*. Englewood Cliffs, NJ: Prentice-Hall, 1983.

Rose, H., Pinkerton, A. *The Energy Crisis, Conservation and Solar*. Ann Arbor, MI: Ann Arbor Science, 1981.

3

Fundamentals of Electric Energy

3.1 Electric Energy Engineering

In Chapter 2, we reviewed the basic definitions of energy and power. The more important nonelectrical forms of energy and energy transformations were discussed. We now focus our attention on electrical energy and the various means used to generate and consume it. At a first glance it would seem unavoidable that a study of electric energy would involve the varied aspects of electrical engineering. However, it is possible to divide (in a broad sense) electrical engineering into two major subareas: (1) communication and data technology, (2) energy conversion and control.

In the first subarea we include all applications that have as their collective primary function the processing, transmission, and general handling of "information." To most electrical engineers the single term "electronics" suffices to adequately describe this entire spectrum of important electrical engineering activities. As a general statement it can be said that electronic components operate at relatively low electric power levels, usually in the range 10^{-6} to 10^{2} W.

In this book, we concentrate on the second subarea of electrical engineering—energy conversion and the techniques for its control. Generally the electric "energy" or "power" engineer is concerned with electricity in the power range of 10^{3} to 10^{9} W. The electric power engineer designs and operates the electric power stations where electric energy is generated, the transmission systems over which it is routed, and the distribution systems from which it is retailed to the users. Generators, transformers, transmission, and distribution lines constitute a power system or *grid*, the proper operation of which requires a thorough knowledge of systems and control theory. The electric energy engineer's interests also includes design and operation of motors and transducers, which transform electric energy into mechanical and other forms of energy.

Our goal is to present the material in this book in a manner that makes it "digestible" to all engineering students. Matters electrical have a tendency to appear mysterious to most nonelectrical engineers, the reason being that "electricity" itself is often poorly understood.

Mass is one of the easier concepts to grasp as we first encounter it in the cradle. Everyone learns early in life, to feels quite at ease with the concept of mass, which forms the cornerstone of most of the other subdisciplines of engineering.

Electricity is something quite different. Our first contact with it is usually in the form of an electric shock and we never really lose the respect for this strange medium that we cannot see but the effects of which can be quite dramatic.

Some of the mystery of electricity can be removed by the utilization of mechanical analog for the explanation of observed electrical phenomena. This technique is often possible and is used extensively in this text. However, the reader is cautioned that this is not always possible. After all, electricity is a unique medium with unique features.

An electric power engineer is interested in electric and magnetic phenomena for reasons quite different from those of an electronics engineer. In this chapter, we shall stress phenomena that are of particular importance for later discussions. Although we run a certain risk of duplicating material from other engineering and physics courses, we think it is worthwhile to have all the necessary electric energy theory summarized in this chapter.

3.2 Physical Nature of Electricity: Electric Charge

Electrical science and engineering are based on the physical characteristics and behavior of the medium we call "electricity." The ancients knew that rubbing the fossil resin called amber (Greek *elektron*) would exert attraction on certain lightweight materials. By the end of the eighteenth century it was established that there were actually two kinds of electricity in nature. Benjamin Franklin introduced the names "positive" and "negative," which are still in use.

Today we know that electricity is closely interrelated with the microstructure of all matter, the basic building block of which is the atom. The modern view depicts the atom as consisting of a core or a *nucleus* surrounded by a "cloud" of orbiting *electrons*. The nucleus of a stable atom is made up of *protons* and *neutrons*. To each proton is ascribed an elementary positive *charge* of electricity, to each electron an elementary negative charge of equal magnitude. The neutrons are electrically neutral. These elementary charges are the smallest electrical quanta found in nature.

An undisturbed atom contains exactly as many electrons as protons, and the positive and negative charges balance or neutralize each other. If, due to external influences, atoms are robbed of some of their electrons, they will hold a surplus of positive charge and are referred to as *positive ions*. An atom may also acquire more than its normal complement of electrons, thus becoming a *negative ion*.

Particular combinations of protons, neutrons, and electrons form the atoms of about 100 known chemical elements. The number of protons in the nucleus identifies the element. Elements with equal numbers of protons, but different numbers of neutrons, constitute *isotopes* of the element in question. Protons, neutrons, and electrons have physical mass. Furthermore, they all perform extremely complicated spins and orbital motion.

In all engineering applications, we shall be concerned with the macrostructure of matter. Even the tiniest microcircuit used in modern electronics contains trillions of atoms. A typical solid contains on the order of 10^{20} atoms per cubic millimeter. Should we charge, or electrify, this material negatively by adding an electron to only every billionth atom, the net charge of this single small cube will still contain the incredible number of 10^{11} charge quanta.

Electricity, as we encounter it in most engineering applications, involves myriad elementary charge quanta. To picture electricity, we may think of it as an extremely fine-grained, or fluidic, substance. Perhaps a useful definition would be "an invisible fluid that shocks on touch."

The fine-grained character of electricity is best demonstrated by the units in which it is measured. The SI unit of charge, coulomb [C], is defined as

$$1\ \mathrm{C} = 6.242 \cdot 10^{18} \text{ charge quanta}$$

Clearly, the quantum of charge of a proton is $+1.602 \cdot 10^{-19}$ C; that of an electron, $-1.602 \cdot 10^{-19}$ C.

It may seem impractical to define the basic unit of charge in terms of an uneven number of quanta, but that choice was made long before we knew the quantized character of electricity.

3.3 Coulomb's Law: The Gravity Analog

The natural takeoff point in introducing electric energy is Coulomb's law, which describes the electrostatic force between static[1] charges. The electrostatic forces have close similarities with gravity forces between masses. We can make a better use of analogs between the two if we present Coulomb's law for a system of charges that have the same spherical and symmetry features as the system of masses that we discussed in Chapter 2.

[1] Although as we have said electricity in a "microsense" is never at rest, it is still possible to ascribe to macroconglomerations of electricity static characteristics. Consider the following analogy: All gases are made up of individual gas molecules, each of which is in a state of perpetual and complex motion. As we consider a tank filled with gas under certain pressure, we can still derive certain static relations between the volume, pressure, and temperature of this gas. For example, what we measure as a constant "pressure" is a result of a statistical averaging of countless collisions of billions of molecules against the walls of the tank.

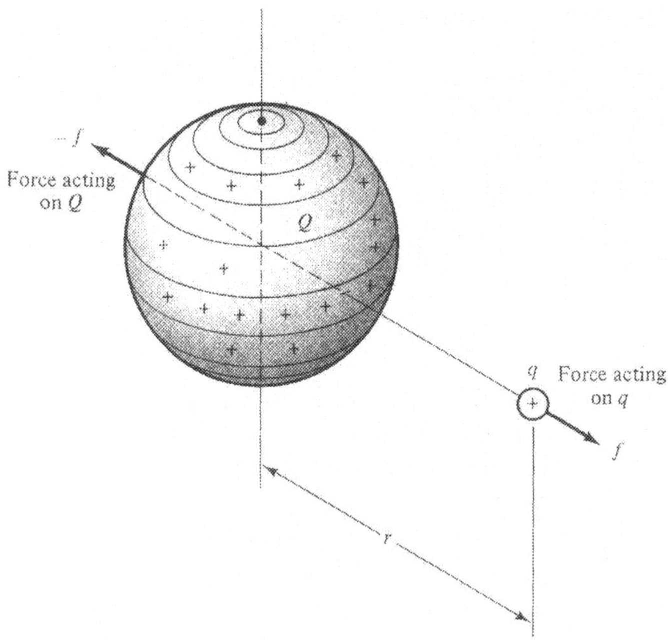

Figure 3.1

Consider the large spherical ball shown in Figure 3.1 and assume that it contains the electric charge Q coulombs.[2] At a distance r meters we place a small test charge, q coulombs. We can now observe an *electrostatic* force on the q charge. (An equal and opposite force is also felt by the charge Q.) Careful experiments established the following equation for the magnitude of the force, which is directed radially:

$$f = \frac{1}{4\pi\varepsilon\varepsilon_0}\frac{Qq}{r^2} \quad [\text{N}], \tag{3.1}$$

where $\varepsilon_0 = 8.854 \cdot 10^{-12}$ (the dielectric constant of vacuum), and ε = constant that depends upon the material surrounding the charges. For air (or vacuum), $\varepsilon = 1$.

[2] Where in the ball would this charge reside? If the sphere were made of metal or some other good *conductor*, that is, material in which electricity can travel or move freely, then we would find the total charge Q spread out uniformly over the surface of the sphere in the form of a *surface charge*. The reason for this is that the charges repel each other and being free to move they would travel to the surface and distribute themselves uniformly around it. (Compare this to the shape a spherical balloon will take when subjected to a uniform internal pressure.) If the ball were made of an insulating material, such as amber, the internal charge distribution would be more complex. In general, charges would be found throughout the interior as well as on the surface.

We note the close similarity with the gravity equations in Chapter 2. The gravity force between two positive masses is attractive, and the electrostatic force between two positive (or two negative) charges is repulsive. If the charges are of different signs the force is attractive.

3.4 The Electric Field

We can write equation (3.1) in the alternate form:

$$f = q\left(\frac{1}{4\pi\varepsilon\varepsilon_0}\frac{Q}{r^2}\right) \qquad [\text{N}]. \tag{3.2}$$

In a manner analogous to the introduction of the gravity vector, g, we identify the expression within the parentheses as a new vector—the *electric field vector*. It represents the force-per-unit charge q and has the magnitude

$$E \equiv \frac{1}{4\pi\varepsilon\varepsilon_0}\frac{Q}{r^2} \qquad [\text{N/C}]. \tag{3.3}$$

Coulomb's law can be written as

$$f = qE \qquad [\text{N}]. \tag{3.4}$$

The electric field is a vector that evidently "radiates" from the Q charge in a manner analogous to the g vector around the earth. However, they are in opposite directions. We can view the electric field as a feature associated with the Q charge, just as the gravity field around the earth can be said to be associated with the mass of the earth, M. The radial nature of E is shown in Figure 3.2 (cf. Figure 2.2).

3.5 Electrostatic Energy

If we move a mass m against the gravity field, we store a potential gravitational energy in it. Similarly, as we move the charge q against the electric field, we expend energy, which will be stored in the charge or in the field (depending on what philosophical view we take). As we move the charge radially from the initial radial distance r_1, to the final position r (Figure 3.3) we must clearly impart the energy

$$w_e = \int_{r_1}^{r} qE(-dr) = q\int_{r_1}^{r} E(-dr) \qquad [\text{J}] \tag{3.5}$$

on the charge. (Note that the negative sign for dr results from our definition of r to be positive in an outward sense.)

By using equation (3.3) we get the following expression for the energy

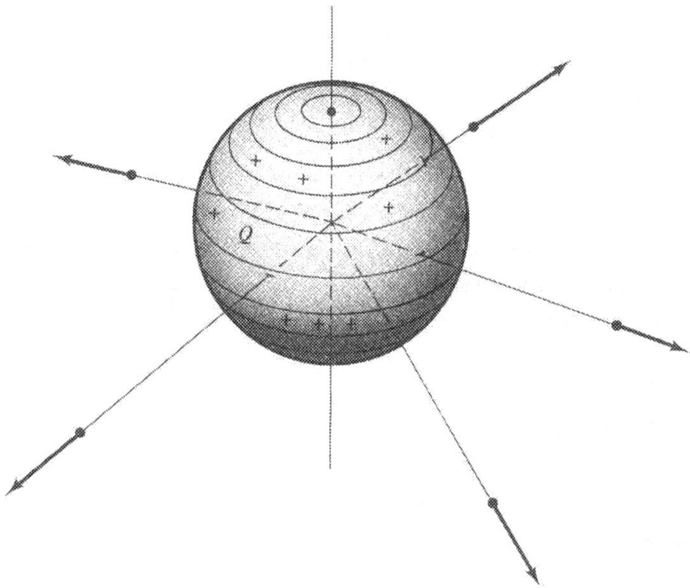

Figure 3.2

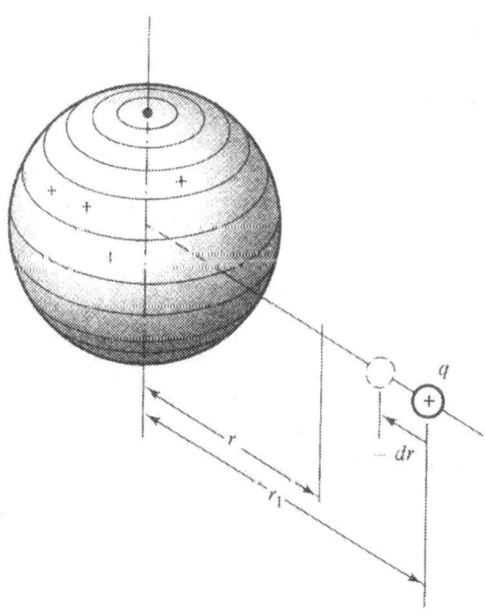

Figure 3.3

$$w_e = -\int_{r_1}^{r} q \frac{1}{4\pi\varepsilon\varepsilon_0} \frac{Q}{r^2} dr \qquad [\mathrm{J}], \tag{3.6}$$

which, on integration, yields

$$w_e = \frac{qQ}{4\pi\varepsilon\varepsilon_0}\left(\frac{1}{r} - \frac{1}{r_1}\right) \qquad [\mathrm{J}]. \tag{3.7}$$

In the gravitational field the stored energy increases as we move away from the earth. In the electric field the stored energy increases as we move toward the Q charge (assuming, of course, that q is positive).

Example 3.1

Find the energy stored in a system consisting of two charges Q and q when one of them is moved toward the other from a very large distance!

Solution: If we set $r_1 = \infty$ in equation (3.7) we obtain the energy required to move a charge from "infinity":

$$w_{e\infty} = \frac{qQ}{4\pi\varepsilon\varepsilon_0}\frac{1}{r} \qquad [\mathrm{J}]. \tag{3.1.1}$$

3.6 Electric Potential

We consider the energy stored per unit charge:

$$v \equiv \frac{w_e}{q} = -\int_{r_1}^{r} E\, dr \qquad [\mathrm{J/C}]. \tag{3.8}$$

Using gravity as an analog, we refer to this new "per unit energy" as the *electric potential.* It obviously has the unit of joules per coulomb. In electrical engineering this unit is of such great importance that we give it a very special name—volt [V]. The unit for electric field intensity was earlier given as newton/coulomb. In view of equation (3.8) we note that its unit can be expressed in volts per meter. This unit is more popular among electrical engineers.

Similar to the case of gravitational potential we are free to choose zero potential point wherever we wish. In this case we find it practical to set the zero potential at a great distance (i.e., at infinity).[3] Combination of expressions (3.1.1) and (3.8) gives the potential around the spherical Q charge as

[3] In most practical applications we define "ground" to have zero potential. Electrical apparatus usually have their chassis connected to ground (grounded) to avoid the possibility of charge buildup, which might result in a shock.

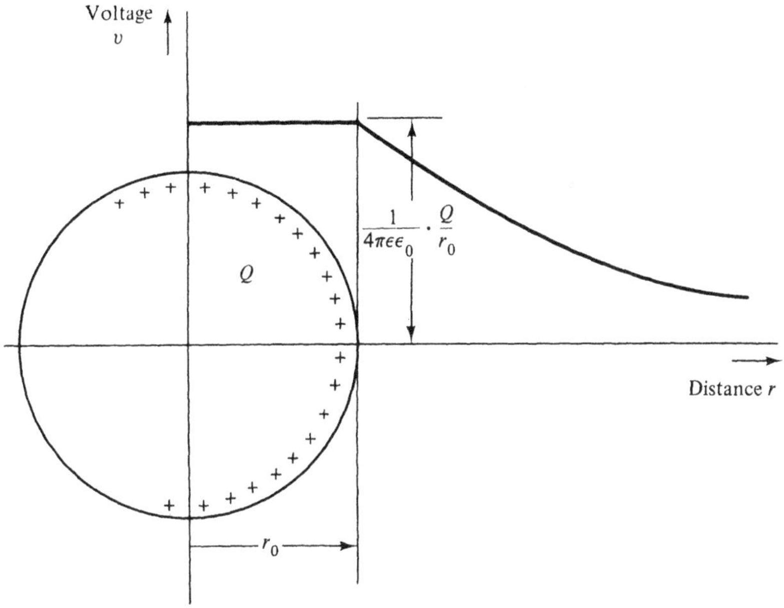

Figure 3.4

$$v = \frac{1}{4\pi\varepsilon\varepsilon_0} \frac{Q}{r} \quad \text{[V]} \tag{3.9}$$

Figure 3.4 shows the hyperbolic variation of the electric potential outside the Q charge. Note that the "potential hill" now slopes in the opposite direction as compared with the gravitational case (cf. Figure 2.5).

The potential is equal at all radially equidistant points and the *equipotential* surfaces are therefore spherical shells, as in the gravitational case. Equations (3.1) and (3.9) are valid outside the large sphere. The variation of the potential inside the large sphere depends on the inside charge distribution. If the sphere is made of a conducting material, and the Q charge is a surface charge, then the inside potential is constant and equal to the surface value (this has been assumed in Figure 3.4).

Example 3.2

Electric power engineers often work with very high voltages. There is a physical limit to how much voltage a given conductor can withstand. When the field strength E reaches a value of about 3 kV/mm ($3 \cdot 10^6$ V/m) in dry air the air molecules around the conductor become ionized and *corona discharge* will occur, accompanied by a hissing sound and a bluish glow around the conductor. (In vacuum or for gases under high pressure this dielectric breakdown will take place at much higher field strengths.)

Find the highest voltage that can be tolerated on a 20-cm spherical conductor in air without corona discharge occurring.

Solution: The E field has the largest value according to equation (3.3) at the surface of the ball, that is, for $r = r_0 = 0.1$ m. The largest possible charge, Q_{max}, is obtained from

$$E_{max} = 3 \times 10^6 = \frac{1}{4\pi\varepsilon\varepsilon_0} \frac{Q_{max}}{(0.1)^2} \quad [\text{V/m}]. \tag{3.2.1}$$

Solving for Q_{max} yields

$$Q_{max} = 3.338 \times 10^{-6}\ [\text{C}] \quad \text{or} \quad 3.338 \quad [\mu\text{C}]. \tag{3.2.2}$$

From equation (3.9) we then solve for the corresponding voltage:

$$v_{max} = \frac{1}{4\pi\varepsilon\varepsilon_0} \frac{Q_{max}}{0.1} = 300{,}000 \quad [\text{V}]. \tag{3.2.3}$$

Example 3.3

What would be the electrostatic force between two 20-cm balls placed with their centers 1 [m] apart and charged to the maximum voltage?

Solution: Equation (3.1) yields the repulsive force:

$$f = \frac{1}{4\pi\varepsilon\varepsilon_0} \frac{Q_{max}}{(0.1)^2} = 0.10 \quad [\text{N}] \tag{3.3.1}$$

The two examples above teach us a couple of important facts:

1. Even very small static charges result in very high voltages.
2. The highest attainable electrostatic forces are small. Electrostatic motors are therefore very weak and of little practical interest.

3.7 General Field Configurations

The electric field in the vicinity of a spherical charged conductor takes a simple geometrical form. A small-charge q placed in this field will distort the field pattern, but only slightly.

The field pattern around arbitrary conductor configurations is more complex. In Figure 3.5 we have sketched the two-dimensional field pattern around two irregularly shaped conductors, carrying equal charges of opposite sign.

For more general conductor geometries, finding the field and potential distributions can be very difficult. In fact, if the conductors have no simple geometric

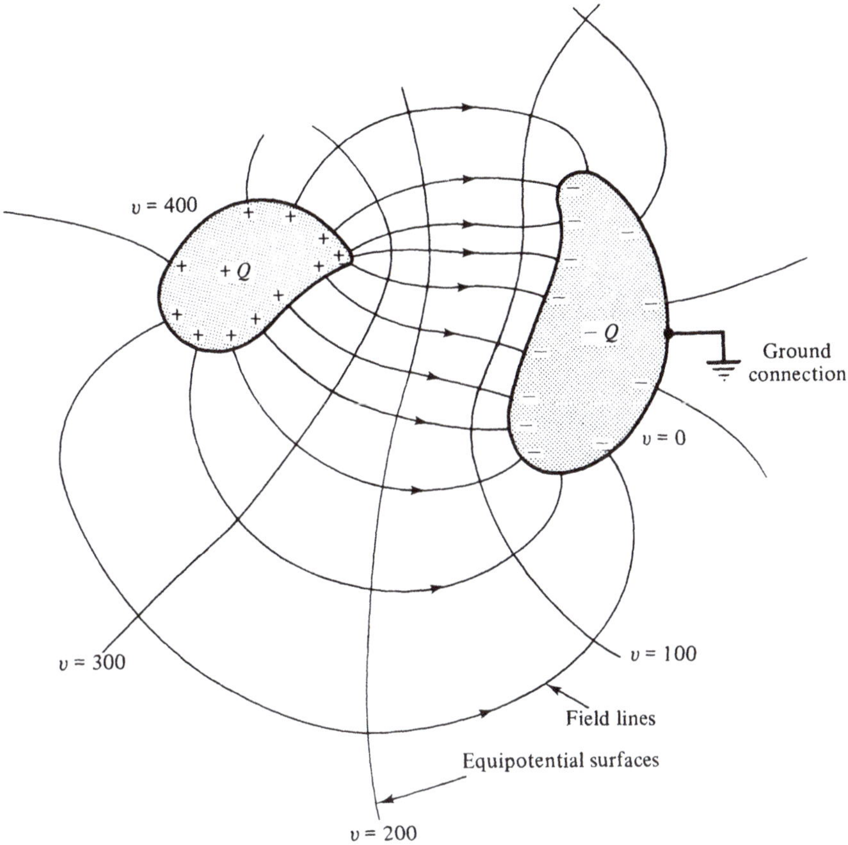

Figure 3.5

character, the field pattern can be obtained only after tedious computational processes. We therefore give some generalized characteristics:

1. The electric field lines always cross the equipotential surfaces and terminate on the conductor surfaces at right angles.
2. High field concentrations occur at sharp corners on the conductor surface.[4]

The effect of the electric field in the vicinity of electric overhead transmission line conductors is of very great importance in electric power engineering. Figure 3.6a shows the electric field in the vicinity of a single cylindrical conductor. The electric field is radial and most intense at the conductor surface. (Note that the closeness of the E lines is a measure of the field intensity.)

[4] This explains why electrostatic discharges ("corona") always start at perturbations. In high-voltage technology all conductors must have rounded forms.

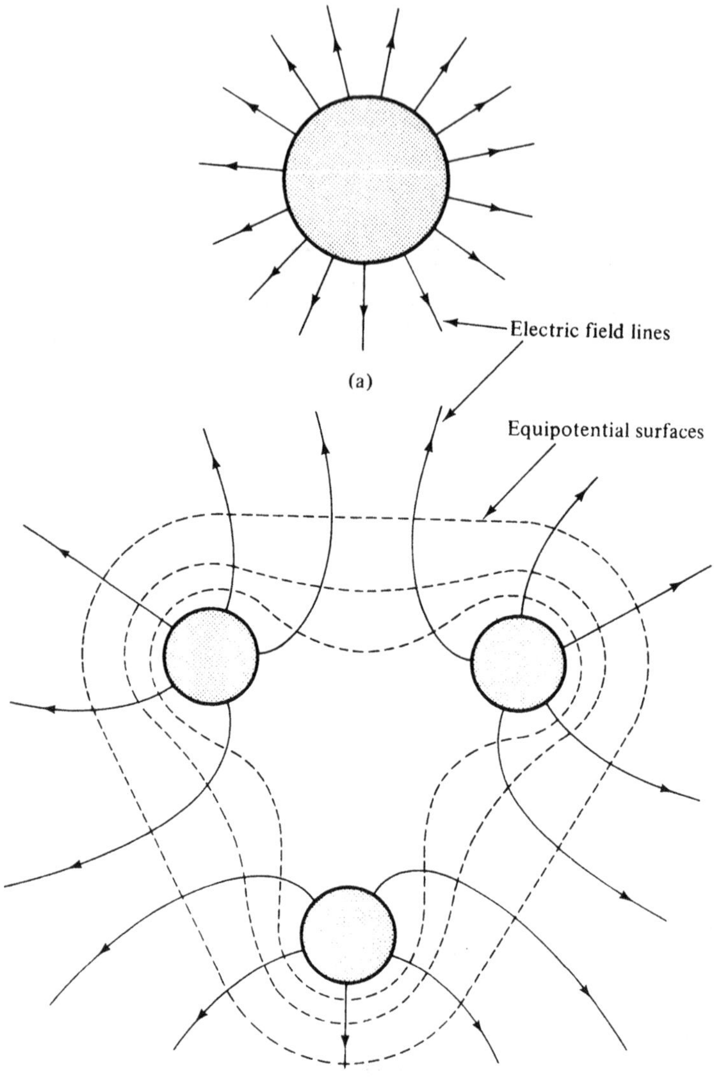

Figure 3.6

For extra-high-voltage (EHV) lines, usually in excess of 250 kV, the field concentration becomes so intense that the single conductor must be replaced by bundled conductors. Figure 3.6b shows a *triple bundling* and the corresponding field pattern. The three conductors are kept apart by spacers about 20–30 cm long. It is clear that the field intensity at each conductor has been reduced because the total charge is now divided equally among the three conductors. (Note that in order to make a meaningful comparison, the two cases must have equal total cross-sectional area.)

3.8 Electrostatic Energy Storage: Capacitance

From equation (3.9) we note that the charge is proportional to the voltage of the spherical conductor. By setting $r = r_0$,

$$Q = 4\pi\varepsilon\varepsilon_0 r_0 v \qquad [\mathrm{C}], \tag{3.10}$$

hence

$$Q = Cv \qquad [\mathrm{C}], \tag{3.11}$$

where

$$C \equiv 4\pi\varepsilon\varepsilon_0 r_0 \qquad [\mathrm{F}] \tag{3.12}$$

is referred to as the *electric capacitance* of the sphere. It has the physical unit coulomb/volt, which is given the name *farad* [F].

As we move a small charge dQ "from infinity" to the sphere, we add to its charge and also to its potential. The potential increase dv follows from equation (3.11) and is equal to

$$dv = \frac{1}{C}\,dQ \qquad [\mathrm{V}]. \tag{3.13}$$

The addition of the charge dQ increases the energy of the sphere [according to equation (3.8)] by

$$dw_e = v\,dQ \qquad [\mathrm{J}]. \tag{3.14}$$

Combining equations (3.13) and (3.14) we have

$$dw_e = Cv\,dv \qquad [\mathrm{J}]. \tag{3.15}$$

As we "charge" the sphere, thus increasing its potential from 0 to v, we increase its stored electric energy by the amount:

$$w_e = \int dw_e = \int_0^v Cv\,dv = \frac{1}{2}Cv^2 \qquad [\mathrm{J}] \tag{3.16}$$

Example 3.4

Compute the stored energy of the charged sphere in Example 3.2.

Solution: We have

$$C = 4\pi \cdot 8.854 \cdot 10^{-12} \cdot 0.1 = 11.13 \cdot 10^{-12} \qquad [\mathrm{F}] \tag{3.4.1}$$

or

$$C = 11.13 \text{ [pF] (picofarad).} \qquad (3.4.2)$$

As the voltage is 300 kV, we get for the energy stored:

$$w_e = \tfrac{1}{2} \cdot 11.13 \cdot 10^{-12} \cdot (300{,}000)^2 = 0.501 \qquad \text{[J].} \qquad (3.4.3)$$

This is not a very impressive amount of energy. Capacitor storage of large amounts of electric energy does not look promising.

Example 3.5

The two vessels shown in Figure 3.7 are connected by a pipe to form a *hydraulic capacitor*. The pump is used to transfer the fluid between the vessels. The figure shows a fluid of mass Q "stored" in the left tank. Find the "hydraulic capacitance" of the storage system and the energy stored! Assume that the cross-sectional area of each vessel is A [m^2].

Solution: If A represents the vessel area and ρ the density of the fluid, we have for the stored mass:

$$Q = \rho A h \qquad \text{[kg].} \qquad (3.5.1)$$

The head $2h$ represents a gravitational potential increase of

$$v_g = g_0 2h \qquad \text{[J/kg].} \qquad (3.5.2)$$

In terms of this potential difference we can write for the stored mass:

$$Q = \frac{\rho A}{2g_0} v_g \qquad \text{[kg].} \qquad (3.5.3)$$

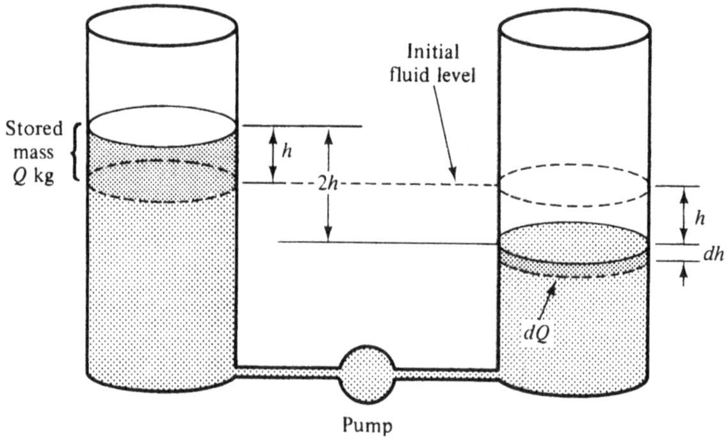

Figure 3.7

We introduce the hydraulic capacitance,

$$C_h \equiv \frac{\rho A}{2g_0}, \tag{3.5.4}$$

and can then write for the mass of the stored fluid,

$$Q = C_h v_g \qquad [\text{kg}]. \tag{3.5.5}$$

The stored mass, Q evidently represents a certain hydraulic energy w_h. If we were to release it, it could drive a turbine, which can perform work. Let us compute how much energy is, in fact, stored. As the fluid is pumped into the left vessel, clearly the hydraulic head is increasing. We can therefore expect that the pump effort, or *power*, will have to increase as more fluid is transferred.

Consider the incremental energy dw_h required to move the elemental mass dQ against the head $2h$ (see Figure 3.7). Using equation (2.12), we have

$$dw_h = g_0 2h\, dQ = g_0 2h\rho A\, dh \qquad [\text{J}]. \tag{3.5.6}$$

The total energy needed is obtained by integration:

$$w_h = 2g_0\rho A \int_0^h h\, dh = g_0 \rho A h^2 \qquad [\text{J}]; \tag{3.5.7}$$

$$w_h = \frac{1}{4}\frac{\rho A}{g_0} v_g^2 \qquad [\text{J}]. \tag{3.5.8}$$

In view of expession (3.5.4) we get

$$w_h = \tfrac{1}{2} C_h v_g^2 \qquad [\text{J}]. \tag{3.5.9}$$

Note the analogy with equation (3.16).

Consider the following numerical example:

$$A = 1\ \text{m}^2,$$

$$\rho = 1000\ \text{kg/m}^3,$$

$$h = 1\ \text{m}.$$

We have

$$C_h = \frac{1000 \times 1}{2 \times 9.81} = 50.97. \tag{3.5.10}$$

$$v_g = 9.81 \cdot 2 = 19.62. \tag{3.5.11}$$

From equation (3.5.9) we compute the hydraulic energy stored:

$$w_h = \tfrac{1}{2} \cdot 50.97 \cdot 19.62^2 = 9810 \qquad [\text{J}]. \tag{3.5.12}$$

Note the difference in magnitude compared to that of Example 3.4. It is much easier to store large quantities of energy hydraulically than electrically. Several examples of this phenomenon can be found in practice. (Recall the energy storage facilities discussed in Chapter 2.)

3.9 Practical Electric Capacitors

A single conductor capacitor (like the sphere in Section 3.8, with the charge taken "from infinity") is not a very practical electric storage device. A more practical element is an electric capacitor made up of two adjacent conductors separated by an insulator or dielectric. For example, the charge $+Q$ on the positive conductor in Figure 3.5 can be "pumped" from the opposite conductor, which then will have the charge deficiency $-Q$.

The potential difference

$$v = v_1 - v_2 \qquad [\mathrm{V}] \tag{3.17}$$

between the two conductors increases linearly with the charge transfer Q, and again we can express this by

$$Q = C(v_1 - v_2) = Cv \qquad [\mathrm{C}]. \tag{3.18}$$

The capacitance C depends upon the conductor geometry, size, and dielectric constant of the insulating material. We give without proof the following formula for the capacitance of a plate capacitor (with d and A as shown in Figure 3.8a):

$$C = \varepsilon\varepsilon_0 \frac{A}{d} \qquad [\mathrm{F}]. \tag{3.19}$$

The electric field is uniform between the plates (except for the "fringe effect" shown in Figure 3.8b), and it is equal to

$$E = \frac{v}{d} \qquad [\mathrm{V/m}], \tag{3.20}$$

where v is the voltage between the plates. The equipotentials are parallel planes between the plates and show the effect of fringing. The capacitance increases with increased area A and decreased plate separation d. Commercial capacitors are made of layered foils (Figure 3.9). They are rolled into bundles to make them compact.

Example 3.6

The plate capacitor shown in Figure 3.8 is connected to a 12-V automobile battery, that is, a voltage differential of 12 V exists between the two plates. The plates are parallel and 1 mm apart, therefore the electric field between the plates is

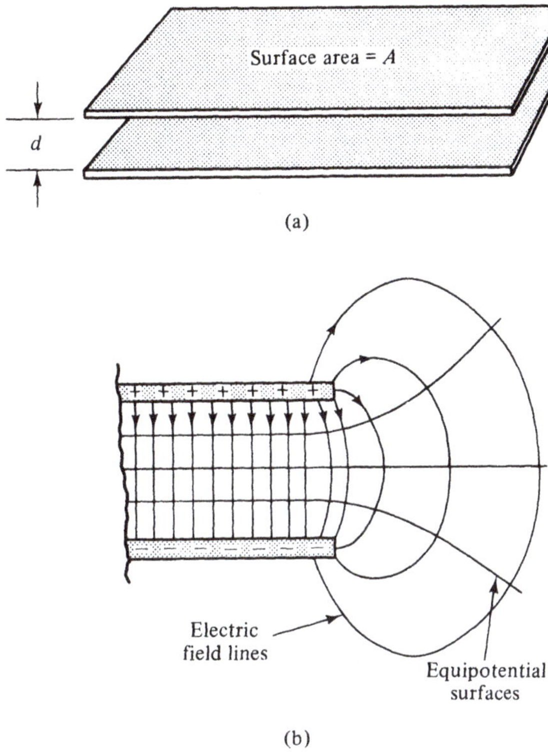

Figure 3.8

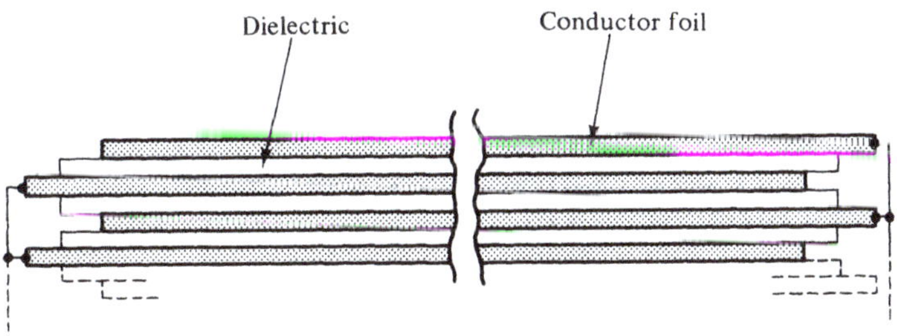

Figure 3.9

$$E = \frac{12}{0.001} = 1.2 \times 10^4 \qquad [\text{V/m}]. \tag{3.6.1}$$

For all practical purposes the electric field is of uniform strength. An electron is placed on the negative plate. Under the influence of the electrostatic force the electron will accelerate toward the positive plate. Assume a vacuum between the plates and describe the motion of the electron.

Solution: The mass of the electron is $m = 9.11 \cdot 10^{-31}$ kg and its charge $q = -1.602 \cdot 10^{-19}$ C.

If we neglect all relativistic effects (which we are permitted to do because the velocity will be much less than that of light) the electron acceleration a will be

$$a = \frac{\text{force}}{\text{mass}} = \frac{qE}{m} = \frac{(-1.602 \times 10^{-19}) \times 1.2 \times 10^4}{9.11 \times 10^{-31}} = 2.11 \times 10^{15} \quad [\text{m/s}^2]. \tag{3.6.2}$$

Considering that the electron acceleration due to the earth's gravity is only 9.81 m/s^2, we conclude that the electric force is overwhelmingly dominant.

Assuming that the distance d between the plates is covered in t_0 seconds, and because the acceleration is constant, we can find t_0 from

$$d = \frac{a}{2} t_0^2. \tag{3.6.3}$$

Thus,

$$t_0 = \sqrt{\frac{2d}{a}} = \sqrt{\frac{0.002}{2.11 \times 10^{15}}} = 0.974 \times 10^{-9}\,[\text{s}] = 0.974 \quad [\text{ns}]. \tag{3.6.4}$$

The electron, when it hits the positive plate, will have attained the velocity

$$s = at_0 = 2.11 \cdot 10^{15} \cdot 0.974 \cdot 10^{-9} = 2.06 \cdot 10^6 \quad [\text{m/s}]. \tag{3.6.5}$$

In words: The electron, under the influence of the relatively modest electric field caused by a car battery, will, in the short distance of 1 mm, accelerate to a velocity over 2000 km/s in the time span of about 1 ns. Electricity is certainly a volatile medium!

Example 3.7

In this example we estimate the electric energy involved in a typical thunderstorm. We hasten to explain that our model of the storm cloud is a highly simplistic one, but it should give us rough estimates of the magnitude of the voltage, current, and energy involved.

Figure 3.10 depicts a thundercloud, having its negatively charged base at an altitude of 1500 m above the earth. The base area of the cloud is 90 km^2. The cloud is about to discharge a bolt of lightning. Experimental measurements indicate that discharge occurs when the electric field reaches a value of about $4 \cdot 10^5$ V/m. (Note that this is considerably lower than the dry-air breakdown in Example 3.2.)

1. Compute the voltage between cloud and earth before discharge.
2. If you were to model the cloud-earth system as a parallel-plate capacitor what would be its capacitance?
3. Compute the negative charge on the cloud just before the discharge.

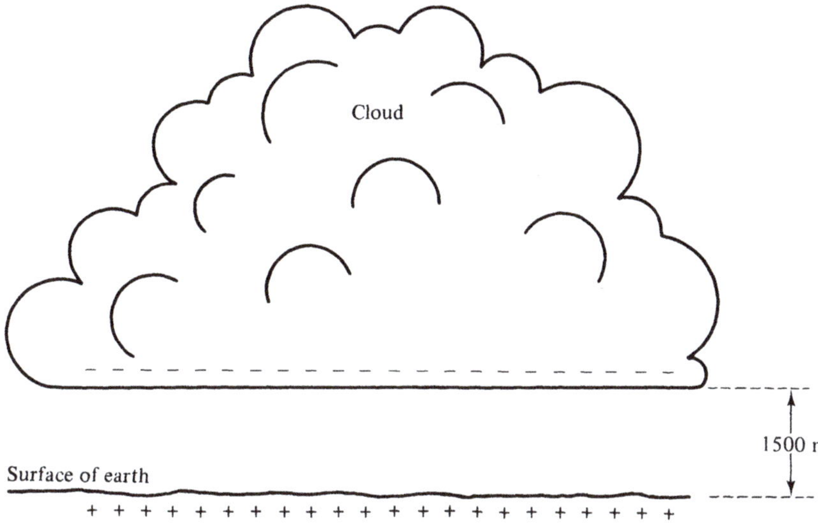

Figure 3.10

4. Compute the total electric energy stored before the lightning stroke.
5. If the discharge lasts about 100 ms, compute the average power developed during the stroke.
6. The cloud has a lifetime of about 1 hour and delivers 100 strokes of lightning during this time. What average power does this correspond to?

Solution

1. If we consider the electric field to be uniform between the "plates" then we obtain for the cloud-to-earth voltage:

$$v = -1500 \cdot 4 \cdot 10^5 = -6 \cdot 10^8 \qquad [\mathrm{V}]. \tag{3.7.1}$$

2. From equation (3.19) we get the capacitance:

$$C = 8.854 \times 10^{-12} \frac{90 \times 10^6}{1500} = 0.531 \times 10^{-6}\,[\mathrm{F}] = 0.531 \qquad [\mu\mathrm{F}]. \tag{3.7.2}$$

3. Equation (3.11) gives the charge stored in the cloud as

$$Q = -0.531 \cdot 10^{-6} \cdot 6 \cdot 10^8 = -319 \qquad [\mathrm{C}]. \tag{3.7.3}$$

4. The energy stored from equation (3.16) is

$$w_e = \tfrac{1}{2} \cdot 0.531 \cdot 10^{-6} \cdot (-6 \cdot 10^8)^2 = 9.56 \cdot 10^{10} \qquad [\mathrm{J}]. \tag{3.7.4}$$

5. If this energy is discharged in 10^{-1} seconds then the rate of release of energy will be

$$p_{\text{stroke}} = \frac{9.56 \times 10^{10}}{0.1} = 9.56 \times 10^{11} \qquad [\mathrm{W}]. \tag{3.7.5}$$

Expressed in the large power unit, gigawatt [GW] (10^9), the discharge power is 956 GW. (For comparison, the total electric generating capacity of the United States in 1994 was about 755 GW.)

6. If the lightning stroke is of average size, then the total energy discharged during the lifetime of the cloud is

$$w_{tot} = 100 \cdot 9.56 \cdot 10^{10} = 9.56 \cdot 10^{12} \qquad [J]. \tag{3.7.6}$$

The average discharge power is obtained by dividing the energy into the lifetime of the cloud, which is 3600 seconds:

$$p_{avg} = \frac{9.56 \times 10^{12}}{3600} = 2.6 \times 10^9 \qquad [W] \tag{3.7.7}$$

or

$$2600 \qquad [MW].$$

This power would be enough for a city of about two million people, but, we do not know how to harness it. The example demonstrates the vast electrical energy (as well as potential and kinetic energy stored in rain and wind) at play in the atmosphere.

3.10 Electrodynamics: Electric Current

So far we have concerned ourselves exclusively with static electricity and its associated forms of energy. The most important electric energy conversion phenomena involve electrons in motion or *electric current*.

If a charge ΔQ coulombs passes through a surface S (Figure 3.11) in the time interval Δt then there exists an electric current of magnitude

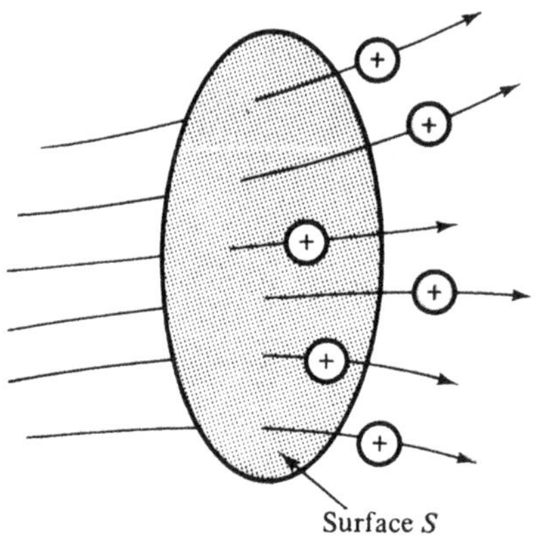

Figure 3.11

$$i \equiv \frac{\Delta Q}{\Delta t} \qquad [\text{C/s}] \quad \text{or} \quad [\text{A}] \tag{3.21}$$

through the surface in question.

Example 3.8

The charge on the cloud, −319 C in Example 3.7 discharges to ground in 100 ms. What is the corresponding (average) current between the cloud and the ground?

Solution: From the definition (3.21), we get[5]

$$i_{\text{ave}} = \frac{-319}{0.1} = -3190 \qquad [\text{A}]. \tag{3.8.1}$$

Example 3.9

A positive charge of 10^{-6} C is placed on a toroidal wheel, as shown in Figure 3.12, the spokes of which are insulators. This charge is the maximum that the conductor can withstand (see Example 3.2).

The wheel is spun at the speed of 1200 rpm. What is the current through the surface S?

Solution: In 1 second the wheel makes 20 full revolutions. The current will be

$$i = 20 \cdot 10^{-6} = 2 \cdot 10^{-5} \qquad [\text{A}] \quad (\text{or } 20 \qquad [\mu\text{A}]). \tag{3.9.1}$$

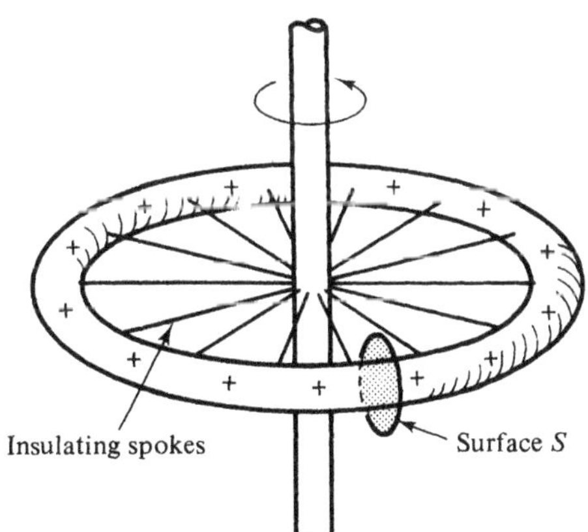

Figure 3.12

[5] The negative sign means that the current is directed from the ground to the cloud. Why?

The two examples above involve very large and very small electric currents. Neither has important applications.

3.11 Currents in Electric Conductors

Electric currents take on practical significance when they occur in *electric conductors*. The basic feature of an electric conductor is the presence within its atomic lattice of *free electrons*. In nonconducting material all electrons are bound to their "home" atoms. In conducting material, however, the free electrons are not bound to any particular atom but are free to drift throughout the material.

For example, copper has $0.849 \cdot 10^{20}$ atoms per cubic millimeter. If we assume the existence of at least one free electron per atom, then the total negative charge of all the free electrons in 1 mm^3 is

$$(0.849 \cdot 10^{20}) \cdot (-1.602 \cdot 10^{-19}) = -13.6 \text{ C}.$$

Each copper atom has 29 electrons, and therefore every cubic millimeter contains a total negative charge of about -394 C. The protons carry an equal positive charge.

Example 3.10

The charges calculated above are enormous. Note that the total negative charge in one single cubic millimeter of copper exceeds the total charge in a thunder cloud (cf. Example 3.7).

Assume that you could somehow separate the positive and negative charges in 1 mm^3 of copper and place these "blobs" of charge 1600 m apart. With what force would they attract each other?

Solution: Equation (3.1) gives

$$f = \frac{1}{4\pi\varepsilon_0}\frac{(394)^2}{(1600)^2} = 5.39 \times 10^8 \qquad [\text{N}] \tag{3.10.1}$$

or 55,000 tonnes.

These figures give us some idea of the cohesive forces that hold matter together.

Because within each atom there is an equal amount of positive charge as there is negative charge, on average, the atom has complete charge neutrality. The free electrons can drift around "freely" within the atomic lattice but only so as to preserve this charge neutrality. If, for example, a local "lumping" of free electrons were to occur, then there would be a negative charge concentration. Due to the repulsive force between charges of equal sign this concentration would instantly dissipate.

The free electrons can be thought of as a "cloud" or a "fluid" drifting around within the conductor, uniformly distributed to preserve overall charge neutrality.

As the internal electric forces prevent any deviation from this uniformity, the electron fluid is in effect completely incompressible. Also, because the electric forces completely dominate other forces, the electric fluid is essentially inertialess. (See Example 3.6.)

Each individual free electron performs a random motion as it bounces around within the atomic lattice. The overall velocity of the total "electron fluid" is, however, zero. A gas-filled balloon serves as a good analogy.[6] Each individual gas molecule performs random motion inside the balloon, but the total gas volume has zero velocity.

When an external *E* field is superimposed on this "electron fluid" every electron will be subjected to a force in the negative *E* direction, and the whole "fluid" will start a collective drift in a direction opposite the *E* field. But the uniform electron density, and hence the charge balance within the atomic lattice, must be preserved. (We assume, of course, that the conductor is part of a closed loop circuit so that the electron motion is not impeded.)

From the moment the *E* field is applied until a steady-state drift velocity is achieved, a certain startup time will have elapsed. However, due to the very small inertia of the electrons (see Example 3.6) this acceleration period is of the order of nanoseconds (10^{-9} seconds). For practical purposes we can assume an "instantaneous" or inertialess response. The drift velocities are normally very small. Consider a current of 1 [A] in a copper wire with a 1-mm^2 cross-sectional area. (A current of 1 [A] represents a charge flow of 1 C/s.) Earlier we had concluded that copper contains approximately 14 C of free electrons per cubic millimeter (assuming one free electron per atom). If all these free electrons would drift at the rate of 1 [mm/s] in the 1-mm^2 wire, a current of 14 [A] would flow. Consequently, a 1-[A] current corresponds to a drift velocity of only 1/14 [mm/s].

3.12 Ohm's Law

For most conductor materials the drift velocity (and hence the current, i) can be found experimentally to be proportional to the *E* field. As the field in turn is proportional to the voltage v applied across the conductor, we can write:

$$i = Gv \qquad [\mathrm{A}], \tag{3.22}$$

where G, a constant, is called *conductance*,
or

$$v = \frac{1}{G} i = Ri \qquad [\mathrm{V}]. \tag{3.23}$$

[6] In one important aspect the analogy fails. The electron cloud is incompressible; this is certainly not true for the volume of gas.

This is *Ohm's law*. The constant, R, is referred to as the *resistance* of the conductor, measured in ohms [Ω]. Its inverse, G, the conductance, is measured in siemens [Ω^{-1}]. The resistance R varies with the length of the conductor l and cross-sectional area A in accordance with:

$$R = \rho \frac{l}{A} \qquad [\Omega], \tag{3.24}$$

where ρ is the resistivity. (For copper $\rho \approx 1.75 \cdot 10^{-8}\ \Omega \cdot \text{m}$.) ρ increases slightly with the temperature of the conductor.

3.13 Basics of Electric Power

The potential energy stored in water behind a dam, for example, can be transformed into other forms of energy. In a free fall followed by impact, the energy is first transformed into the kinetic and then, suddenly, into the caloric form. In a controlled fall, for example, in a Pelton turbine (Example 2.4), most of the energy is converted into "useful work" performed by the turbine. The turbine will typically supply power to some device, such as a mill or electric generator.

Some of the energy never reaches the turbine but is transformed in the penstock into heat due to friction between the water and the tube walls and also within the fluid itself. This *resistive heat loss*, p_{res}, increases with the fluid flow, i, and also with increased tube resistance, R. The latter parameter is a function of tube dimensions and tube surface characteristics.

The potential energy stored in an electric charge can likewise be transformed into other forms of energy. If we let the charge "fall freely" to a lower potential level (or to higher potential levels if the charge is negative) the energy is transformed into the caloric form. If we control the "fall" of the charge by guiding it via conductors through an electric energy converter, for example, a motor, then the potential energy of the electric charge will be transformed into useful mechanical energy.

We can derive a simple and useful equation for the electric power involved in such a transformation. Assume that the incremental charge element ΔQ experiences a potential drop of v volts. In view of equation (3.14), it will release energy:

$$\Delta w_e = v \Delta Q \qquad [\text{J}]. \tag{3.25}$$

Assume that this energy transfer takes place in the short time interval Δt. The rate of energy transfer, or *electric power*, will be equal to

$$p = \frac{\Delta w_e}{\Delta t} = v \frac{\Delta Q}{\Delta t} = vi \qquad [\text{W}]. \tag{3.26}$$

This is the basic electric power. Compare it to the hydraulic power expressed in equation (2.4.2).

3.14 Resistive or Ohmic Power Dissipation

When an electric current i flows through a conductor of resistance R, electric energy will be dissipated in the conductor in the form of heat called *ohmic heat dissipation*. If the voltage drop across the conductor is v, then, according to equation (3.26), electric power,

$$p_\Omega = vi \qquad [\mathrm{W}], \tag{3.27}$$

will be lost in heat. According to Ohm's law, v and i are related by

$$v = Ri \qquad [\mathrm{V}], \tag{3.28}$$

and we have for the ohmic heat loss,

$$p_\Omega = i^2 R \qquad [\mathrm{W}]. \tag{3.29}$$

3.15 Electric Power Transmission

Figure 3.13 shows the simplest possible electric energy transmission system. A "generator" delivers energy to a "load" via a transmission line. If we assume that the generator voltage v_g is only slightly higher than the load voltage v_l due to a low voltage drop along the line, then we can let v represent the average of the two voltages.

According to equation (3.26) the transmitted power

$$p_{\mathrm{trans}} = vi \qquad [\mathrm{W}]. \tag{3.30}$$

If R represents the total line resistance (both leads), then from equation (3.29) we have the total transmission loss,

$$p_\Omega = i^2 R, \tag{3.31}$$

or, in view of equation (3.30),

$$p_\Omega = R\left(\frac{p_{\mathrm{trans}}}{v}\right)^2 \qquad [\mathrm{W}]. \tag{3.32}$$

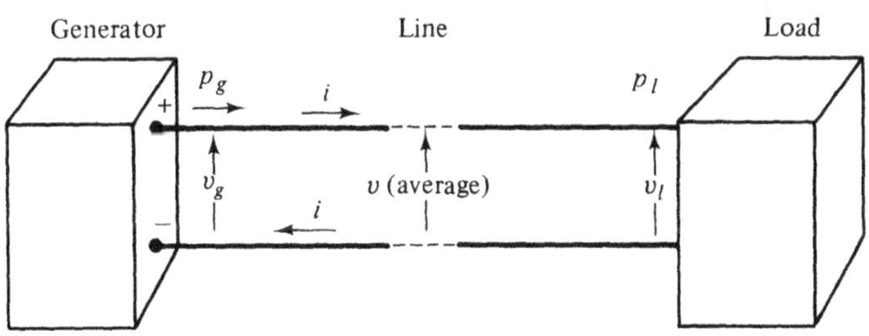

Figure 3.13

The ratio of the power loss in the transmission line to the power transmitted gives

$$\frac{p_\Omega}{p_{\text{trans}}} = R\,\frac{p_{\text{trans}}}{v^2}. \tag{3.33}$$

This expression tells us that the transmission loss is proportional to the inverse of the square of the transmission voltage. Thus the higher the voltage at which we transmit power the lower the power loss in the transmission line. In North America, the highest transmission voltages are close to one million volts (discussed further in Chapter 6).

How fast will the energy travel in the system shown in Figure 3.13? As the voltage is applied to the sending end, we, in effect, give the electron fluid a "push," which will be felt as a "pressure wave" propagating along the "incompressible" electron "fluid" of the conductor. Because of the extremely low inertia of the electrons the wave will travel extremely fast, with a velocity slightly less than the speed of light. For most practical situations this is "instantaneous." (In a hydraulic transmission system, the velocity of the pressure wave would be equal to the velocity of sound in the fluid in question.)

3.16 Electric Sources

We have on several occasions referred to "charge reservoirs," "sources," and "generators." In electrical engineering, a variety of electric energy sources is available. The range varies from giant synchronous generators that can deliver, continuously, power in excess of 1000 MW to small capacitors that can supply "one-shot" energy pulses of the order of microjoules.

A direct current (dc) source delivers a unidirectional current and power. An electric storage battery is probably the most common and versatile dc source in use. It may supply power to a hearing aid, start an automobile engine, or serve as a hospital power emergency source.

In an electric storage battery, positive and negative charges are "separated" from each other in an electrochemical process. The charges are forced to the two battery plates, and, if these are open-circuited, the charges will accumulate there. The result will be a buildup of terminal potential and an electrostatic field. As this field increases, the charge separation process within the battery decreases, and a balance is reached when no new charge separation takes place and a steady potential difference, the *electromotive force* (emf), exists between the terminals.

When an external load, in the form of a resistor, for example, is connected across the terminals, the emf causes a current to flow. This current, in effect, represents a "charge leakage" between the terminals that will upset the previous balance. But immediately, new charges are separated so as to maintain the emf. If the external current drain persists, then we clearly have a continuous process. The process cannot go on *ad infinitum*, as a battery can discharge only a given maximum charge. Usually a battery is rated in ampere-hours [A · h]. A rating of 1 A · h represents 3600 ampere-seconds [A · s] or 3600 C, and this means that the battery can deliver 1 [A] for 1 hour, or 2 [A] for a half hour, before the need for recharging.

During discharge, the terminal voltage v_t (Figure 3.14a) varies with the current drain in a linear fashion, hence

$$v_t = e - R_i i \qquad [V], \tag{3.34}$$

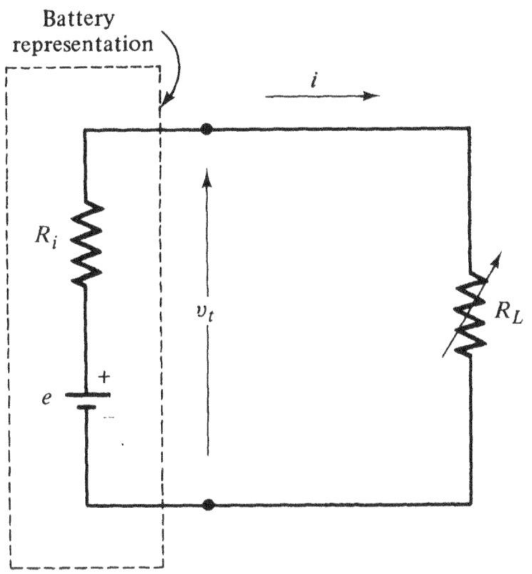

(a)

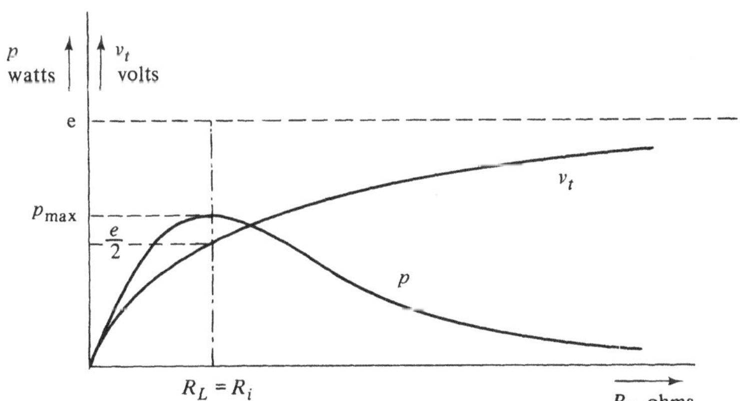

(b)

Figure 3.14

where

$$e = \text{emf in volts;}$$

$$R_i = \text{internal resistance in ohms.}$$

To charge the battery, positive charges are "pumped" into the positive terminal. The current is now reversed, and the terminal voltage is given by

$$v_t = e + R_i i \qquad [\text{V}]. \tag{3.35}$$

During the discharge process, the chemical composition of the cell changes such that the charge separation process is gradually "destroyed." By charging the battery, the proper chemical conditions are restored. The electrical energy supplied to the battery during the charge process is used to build up the *chemical potential energy* of the electrolyte. The potential energy decreases during the discharge.

Example 3.11

Let the transmission system in Figure 3.13 represent the dc supply system in a mobile medical unit. The system is characterized by the following data:

Generator: Consists of a storage battery having an emf $e = 200.0$ V and internal resistance $R_i = 0.031\ \Omega$.

Line: Consists of 100 m (both leads) copper cable of cross-sectional area $A = 20$ mm^2. From equation (3.24) we find its line resistance:

$$R = 1.75 \times 10^{-8} \frac{100}{20 \times 10^{-6}} = 0.0875\ [\Omega]. \tag{3.11.1}$$

Load: An assortment of equipment drawing a total current of $i = 95.0$ A from the battery.

Find the voltages v_g and v_l, the ohmic transmission losses and the total power drained from the battery.

Solution: We find the battery terminal voltage by using equation (3.34):

$$v_g = 200.0 - 0.031 \cdot 95.0 = 197.1 \qquad [\text{V}]. \tag{3.11.2}$$

The voltage drop along the cable is

$$\Delta v = iR = 95.0 \cdot 0.0875 = 8.3 \qquad [\text{V}]. \tag{3.11.3}$$

The load voltage is

$$v_l = 197.1 - 8.3 = 188.8 \qquad [\text{V}]. \tag{3.11.4}$$

Power supplied by the battery is

$$p_g = 197.1 \cdot 95.0 = 18.72 \qquad [\text{kW}]. \tag{3.11.5}$$

Power lost in the cable is

$$p_\Omega = 0.0875 \cdot 95^2 = 0.79 \qquad \text{[kW]}. \tag{3.11.6}$$

Power supplied to the load is

$$p_l = 18.72 - 0.79 = 17.93 \qquad \text{[kW]}. \tag{3.11.7}$$

Example 3.12

A battery (Figure 3.14a) has an emf $e = 100$ V and an internal resistance of $R_i = 1\ \Omega$. Determine the maximum power that can be supplied by the battery to a variable load resistance R_L.

Solution: The current drawn from the battery is

$$i = \frac{e}{R_i + R_L} \qquad \text{[A]}. \tag{3.12.1}$$

The terminal voltage is

$$v_t = e - R_i i = e - R_i \frac{e}{R_i + R_L} = \frac{R_L}{R_i + R_L} e \qquad \text{[V]}. \tag{3.12.2}$$

The power supplied by the battery is

$$p = v_t i = \frac{R_L}{(R_i + R_L)^2} e^2 \qquad \text{[W]}. \tag{3.12.3}$$

Figure 3.14b indicates that the power will approach zero for both $R_L = 0$ and $R_L = \infty$. Clearly, for some finite value of R_L, we must have maximum power supplied to the load. We can find this p_{max} by setting

$$\frac{dp}{dR_L} = 0. \tag{3.12.4}$$

A simple analysis reveals that maximum power will occur when $R_L = R_i$, and it is equal to

$$p_{max} = \frac{1}{4}\frac{e^2}{R_i} \qquad \text{[W]}. \tag{3.12.5}$$

Substituting numerical values, we get

$$p_{max} = \frac{1}{4}\frac{100^2}{1} = 2.5 \qquad \text{[kW]}. \tag{3.12.6}$$

Note that the terminal voltage drops to half its open-circuit value, and note also that the power dissipation in the internal resistance R_i is equal to the power dissipation in the load, that is, 2.5 kW. The battery would overheat in a short time, but this type of optimum battery discharge can be tolerated only for short periods, for example, when starting an automobile engine. In normal battery operation the load resistance, R_L, is typically much larger than the internal resistance of the battery, R_i. In Figure 3.14b this corresponds to points far to the right of p_{max}.

A fully charged automobile battery can deliver a current in the 1-A range for hours at a relatively constant terminal voltage. As 1 A · h is equal to 3600 C, a storage battery may store tens of thousands of coulombs. This charge must not be thought of as a free charge existing on the electrolytic plates within the battery, similar to a charged capacitor. Rather, the charges exist in neutralized form ready to be separated at a slow and controlled rate to match the external "leakage" from the terminals.

A capacitor, on the contrary, has much less charge storage capacity (Example 3.4)—usually much less than 1 C, and often only microcoulombs. As a capacitor is discharged, it is to be expected that its terminal voltage will diminish more rapidly than a battery.

Example 3.13

Consider the *RC* circuit shown in Figure 3.15. The capacitor is initially charged to a voltage v_0. Derive an expression for the voltage as a function of time during discharge!

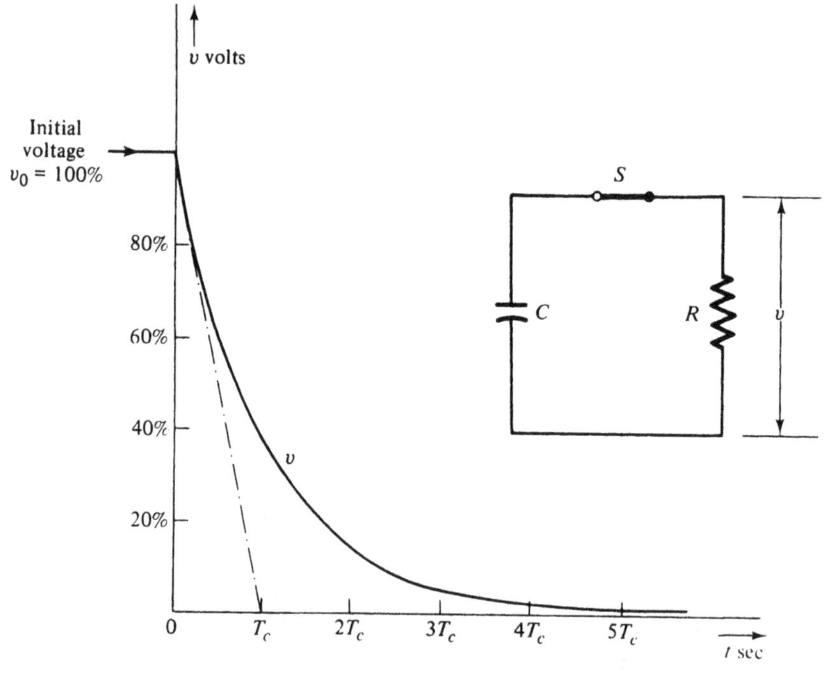

Figure 3.15

Solution: When the switch S is closed, charges will flow through the resistor R where energy will be dissipated in the form of heat. The power dissipated in the resistor, v^2/R, must be equal to the rate of change of the energy stored in the capacitor, therefore,

$$-\frac{d}{dt}\left(\frac{1}{2}Cv^2\right) = \frac{v^2}{R} \qquad [\mathrm{W}]. \tag{3.13.1}$$

(The negative sign indicates that the capacitor is losing energy.) Differentiating the expression, we obtain the differential equation

$$\frac{dv}{dt} + \frac{1}{RC}v = 0. \tag{3.13.2}$$

The reader can verify that this equation has the solution:

$$v = v_0 e^{-(t/T_c)} \qquad [\mathrm{V}], \tag{3.13.3}$$

where $T_c \equiv RC$ is referred to as the *time constant* of the circuit. The voltage is plotted against time in Figure 3.15. We note that the capacitor is essentially completely discharged after about $5T_c$ seconds. (If we arbitrarily define the capacitor to be discharged when the voltage has dropped to 1% of its original value, the discharge time will be 4.605 T_c seconds.)

Substituting the following numerical values:

$$C = 1.0\ \mu\mathrm{F},$$
$$R = 1000\ \Omega,$$
$$v_0 = 100\ \mathrm{V},$$

we get

$$T_c = 10^3 \cdot 10^{-6} = 10^{-3} \qquad [\mathrm{s}] \quad \text{or} \quad 1 \qquad [\mathrm{ms}]. \tag{3.13.4}$$

The capacitor would discharge in about 5 ms.

3.17 The Magnetic Field

One of the more interesting phenomena of electrical science, and certainly the most important one from a technological point of view, is the ability of moving charges to produce a *magnetic field*. Similar to a gravitational or electrostatic field, the magnetic field can be detected, mapped, and measured. This is the basis of the operation of the most important electrical energy conversion apparatus. In this section we present the character of the magnetic field, and in the sections that follow we describe its effects and applications.

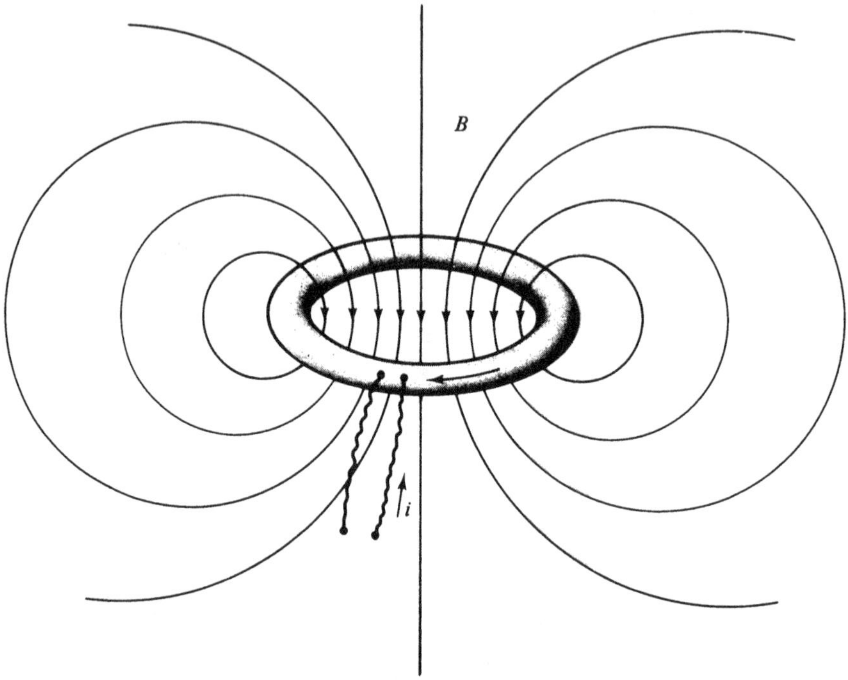

Figure 3.16

Consider the toroidal coil shown in Figure 3.16. When a current of magnitude i amperes circulates in the toroid, a magnetic field is created having the geometry shown. We note some important characteristics of this field.

1. The field, B, has the characteristics of a vector.
2. The field lines are closed (compare this to the open-ended character of electrostatic fields). They always enclose the current from which they originate.
3. The field exists both outside and inside the conductor.
4. The magnitude of B at each point in space is directly proportional to its causative current i [except in ferromagnetic materials (Section 3.26) where the relationship between B and i is nonlinear].
5. The magnitude of B is high at points close to the conductor and decreases as the distance from the conductor increases (this is similar to contour lines on a geographic map; the magnitude of the B field is indicated by how close the field lines are drawn).
6. The direction of the B field can be obtained by the "right-hand rule": Hold the coil with your right hand, with your fingers pointing in the direction of the current. The B field vector will be in the direction of your thumb.

The SI unit for B is tesla [T], or webers per square meter [Wb/m^2]; the latter unit reveals the nature of B as a flux density (*magnetic flux density* is the proper name for B).

Once the conductor geometry is known and the magnitude of the current given, the magnitude and direction of the magnetic field can be computed. However, the computation for arbitrary conductor geometries is not simple. We consider here only the simplest possible cases. For example, the B field around a long straight wire (Figure 3.17) consists of concentric circles, and the magnitude of B decreases inversely with the radial distance r from the center of the wire in accordance with

$$B = \mu_0 \frac{i}{2\pi r} \qquad [\text{T}], \tag{3.36}$$

where $\mu_o = 4\pi \cdot 10^{-7}$ is a constant, called the *permeability* of vacuum.

Another conductor geometry, which is of great practical interest, is the *coil* (Figure 3.18). The usefulness of this device is its ability to amplify the B field. To explain this, consider the straight conductor shown in Figure 3.17. If we double the current to $2i$ the B field will double at every point in space. However, the same result would be obtained if, instead of one conductor carrying $2i$, we employ two adjacent conductors, each carrying the current i. However, it is not necessary to provide two separate currents, as the same current can be made to flow past a

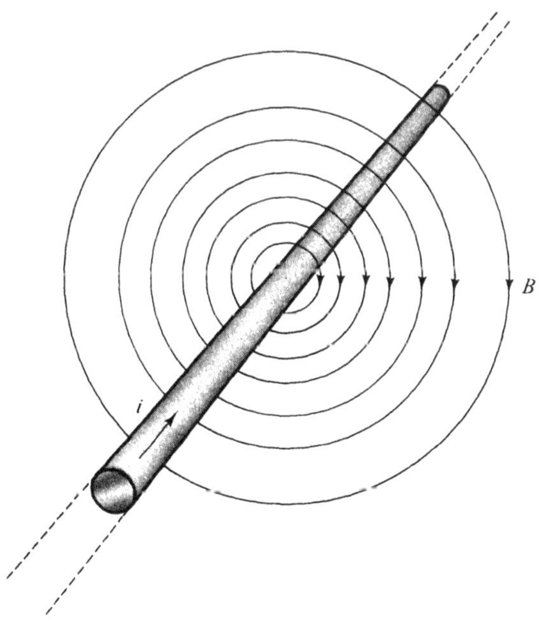

Figure 3.17

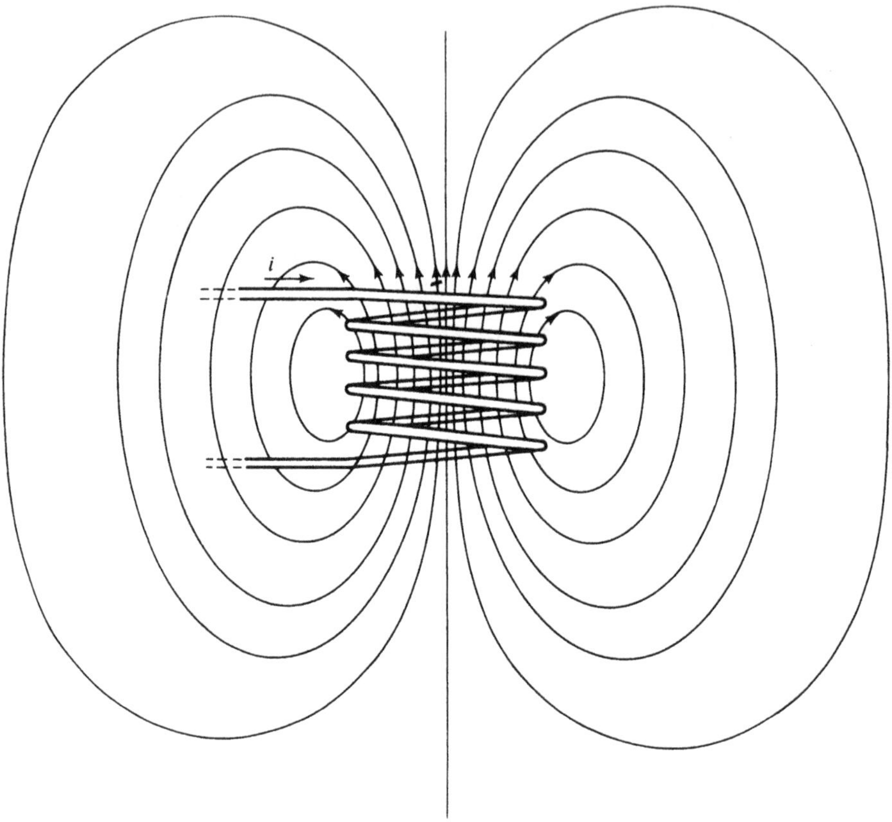

Figure 3.18

given point twice instead of once by using a coil with two turns. Using an n-turn coil will produce an n-multiplication of the field.

3.18 Magnetic Flux

Consider a magnetic field penetrating a surface S, as shown in Figure 3.19. The field that penetrates the elemental surface dS (considered to be constant over the elemental surface) has the component $B \cos\beta$ measured perpendicularly to the surface of the element. The differential magnetic flux, $d\Phi$ penetrating the elemental surface is

$$d\Phi \equiv B \cos\beta \, dS \qquad [\text{Wb}]. \tag{3.37}$$

The total magnetic flux penetrating the finite surface S is then obtained by integrating over the total surface:

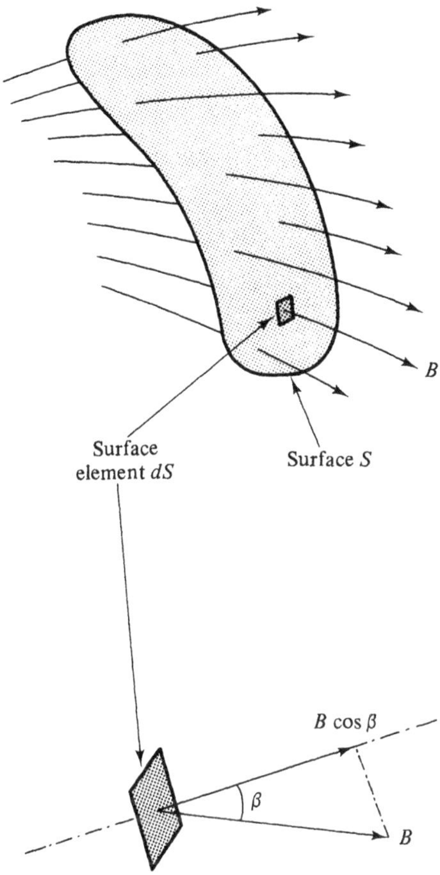

Figure 3.19

$$\Phi = \int\int d\Phi = \int\int B \cos\beta \, dS \qquad \text{[Wb]}. \tag{3.38}$$

(We could have defined gravitational and electric fluxes in terms of the vectors g and E, respectively. However, they were not needed in our elementary discussion of energy. For a discussion of the magnetic flux, the use of vectors is inevitable.)

Consider the situation shown in Figure 3.20 in which a long, straight conductor is carrying a current i. We can derive an expression for the total magnetic flux passing through the rectangular surface located in the same plane as the conductor. We note that the flux density is perpendicular to the surface, that is, $\cos \beta = 1$ over the total surface, and the flux density given by equation (3.36) is constant along the differential surface strip of width dr. The differential flux through this strip is

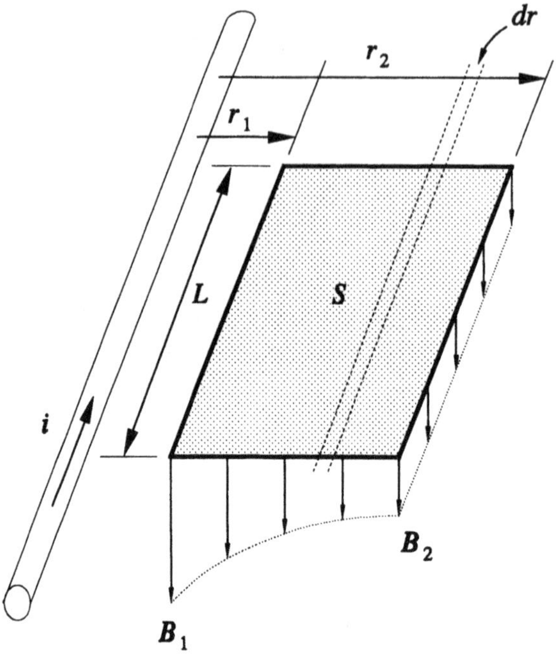

Figure 3.20

$$d\Phi = BL\,dr = \mu_0 \frac{i}{2\pi r} L\,dr \qquad \text{[Wb]}. \tag{3.39}$$

By integration from r_1 to r_2 we obtain the total flux:

$$\Phi = \int_{r_1}^{r_2} \mu_0 \frac{iL}{2\pi r}\,dr = \mu_0 \frac{iL}{2\pi} \ln\left(\frac{r_2}{r_1}\right) \qquad \text{[Wb]}. \tag{3.40}$$

3.19 Electromagnetic Induction: Faraday's and Lenz's Laws

Electromagnetic induction (*induction* for short) is an extremely important phenomenon to electrical engineers. It can be described as follows:

> When the magnetic flux Φ passing through a surface S changes its magnitude, an electric field is induced along the contour of S.

Faraday's law gives the quantitative relationship between the flux and the electric field. It states:

When the magnetic flux Φ passing through a surface S changes its magnitude, if a thin N-turn coil is placed along the contour of S, a voltage v can be measured across its terminals of magnitude:

$$v = N\frac{d\Phi}{dt} \qquad [\mathrm{V}]. \tag{3.41}$$

(This equation is often written in the form $v = (d/dt)(N\Phi) = (d/dt)(\Phi_c)$, where $\Phi_c \equiv N\Phi$ is referred to as the coupled or linked flux, or the flux that effectively induces the voltage.)

The polarity of the voltage follows Lenz's law, which states:

> If the induced voltage is permitted to produce a current (by closing the loop), this current will have such a direction as to oppose the change of the flux.

Let us examine the nature of magnetic induction by studying the coil shown in Figure 3.21. It is placed so that it coincides with the contour of the surface S shown in Figure 3.20. The flux Φ, which was computed earlier [equation (3.40)] can be changed in at least three different ways:

1. By changing the current i in the straight wire
2. By moving the coil radially without changing its plane (as the coil moves from a high-B area to a low-B area, the total flux penetrating the coil must decrease)
3. By rotating the coil along its axis

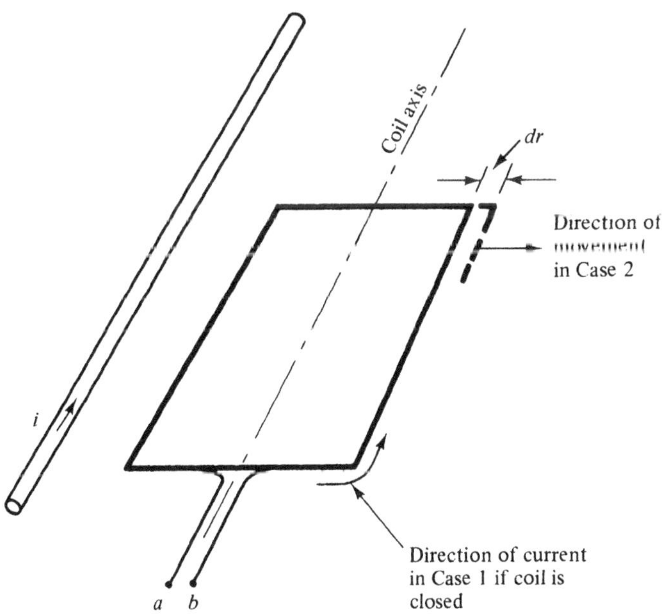

Figure 3.21

CASE 1. The flux change is accomplished by keeping the coil position fixed but increasing the current i at the rate di/dt.

From equation (3.40) the flux change is given by

$$\frac{d\Phi}{dt} = \mu_0 \frac{L}{2\pi} \ln\left(\frac{r_2}{r_1}\right) \frac{di}{dt} \qquad \text{[Wb/s]}. \tag{3.42}$$

Using Faraday's law, the induced voltage is

$$v = N\frac{d\Phi}{dt} = N\mu_0 \frac{L}{2\pi} \ln\left(\frac{r_2}{r_1}\right) \frac{di}{dt} \qquad \text{[Wb/s]}. \tag{3.43}$$

The voltage would be constant as long as the current continues to change at the same rate, that is, di/dt = constant.

Example 3.14

Let us explore Case 1 further by using the following numerical data: $N = 100$ [T]; $i = 100$ [A]; $r_1 = 0.1$ [m]; $r_2 = 1$ [m]; $L = 1$ [m]; and $di/dt = 10^5$ [A/s].

Solution: From equation (3.42),

$$\frac{d\Phi}{dt} = 4\pi \times 10^{-7} \times \frac{1}{2\pi} \ln\left(\frac{1}{0.1}\right) \times 10^5 = 0.04605 \qquad \text{[Wb/s]}. \tag{3.14.1}$$

From equation (3.43) the voltage induced in the coil is

$$v = N\frac{d\Phi}{dt} = 100 \times 0.04605 = 4.605 \qquad \text{[V]}. \tag{3.14.2}$$

What would be the polarity of this voltage?

According to Lenz's law, if we close the loop, by placing a short circuit across the terminals a and b, the voltage would give rise to a coil current, which in turn produces a flux in a direction such as to oppose the change of the flux. With the given direction for the conductor current i, the coil flux is directed downward. This means that the induced coil current would try to build up a coil flux directed upward. According to the "right-hand rule" (see Section 3.17) the coil current must have the direction indicated in Figure 3.21, that is, the current flows out of terminal a and into terminal b. Terminal a is then positive relative to terminal b—as one would expect from a voltage source.

CASE 2. In this case we accomplish the flux change by moving the coil radially (i.e., in the increasing r-direction).

If B_1 and B_2 are the flux densities at the radial distances r_1 and r_2 respectively (see Figure 3.20). The change dr means that the coil flux decreases by $B_1 L\, dr$ at its left side and increases by $B_2 L\,dr$ at its right side. The total flux change experienced by the coil will be

$$d\Phi = B_2 L\,dr - B_1 L\,dr \qquad \text{[Wb]}. \tag{3.44}$$

The derivative of the flux with respect to time is

$$\frac{d\Phi}{dt} = L\left(B_2\frac{dr}{dt} - B_1\frac{dr}{dt}\right) = L\frac{dr}{dt}(B_2 - B_1) \qquad \text{[Wb/s]}. \tag{3.45}$$

The induced voltage, according to Faraday's law is given by

$$v = NL\frac{dr}{dt}(B_2 - B_1) \qquad \text{[V]}. \tag{3.46}$$

Example 3.15

Use the numerical values given in Example 3.14, to calculate the voltage induced in the coil, assuming that the coil is moving from its original position with a velocity of 10 [m/s], that is,

$$\frac{dr}{dt} = 10 \qquad \text{[m/s]} \tag{3.15.1}$$

(dr is shown in Figure 3.21).

Solution: From equation (3.36) we calculate the flux densities:

$$B_1 = 4\pi \times 10^{-7} \times \frac{100}{2\pi \times 0.1} = 2 \times 10^{-4} \qquad \text{[T]} \tag{3.15.2}$$

and

$$B_2 = 4\pi \times 10^{-7} \times \frac{100}{2\pi \times 1.0} = 0.2 \times 10^{-4} \qquad \text{[T]}. \tag{3.15.3}$$

The induced voltage[7] will be (neglecting the polarity)

$$v = N\frac{d\Phi}{dt} = 100 \times 1 \times 10 \times (2 - 0.2) \times 10^{-4} = 0.18 \qquad \text{[V]}. \tag{3.15.4}$$

What is the polarity of the voltage in this case?

Evidently, the movement of the coil away from the current-carrying conductor results in a decrease of the total flux penetrating it. The coil flux must therefore be

[7] This voltage will *not* be constant as in Case 1. The computed value (0.18 V) applies *only* to the instant when the coil occupies the assumed position. As the coil travels on, the induced voltage approaches zero.

in the direction in which it opposes the decreasing flux due to the current i, that is, the coil flux must be directed downward. The polarity of the induced voltage is therefore the opposite of the one in Case 1, that is, terminal b will be positive relative to a.

Let us make one comment on equation (3.45). The two terms in the expression for flux change can be interpreted as the voltages induced in each side of the coil of length L. In a conductor moving perpendicularly to a flux B with velocity s a voltage of magnitude,

$$v = sB \qquad [\text{V/m}] \tag{3.47}$$

is induced.

Note that

1. The value of B at each conductor of length L will be different.
2. The two remaining coil sides do not "cut" any flux and will therefore have no voltage induced in them.

CASE 3. The rotation of a coil in the nonuniform magnetic field, resulting from the current in the long straight conductor shown in Figure 3.21, is more complicated and would not serve a useful function at this point. Instead we consider the rotation of a coil in a uniform magnetic field.

3.19.1 Voltage Induced in a Coil Rotating in a Uniform Magnetic Field

The coil shown in Figure 3.22a has a cross-sectional area equal to A [m^2]. It has N turns and rotates at n [rpm] in a uniform magnetic field of density B [T]. In order to derive an expression for the induced voltage, we consider an instant when the coil is inclined at an angle α to the plane orthogonal to the direction of the flux.

The total magnetic flux passing through the coil is

$$\Phi = BA\cos\alpha \qquad [\text{Wb}]. \tag{3.48}$$

The coil rotates through $n/60$ full revolutions per second, that is, it rotates at the angular velocity

$$\omega = \frac{2\pi n}{60} \qquad [\text{rad/s}]. \tag{3.49}$$

As the angular velocity is constant, the angle α will grow in direct proportion to time t, and hence

$$\alpha = \omega t \qquad [\text{rad}]. \tag{3.50}$$

For the flux as a function of time, we have

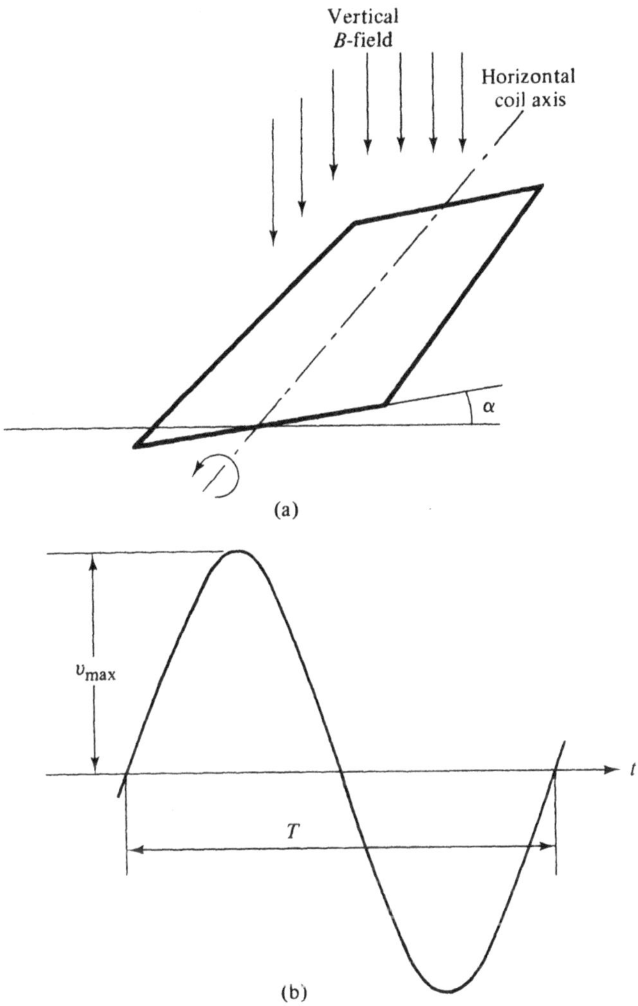

Figure 3.22

$$\Phi = BA \cos \omega t \qquad [\text{Wb}]. \tag{3.51}$$

By applying Faraday's law we obtain

$$v = N\frac{d\Phi}{dt} = N\frac{d}{dt}(BA \cos \omega t) = -N\omega BA \sin \omega t \qquad [\text{V}]. \tag{3.52}$$

We conclude that the induced voltage in a coil spinning at a constant angular velocity in a uniform magnetic field is sinusoidal, that is, it is *periodic*. A voltage of the alternating type, plotted in Figure 3.22b, is called a sinusoidal voltage. According to equation (3.52), the voltage wave has a peak value

$$v_{\max} = N\omega BA \qquad [\mathrm{V}]. \tag{3.53}$$

The voltage wave completes $n/60$ full cycles per second [cps], that is, its *frequency, f,* is

$$f = \frac{n}{60} \qquad [\mathrm{cps}] \quad \text{or} \quad \text{hertz} \qquad [\mathrm{Hz}]. \tag{3.54}$$

The period T (see Figure 3.22b) is

$$T = \frac{1}{f} \qquad [\mathrm{s}]. \tag{3.55}$$

Example 3.16

A coil having a sectional area of 0.4 [m^2] with $N = 10$ turns, rotates about a horizontal axis with a constant speed $n = 3600$ [rpm] in a vertical magnetic field that is uniform and of density $B = 1$ [T]. The situation is shown in Figure 3.22a. Find the magnitude of the voltage induced in the coil, its frequency, period, and angular frequency.

Solution: From equation (3.53), the induced voltage is

$$v_{\max} = 10 \times 377 \times 1 \times 0.4 = 1508 \qquad [\mathrm{V}]. \tag{3.16.1}$$

From equation (3.54), the frequency is

$$f = \frac{3600}{60} = 60 \qquad [\mathrm{Hz}], \tag{3.16.2}$$

and hence the period is

$$T = \frac{1}{60} = 0.0167 \qquad [\mathrm{s}]. \tag{3.16.3}$$

From equation (3.49), the angular frequency is

$$\omega = 60 \times 2\pi = 377 \qquad [\mathrm{rad/s}]. \tag{3.16.4}$$

The induction processes described in Case 1 and Case 2 result in a dc emf, which, for obvious reasons, cannot be maintained easily. In addition, the magnitudes of the induced voltages are small. The induction process in which a coil is rotated in a uniform magnetic field results in an ac emf, which can be maintained indefinitely, or as long as the coil rotates. In addition, the magnitude of the voltage is considerably higher. In fact, this is the preferred method for the production of most of the world's electricity. We make use of it repeatedly in the rest of this book.

3.20 The Electromagnetic Force Law

Electromagnetic induction explains how a "passive wire" can be transformed into a generator of emf. The *electromagnetic force law* describes the interaction between a magnetic field and a current-carrying conductor, which results in a mechanical force. This is the basis of operation of all electric motors.

Consider the three-dimensional coordinate system shown in Figure 3.23. A uniform magnetic field with a flux density, B [T] is oriented in the z direction. A thin conductor carrying a current, i [A] is held in the x–z plane, forming an angle, α, with the direction of the flux. The force law states that the conductor will be subjected to an electromechanical force, f directed in the y direction, of magnitude:

$$f = BiL \sin \alpha \qquad [\mathrm{N}], \tag{3.56}$$

where L is the length of the conductor. The force evidently reaches a maximum,

$$f_{\max} = BiL \qquad [\mathrm{N}], \tag{3.57}$$

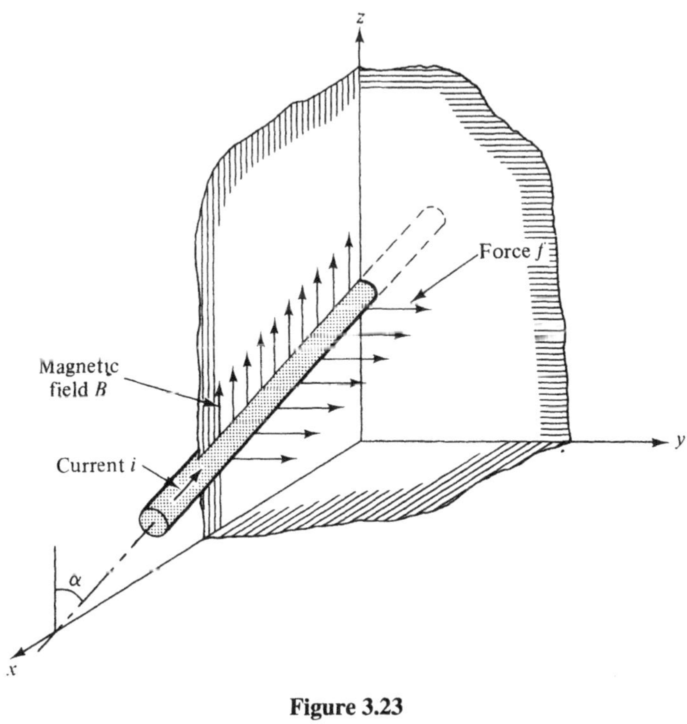

Figure 3.23

when $\alpha = 90°$. In that case, f, i, and B are all orthogonal, and parallel to the y, x, and z axes, respectively. Those readers familiar with the *vector cross-product* realize that we can express the force as

$$\mathbf{f} = \mathbf{i} \times \mathbf{B} \qquad [\text{N/m}]. \tag{3.58}$$

Note that the force will always be perpendicular to both the magnetic field vector and the conductor carrying the current.

An easy method for determining the relative directions of the force, current, and flux is the *left-hand rule*.

> The thumb and first and second fingers of the left hand are positioned to be mutually at right angles with each other.

If

> the **F**irst finger is pointed in the direction of the **F**ield,

and

> the se**C**ond finger is pointed in the direction of the **C**urrent,

then

> the thu**M**b points in the direction of **M**otion, that is, in the direction of the force.

We also make the following important observation: As we place the current-carrying conductor in the field B, the field will change due to the effect of the current i. When we use equation (3.56) the field B must be understood to mean the magnetic field that existed before the presence of the current i.

Example 3.17

Calculate the force on a conductor carrying a current of 100 [A] in a magnetic field of flux density 1.5 [T] orthogonal to the flux.

Solution: According to equation (3.58), the force is given by

$$f = 1.5 \cdot 100 = 150 \qquad [\text{N/m}]. \tag{3.17.1}$$

We can obtain considerable force (and torque) by using this system. Compare this to the weak force in Example 3.3, obtained from the interaction between static charges.

3.20.1 Torque on a Coil in a Magnetic Field

The thin N-turn coil shown in Figure 3.24 is carrying current i, and it is free to rotate about its horizontal axis in a vertical B field. Forces will act on the four coil sides as shown. The forces acting on the coil sides of length b will cancel each other. The forces acting on the coil sides of length a will form a torque ("couple")

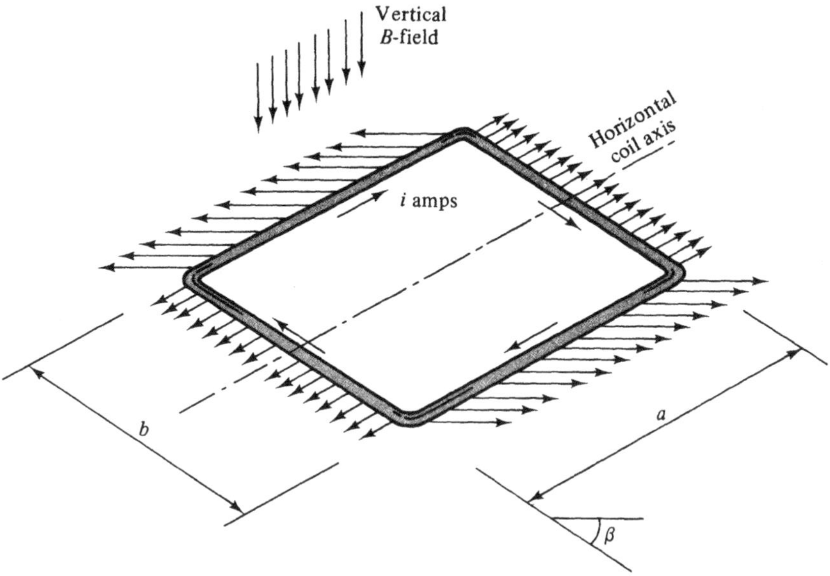

Figure 3.24

trying to rotate the coil so that the magnetic field generated by the coil lines up with the B field (i.e., the angle β will decrease). The forces on each side of length a have a horizontal direction, and each of them is equal to $BNia$ newtons. The torque on the coil will be

$$T = BNiab \sin \beta \qquad [\text{N} \cdot \text{m}]. \tag{3.59}$$

As the coil is free to rotate, it will assume a horizontal position. In the equilibrium position all four coil sides will be subject to equal forces (expressed in newtons per meter)—all acting in the plane of the coil and all trying to expand the coil—and the coil flux will be lined up with the B field.

3.20.2 Force Between Two Long Parallel Conductors

The magnetic field caused by each conductor is a family of concentric circles (Figure 3.17), with a magnitude given by equation (3.36). From the application of the force law, we conclude that, if the currents have the same direction, the conductors will attract each other; if in opposite directions, they will repel each other. Here we are concerned with the latter case. By using equations (3.36)[8] and (3.58) we find the magnitude of the repulsive force that acts on 1 [m] of the conductor to be

[8] Note that we compute the value of B at the conductor 2 as caused by the current in conductor 1 *only*.

$$f = Bi = \left(\mu_0 \frac{i}{2\pi a}\right) i = \mu_0 \frac{i^2}{2\pi a} \qquad \text{[N/m]}, \tag{3.60}$$

where a is the distance between the conductors, and i is the current flowing.

Example 3.18

Bus bars in electric power stations carry normal currents of 10 kA. During short circuits these currents can reach 100 kA before the fault can be isolated. Calculate the force between two bus bars carrying 100 kA each and placed 1 [m] apart.

Solution: According to equation (3.60), the force will be

$$f = 4\pi \times 10^{-7} \frac{(10^5)^2}{2\pi \times 1} = 2000 \qquad \text{[N/m]} \tag{3.18.1}$$

Forces of such a magnitude may cause considerable damage unless the supporting insulators have sufficient strength.

Example 3.19

Determine qualitatively the nature of the forces acting on two parallel toroidal coils carrying currents in the same direction.

Solution: The magnetic field B_1 around coil 1 is shown in Figure 3.25 (cf. Figure 3.16). We now place coil 2 (shown by the dashed lines) in the field B_1. When the current i_2 flows, the force f_2 acting on coil 2 must be perpendicular to both the field B_1 and the current i_2. The direction of f_2 will have to be in the direction indicated in Figure 3.25.

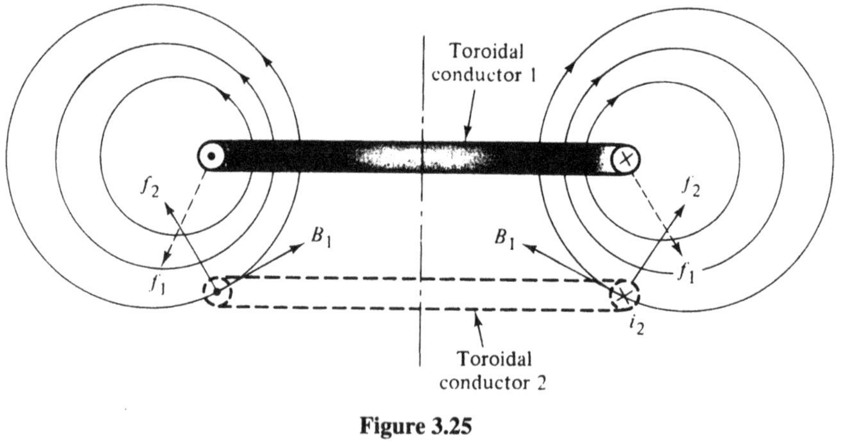

Figure 3.25

Force f_1 acting on coil 1 must be symmetrical as indicated by the dashed lines. The forces tend to attract the coils axially and to expand them radially.

3.21 The Concept of Mutual Inductance

The coil and conductor in Figure 3.21 represent two *coupled magnetic circuits*. A change in the current in circuit 1 (the long conductor) induces a voltage (or emf) in circuit 2 (the coil). Magnetic coupling forms the basis for some of the most important devices in electric power engineering, for example, transformers and induction motors.

It is convenient to account for this coupling by means of a parameter referred to as *mutual inductance*. Using equations (3.41) and (3.42) we obtain the following expression for the coil voltage:

$$v_2 = \frac{d}{dt}\left[\mu_0 \frac{Li_1}{2\pi} N \ln\left(\frac{r_2}{r_1}\right)\right] \quad \text{[V]}. \tag{3.61}$$

We have introduced the subscripts 1 and 2, for the current and voltage, respectively, and the equation then gives us the magnitude of the voltage induced in circuit 2 (the coil), v_2, as a result of a current change in circuit 1 (the straight conductor), i_1.

We now define the mutual inductance between circuits 1 and 2 as

$$M_{12} \equiv \mu_0 \frac{L}{2\pi} N \ln\left(\frac{r_2}{r_1}\right) \quad \text{[H] (henry)} \tag{3.62}$$

and can then write equation (3.61) in the simpler form[9]:

$$v_2 = M_{12} \frac{di_1}{dt} \quad \text{[V]}. \tag{3.63}$$

The units for M_{12} is volt-seconds per ampere and is given the special name *henry* [H]. M_{12} tells us how many volts will be induced in circuit 2 for a current change in circuit 1 of 1 A/s.

In many situations (e.g., in transformers), the opposite situation is of equal importance: What voltage would be induced in circuit 1 as a result of current change in circuit 2? The mutual inductance M_{21} is defined by the equation

[9] Note that we can write the equation in this form *only* if r_2 and r_1 are *constants*, i.e., *if the two circuits are fixed in relation to each other*.

If the coils are moving relative to each other (which is the case in rotating machines), then the expression reads:

$$v_2 = \frac{d}{dt}(M_{12}i_1).$$

$$v_1 = M_{21} \frac{di_2}{dt} \qquad [\text{V}]. \tag{3.64}$$

We will not give the proof here, but it can be shown that the two mutual inductance parameters are equal in magnitude:

$$M_{12} = M_{21} \qquad [\text{H}]. \tag{3.65}$$

As two coupled coils are characterized by one mutual inductance, we prefer the simpler symbol, M.

3.22 The Concept of Self-Inductance

Faraday's law tells us that a voltage is induced in a loop or coil if the magnetic flux linked to it undergoes a change. In the previous examples we calculated these voltages, assuming the flux was due to currents external to the loop or coil in question. We introduced the mutual inductance M to account for the induced voltage.

We must understand, however, that some of the flux linked to a coil can also be attributed to its own current. If the coil has no coupling to other circuits, all of its linked flux must be due to its own current. The voltage induced in a coil is—in part (or possibly entirely)—due to changes in its own current. We call this phenomenon *self-induction*.

For example, consider a toroidal coil of the type shown in Figure 3.16, which depicts the magnetic field due to the current in the coil. The flux is proportional to the current. Specifically, if i is the coil current, then we can write for the magnetic flux, Φ_1, due to one coil turn

$$\Phi_1 = ki \qquad [\text{Wb}], \tag{3.66}$$

where k is a constant. If the coil is relatively thin, then each of the N turns contributes an equal share to the total coil flux, Φ, and we have

$$\Phi = N\Phi_1 = kNi \qquad [\text{Wb}]. \tag{3.67}$$

According to Faraday's law, the induced voltage

$$v = N\frac{d\Phi}{dt} = N\frac{d}{dt}(kNi) = kN^2\frac{di}{dt} \qquad [\text{V}]. \tag{3.68}$$

We now define the self-inductance L of the coil as

$$L \equiv kN^2 \qquad [\text{H}]. \tag{3.69}$$

Accordingly, the self-inductance $L = \Phi_c/i$. In other words, it is a measure of the *coupled flux*, Φ_c, caused by a unit current. (Note the similarity to the definition of capacitance: $C \equiv Q/v$, that is, the charge per unit voltage.)

We can now write for the self-induced voltage:

$$v = L\frac{di}{dt} \qquad [\text{V}]. \tag{3.70}$$

The constant k in equation (3.69) depends on the geometry of the coil. In general, the determination of the value of k, even for a coil with a very simple geometry, such as a toroidal coil, can be quite tedious, and we shall not dwell on this matter. However, it is important to note that the self-inductance increases as the square of the number of turns.

3.23 Electromagnetic Energy Storage

When a mass m (Figure 3.26a) is subjected to a force f, its velocity s can be computed from the following equation, which states that the force f must be equal to the sum of the force of inertia plus the force due to friction f_{fr}:

$$f = m\frac{ds}{dt} + f_{\text{fr}} \qquad [\text{N}]. \tag{3.71}$$

When multiplied by $s\,dt$ and integrated, we obtain

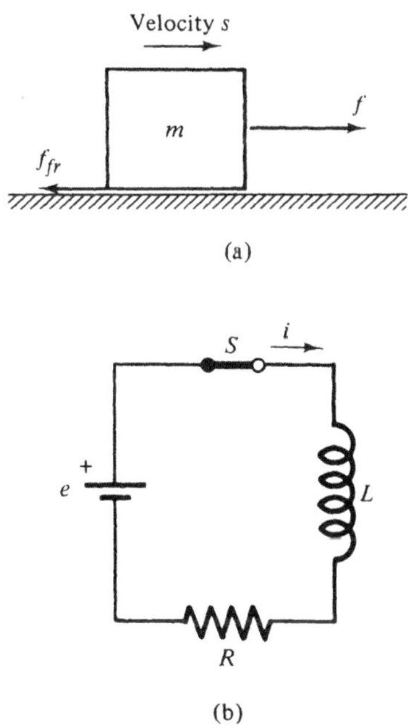

Figure 3.26

$$\int fs\,dt = \int ms\,ds + \int f_{fr}s\,dt \qquad [J]. \tag{3.72}$$

The kinetic energy of the mass is

$$w_{kin} = m\int s\,ds = \frac{1}{2}ms^2 \qquad [J]. \tag{3.73}$$

Equation (3.72) states that the energy supplied by the force f is equal to the sum of the kinetic energy imparted to the mass plus the energy dissipated in heat due to friction ($\int f_{fr}s\,dt$).

When a coil of inductance L and resistance R are connected across a voltage source whose emf is e (Figure 3.26b), the resulting current can be obtained from the voltage equilibrium equation [Kirchoff's voltage law (KVL)] as

$$e = L\frac{di}{dt} + Ri \qquad [V]. \tag{3.74}$$

Multiplication by $i\,dt$ and integration gives

$$\int ei\,dt = \int Li\,di + \int Ri^2\,dt \qquad [J]. \tag{3.75}$$

The first term represents the energy supplied by the voltage source. The third term is the energy dissipated in the resistor. By analogy to the mechanical system above, the second term must represent the *energy*, w_{mag}, stored in the magnetic field:

$$w_{mag} = \int Li\,di = \frac{1}{2}Li^2 \qquad [J]. \tag{3.76}$$

The attention of the reader is drawn to the similarity of equations (3.73) and (3.76) for the kinetic and magnetic energy, respectively. Sometimes the inductance is referred to as *magnetic inertia.*

Example 3.20

The field coil of a synchronous generator (Chapter 4) has a self-inductance of $L = 12$ [H]. How much energy is stored in the coil if the field current, $i = 27$ [A]?

Solution: Equation (3.76) gives

$$w_{mag} = \tfrac{1}{2}\cdot 12\cdot 27^2 = 4374 \qquad [J]. \tag{3.20.1}$$

The possibility of storing large amounts of energy in magnetic coils is an interesting one. The main obstacle is the ohmic loss in the winding.

3.24 Magnetic Energy Storage in Mutually Coupled Circuits

Mutually coupled circuits are of great importance in electric power technology. We shall investigate briefly the nature of energy storage in magnetic circuits.

Consider the two coils shown in Figure 3.27a. The fluxes Φ_1 and Φ_2, resulting from the coil currents i_1 and i_2, respectively, are clearly additive, and therefore the induced voltages caused by the change of these fluxes must also be additive. If each coil is fed from sources with emf's e_1 and e_2, respectively, we obtain the following equations by applying KVL:

$$e_1 = R_1 i_1 + L_1 \frac{di_1}{dt} + M \frac{di_2}{dt} \quad \text{[V]};$$
$$e_2 = R_2 i_2 + L_2 \frac{di_2}{dt} + M \frac{di_1}{dt} \quad \text{[V]}. \tag{3.77}$$

Multiplying the equations by i_1 and i_2, respectively, we obtain

$$e_1 i_1 = R_1 i_1^2 + L_1 i_1 \frac{di_1}{dt} + M i_1 \frac{di_2}{dt} \quad \text{[W]};$$
$$e_2 i_2 = R_2 i_2^2 + L_2 i_2 \frac{di_2}{dt} + M i_2 \frac{di_1}{dt} \quad \text{[W]}. \tag{3.78}$$

The expressions on the left of the equality signs are identified as the powers delivered by the two sources, respectively. The quadratic terms on the right of the equality signs represent the ohmic losses in the windings. The sum of the remaining four terms must therefore represent the rate of change of the energy stored in the magnetic fields in the two-coil system, that is,

$$\frac{d}{dt}(w_{\text{mag}}) = L_1 i_1 \frac{di_1}{dt} + L_2 i_2 \frac{di_2}{dt} + M\left(i_1 \frac{di_2}{dt} + i_2 \frac{di_1}{dt}\right) \quad \text{[W]}. \tag{3.79}$$

After integration, we obtain

$$w_{\text{mag}} = \frac{1}{2} L_1 i_1^2 + \frac{1}{2} L_2 i_2^2 + M i_1 i_2 \quad \text{[J]}. \tag{3.80}$$

Consider the coil system shown in Figure 3.27b. The fluxes Φ_1 and Φ_2 caused by currents i_1 and i_2 are now in opposite directions. The corresponding mutually induced voltages must therefore carry the negative signs. The result would be the same if all equations were to remain unchanged and the mutual inductance M were assigned a negative sign.

Example 3.21

Two coupled coils have the following inductances:

$$L_1 = 10 \quad [\mathrm{H}];$$
$$L_2 = 2 \quad [\mathrm{H}];$$
$$M = \pm 3 \quad [\mathrm{H}] \text{ (sign depending on polarity).}$$

The coil currents are

$$i_1 = 20 \quad [\mathrm{A}];$$
$$i_2 = 10 \quad [\mathrm{A}].$$

Find the energy stored in the magnetic field of the system.

Solution: With the polarity as shown in Figure 3.27a,

$$w_{\text{mag}} = \tfrac{1}{2}\cdot 10\cdot 20^2 + \tfrac{1}{2}\cdot 2\cdot 10^2 + 3\cdot 20\cdot 10 = 2700 \quad [\mathrm{J}]. \quad (3.21.1)$$

With the polarity as shown in Figure 3.27b,

$$w_{\text{mag}} = \tfrac{1}{2}\cdot 10\cdot 20^2 + \tfrac{1}{2}\cdot 2\cdot 10^2 - 3\cdot 20\cdot 10 = 1500 \quad [\mathrm{J}]. \quad (3.21.2)$$

3.25 The Magnetic Moment

The expression for the torque on the rectangular coil shown in Figure 3.24 as given by equation (3.59) can be written as

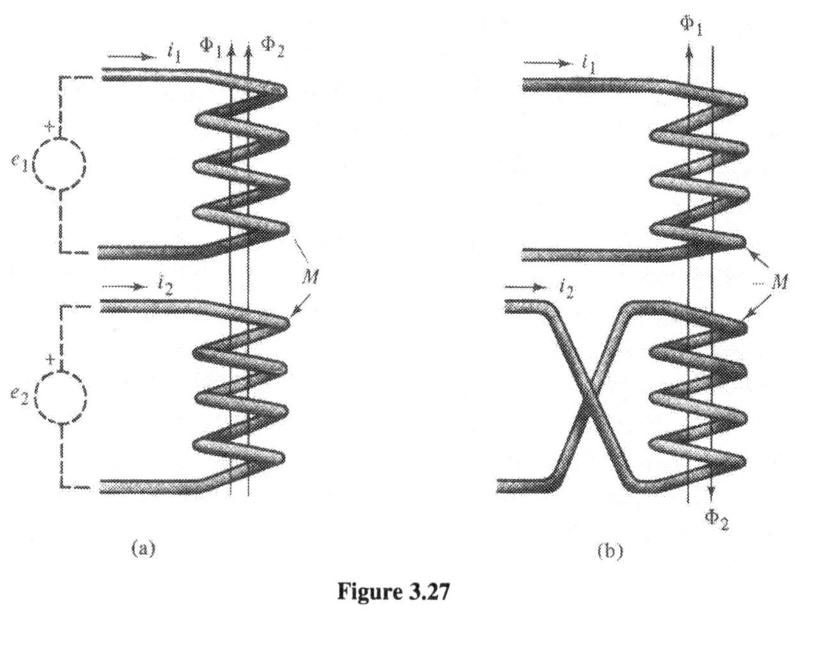

Figure 3.27

$$T = B(Ni)A \sin \beta \qquad [\mathrm{N \cdot m}], \tag{3.81}$$

where A is the area of the coil and Ni is the total current circulating around the contour of A. We introduce the *magnetic moment* defined as

$$m \equiv NiA \qquad [\mathrm{At \cdot m^2}], \tag{3.82}$$

and we can then write equation (3.81) simply as

$$T = Bm \sin \beta \qquad [\mathrm{N \cdot m}]. \tag{3.83}$$

We can think of the magnetic moment as a vector (see Figure 3.28), which is orthogonal to the plane of the coil. When the current flows in a clockwise direction, the vector points in the direction in which a right-hand screw would move if turned in the same direction as the current (i.e., clockwise).

The result obtained in equation (3.59) can now be interpreted as follows:

> Under the influence of the magnetic field B, the coil will assume an equilibrium position where its own magnetic moment vector is lined up with the magnetic field vector B.

We demonstrated this fact for a rectangular coil constrained to rotate around a given axis. However, it is easy to show (and we suggest to the reader to do so) that the rule applies to coils of any shape that are completely free to move in any direction. For example, the circular coil shown in Figure 3.28 will have to turn almost 180° before it comes into the position of equilibrium.

The concept of the magnetic moment vector will prove very useful in interpreting several ferromagnetic phenomena.

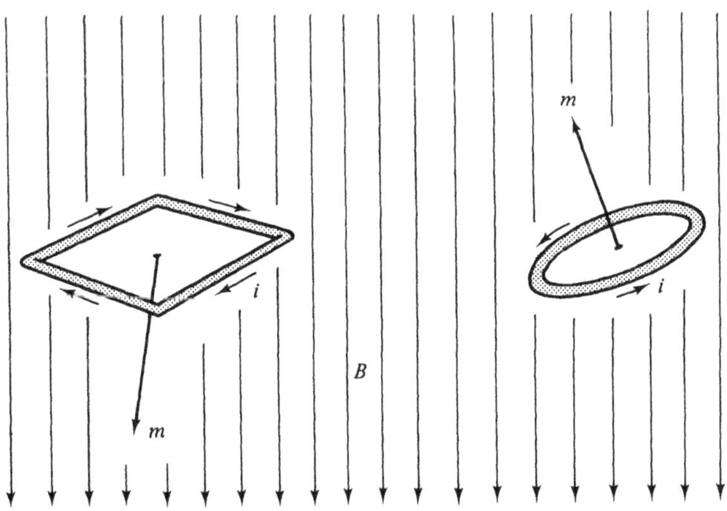

Figure 3.28

3.26 Ferromagnetism

The magnetic phenomena that we have discussed so far apply in vacuum (in the strictest sense). However, for all practical purposes, the expressions developed are valid in air. If we placed a piece of some other "nonmagnetic" material (such as copper, plastic, or wood) in the magnetic field, as shown in Figure 3.16, the field would not change measurably.[10]

However, three elements—iron, cobalt, and nickel (which are adjacent to each other in the periodic table)—have a unique magnetic behavior. We refer to these materials (including some alloys containing small quantities of Al, Cu, and Ti) as "ferromagnetic" (derived from the Latin *ferrum* [iron]). They are important, even crucial, in the design of most electrical power apparatus because they permit us to create magnetic fields of very much greater intensity than are obtainable in air or in vacuum.

3.26.1 Magnetic "Conduction"

We demonstrate the basic feature of ferromagnetism by means of the arrangement shown in Figure 3.29. Around a long straight cable consisting of N insulated

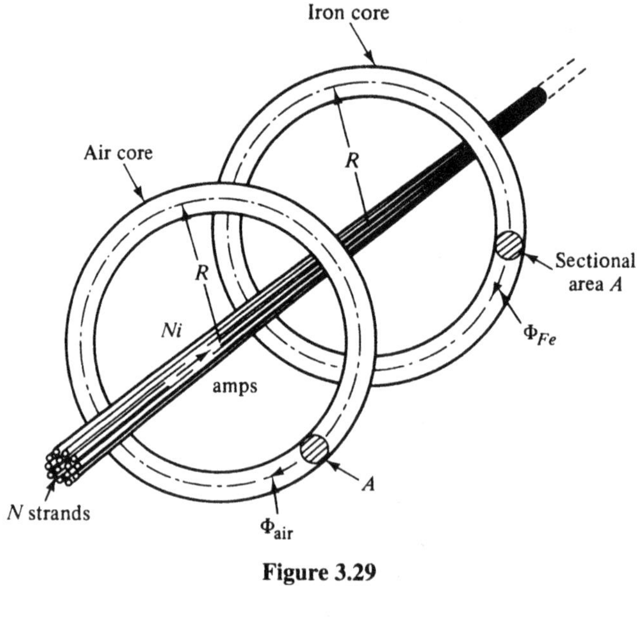

Figure 3.29

[10] If the field is a *variable* one, caused for example, by an ac current, then this statement is not true. Currents would now be induced in the copper due to Faraday's law, and these currents would change the field pattern substantially.

strands, each carrying a current i, we have placed a toroidal core made of a ferromagnetic material. It has the dimensions given in the figure. If we measure the magnetic flux Φ_{Fe} inside the core and compare it with the flux Φ_{air} inside an identical toroidal "core" made up of air, we find that the ferromagnetic core flux density and flux are both larger by a factor μ, that is,

$$\frac{B_{Fe}}{B_{air}} = \frac{\Phi_{Fe}}{\Phi_{air}} = \mu. \tag{3.84}$$

We refer to μ as the *relative permeability* of the core material in question; μ can assume values as large as 10^6. Although μ varies widely among ferrous materials, it is always much greater than unity. As will be explained in Section 3.26.4, μ also varies as a function of flux density—it is not a constant for a given material.

The experiment represented in Figure 3.29 shows that iron is a very good conductor of magnetic flux. The iron path represents a path of least resistance for the flux, and most of the flux will "concentrate" in the small cross-sectional area of the iron, producing μ times higher magnetic flux density in the iron core than in an air core of equal size.

3.26.2 Ohm's Law for a Magnetic Circuit

The magnetic flux density at the center of the air core depicted in Figure 3.29 is given by equation (3.36) as

$$B_{air} = \mu_0 \frac{Ni}{2\pi R} \qquad [\text{T}]. \tag{3.85}$$

Since the flux density in the iron core is μ times larger,

$$B_{Fe} = \mu\mu_0 \frac{Ni}{2\pi R} \qquad [\text{T}], \tag{3.86}$$

and for the iron core flux, we get

$$\Phi_{Fe} = AB_{Fe} = \mu\mu_0 \frac{ANi}{2\pi R} \qquad [\text{W}]. \tag{3.87}$$

By experiment we can confirm a most interesting characteristic of the flux. If we keep the iron core position fixed but move the cable to any position inside the core window, the flux will not change. (The reader should contemplate what happens to the flux in the air core under the same conditions.)

The flux will not change even if we bend the cable into any conceivable shape as long as the total current Ni remains unchanged and as long as it passes through the core window. In fact, the magnetic flux will remain unchanged if we replace the N-strand cable with an N-turn coil carrying a current i as long as the product Ni stays constant.

These experiments prove conclusively that in a ferromagnetic material the core flux is determined solely by the product Ni and is unaffected by the relative positions of the coil and core. The toroidal iron core in Figure 3.29 can be viewed as a magnetic circuit. The magnetic flux Φ_{Fe} in this circuit is "driven" by the product Ni, which we should therefore think of as a *magnetomotive force* (mmf). We can rewrite equation (3.87) as follows:

$$Ni = \Phi_{Fe} \frac{l}{\mu\mu_0 A} \qquad [\text{A} \cdot \text{t}] \text{ (ampere-turns)} \tag{3.88}$$

where $l = 2\pi R$ is the length of the path through the magnetic core.

We introduce the *magnetic resistance* or *reluctance* defined by

$$\mathcal{R} \equiv \frac{l}{\mu\mu_0 A} \qquad [\text{A} \cdot \text{t/Wb}]. \tag{3.89}$$

Equation (3.88) can now be written as

$$Ni = \Phi_{Fe} \cdot \mathcal{R} \qquad [\text{A} \cdot \text{t}], \tag{3.90}$$

which is "Ohm's law for the magnetic circuit." The unit for the reluctance, $\mathcal{R}$, is ampere turns per weber [A · t/Wb]. We shall find this useful in discussing the characteristics of the magnetic circuits in transformers and rotating machines. Note the analogy between the magnetic reluctance $\mathcal{R}$ and the electric resistance R [equation (3.23)].

Example 3.22

Consider the toroidal core shown in Figure 3.30. Find the current needed to produce a flux density of $B_{Fe} = 1.2$ [T] in the following two cases:

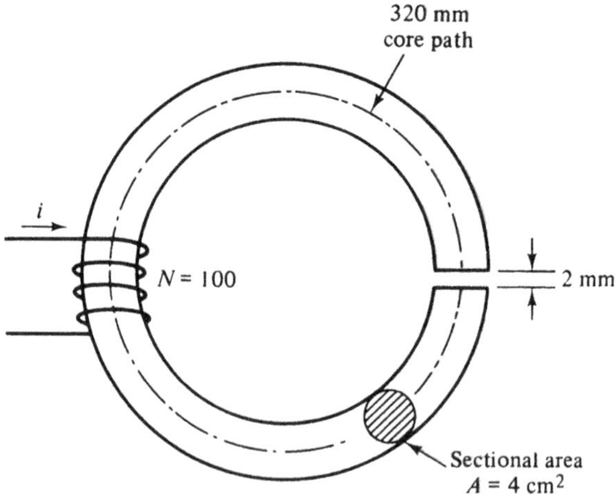

Figure 3.30

1. No air gap
2. With a 2-mm air gap

Let $N = 100$ and $\mu = 4000$ for the iron. The iron cross-sectional area, $A = 4\ \text{cm}^2$.

Solution: The required flux is

$$\Phi_{\text{Fe}} = AB_{\text{Fe}} = 0.0004 \times 1.2 = 48 \times 10^{-5} \qquad [\text{Wb}]. \tag{3.22.1}$$

1. For the reluctance, we have

$$\mathcal{R} = \frac{322 \times 10^{-3}}{4000 \times 4\pi \times 4 \times 10^{-4}} = 1.601 \times 10^5 \qquad [\text{A} \cdot \text{t/Wb}]. \tag{3.22.2}$$

Equation (3.90) gives

$$100i = 48 \times 10^{-5} \times 1.601 \times 10^5. \tag{3.22.3}$$

Therefore,

$$i = 0.769 \qquad [\text{A}]. \tag{3.22.4}$$

2. We must now separate the reluctances for the iron path and the air gap. Equation (3.89) gives

$$\mathcal{R}_{\text{Fe}} = \frac{320 \times 10^{-3}}{4000 \times 4\pi \times 10^{-7} \times 4 \times 10^{-4}} = 1.592 \times 10^5. \tag{3.22.5}$$

$$\mathcal{R}_{\text{air}} = \frac{2 \times 10^{-3}}{4\pi \times 10^{-7} \times 4 \times 10^{-4}} = 39.79 \times 10^5. \tag{3.22.6}$$

The total reluctance is the sum of these two:

$$\mathcal{R}_{\text{tot}} - \mathcal{R}_{\text{Fe}} + \mathcal{R}_{\text{air}} = 41.38 \times 10^5. \tag{3.22.7}$$

From equation (3.90) we get

$$100i = 48 \times 10^{-5} \times 41.38 \times 10^5 \qquad [\text{A} \cdot \text{t}]. \tag{3.22.8}$$

Therefore,

$$i = 19.86 \qquad [\text{A}]. \tag{3.22.9}$$

Note that, although the air gap is only 2 mm in length, its reluctance far exceeds that of the iron. This has great practical significance. For example, in a rotating electric machine, where the iron path for the flux must be interrupted by an air gap (see Figure 3.31). Even a very small air gap increases the need for a "magnetization" current drastically.

The types of magnetic circuits shown in Figure 3.31 are extremely important in electric energy conversion. Consider first the "salient rotor" machine shown in

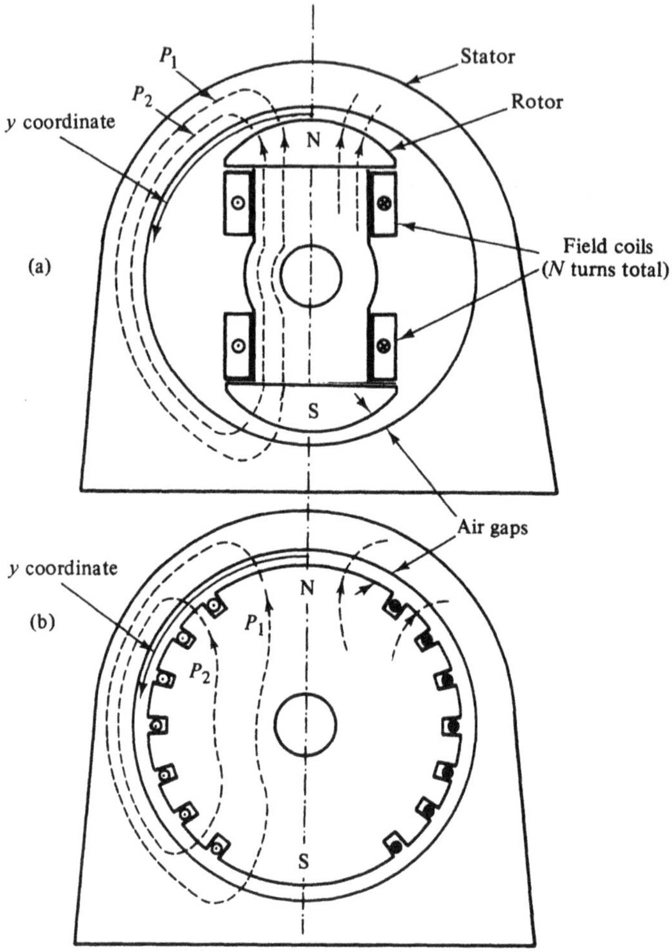

Figure 3.31

Figure 3.31a. The dc[11] current in the rotor *field coils* provides the mmf of magnitude Ni that drives the magnetic flux along the paths indicated. Both paths, P_1 and P_2, enclose the same current, Ni. But because of the tapered poleface, path P_2 contains a wider air gap than path P_1. As a consequence, path P_2 is characterized by a lower magnetic flux density than P_1.

If we plot the flux density against the tangential coordinate y, then we obtain the graph shown in Figure 3.32. The flux density can be made to vary almost sinu-

[11] The term "dc current" is linguistically redundant. However, ac and dc, originally meant as abbreviations, are now commonly used as adjectives in engineering.

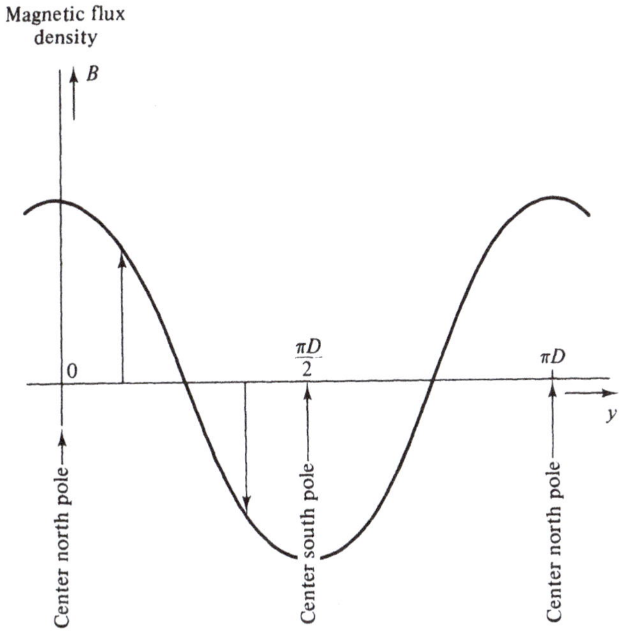

Figure 3.32

soidally by suitably shaping the poleface. The advantage of this is explained in the next chapter.

Consider the "round-rotor" design shown in Figure 3.31b. The dc field winding is placed in the rotor slots. The two flux paths, P_1 and P_2, are now traversing equally wide air gaps. However, path P_1 encloses more current than P_2 and we can therefore expect greater flux densities along the former path. Again we obtain a sinusoidal flux distribution as shown in Figure 3.32 by a suitable distribution of the currents in the slots.

3.26.3 The Magnetic Field Intensity

We return to Ohm's law for electric circuits and combine equations (3.23) and (3.24) to give

$$\frac{v}{l} = \rho \frac{i}{A} \qquad [\text{V/m}]. \tag{3.91}$$

The left-hand side of this equation, having the dimension volts per meter, represents the per meter voltage drop, that is, *the electric field intensity E,* as measured along the current path. The ratio i/A is equal to the current density, measured in amperes per square meter. Equation (3.91) tells us that

$$E = \frac{\text{emf}}{\text{meter}} = \rho \times \text{ current density.} \tag{3.92}$$

Returning to magnetic circuits, according to equation (3.88), we have

$$\frac{Ni}{l} = \frac{1}{\mu\mu_0} B_{\text{Fe}} \tag{3.93}$$

or

$$H \equiv \frac{\text{mmf}}{\text{meter}} = \frac{1}{\mu\mu_0} \text{flux density.} \tag{3.94}$$

We have introduced, by analogy with E in equation (3.92), *the magnetic field intensity H* (also called *magnetizing force*), defined as

$$H \equiv \frac{Ni}{l} \qquad [\text{A} \cdot \text{t/m}]. \tag{3.95}$$

The electric field intensity, E represents a per-meter emf measured along the current path of the electric circuit. It is independent of the electrical characteristics of the conductor.

Similarly, the magnetic field intensity, H represents a per-meter mmf along the flux path of the magnetic circuit. It is also independent of the characteristics of the magnetic conductor. The use of this new quantity is developed in the next section.

3.26.4 Magnetization Curves for Ferrous Materials

We have discovered some far-reaching similarities between electric and magnetic circuits. We now focus our attention on some important dissimilarities.

In general, as we increase the emf in an electric-resistive circuit, the current will increase. Over a very large range of current density, the increase in current will be directly proportional to the increase in emf. This means (cf. equation 3.24) that the resistivity parameter ρ is a constant. The proportionality ceases only when the current density reaches such high values that the conductor heats up excessively.

As we increase the mmf in the magnetic circuit, the flux density will increase but not in direct proportion to mmf. Expressed differently, the μ parameter is not a constant.

The actual relationship between flux density and mmf for a sample of ferromagnetic material is given in a "magnetization curve," a typical example of which is shown in Figure 3.33. In such curves, the flux density B is plotted against H. The advantage of using H rather than mmf is that the data are presented on a per-

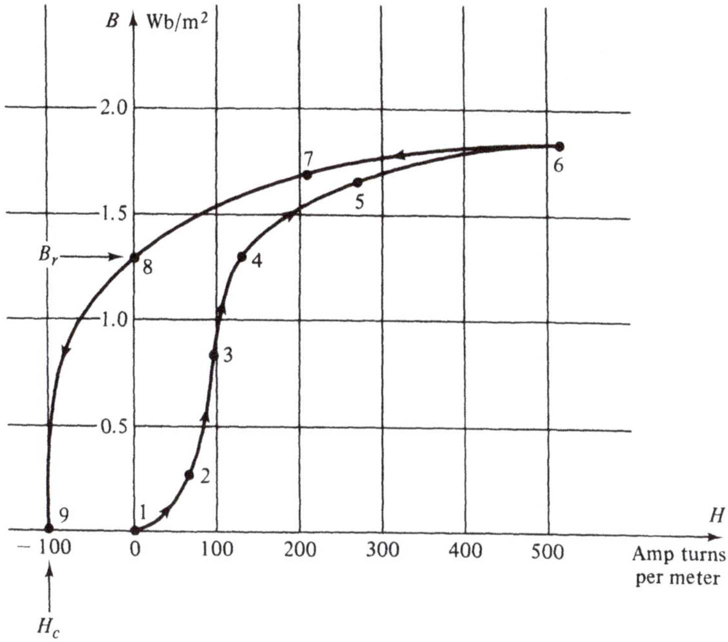

Figure 3.33

unit basis, that is, the data do not refer to a specific core size. Let us explore the main features of ferromagnetic behavior by discussing the curves in Figure 3.33. We assume that ferromagnetic material is initially completely demagnetized.

As we increase the coil current from zero we are moving on a nonlinear curve (points $1 \rightarrow 2$) corresponding to increasing values of μ, maybe in the μ range 50–200. A further increase in coil current takes us to points $3 \rightarrow 4$ characterized by a very rapid growth of the flux, where μ may take on values approximately from 10^4 to 10^5.

A continued increase in mmf through points $5 \rightarrow 6$ produces a decrease in the rate of growth of the flux, that is, a decrease in μ. This decrease in μ will continue and should we apply mmf's corresponding to say, 50,000 [A · t/m] (far beyond the scale of our graph) μ will actually approach unity. This means that the magnetic "conductivity" is reduced to that of vacuum or air. We say that the iron has reached magnetic saturation when the rate of growth of the flux starts to decrease (point $4 \rightarrow 5$).

If we start to reduce the current at point 6 (say) and hence the mmf, the flux density will decrease as expected—but it will now follow a different curve. Should we decrease the mmf to zero (point 8) we note that a given flux density, B_r

called the *residual density*, remains.[12] The core sample has now taken on the characteristics of a permanent magnet. The core will be demagnetized only when we apply a negative mmf, called *coercive intensity* or *force*, $-H_c$.

We should point out that this magnetic behavior applies to iron at normal temperatures. Above 770°C (Curie temperature) iron loses all ferromagnetism and behaves essentially like air.

Example 3.23

In Example 3.22 we studied the magnetic circuit in Figure 3.30 under the assumption of constant permeability. We now solve the problem again under the assumption that the material used in the magnetic circuit has the *B–H* curve shown in Figure 3.33. Specifically, we want to find the current needed to produce a flux density, $B = 1.5$ [T]. As before, the air gap is 2 mm long.

Solution: From the *B–H* curve we find that a flux density of 1.5 [T] corresponds to

$$H_{\mathrm{Fe}} = \frac{Ni}{l} = 195 \qquad [\mathrm{A \cdot t/m}]. \tag{3.23.1}$$

The mmf needed for the iron path (0.320 [m] long) is

$$(Ni)_{\mathrm{Fe}} = 1.195 \times 0.320 = 62 \qquad [\mathrm{A \cdot t}]. \tag{3.23.2}$$

The flux density in the air-gap is equal to that in the iron (if we neglect fringe effects) and from equation (3.94) we get (for $\mu = 1$)

$$H_{\mathrm{air}} = \frac{B_{\mathrm{air}}}{\mu_0} = \frac{1.5}{4\pi \times 10^{-7}} = 1.19 \times 10^6 \qquad [\mathrm{A \cdot t/m}]. \tag{3.23.3}$$

The mmf needed for the air path (0.002 [m] long) is therefore

$$(Ni)_{\mathrm{air}} = 1.19 \times 10^6 \times 0.002 = 2380 \qquad [\mathrm{A \cdot t}]. \tag{3.23.4}$$

Thus the total mmf is

$$(Ni)_{\mathrm{tot}} = 62 + 2380 = 2442 \qquad [\mathrm{A \cdot t}]. \tag{3.23.5}$$

The desired flux density could be obtained, for example, with a 1000-turn coil carrying 2.442 A or a 100-turn coil carrying 24.4 A.

The example teaches us the following important facts. Although the length of the air path is 1% of the total magnetic flux path it requires 97.5% of the total mmf to sustain the required flux density. The iron core, in effect, permits us to concentrate or focus practically all the magnetizing force of the coil on the air gap.

[12] If an electric circuit were to behave in a similar manner, we would be able to measure a circulating current after the removal of the emf.

3.26.5 A Physical Explanation of Ferromagnetism

What we need at this point is a reasonably satisfactory explanation of the ferromagnetic phenomena described above. To present an exhaustive explanation of ferromagnetism would require a thorough knowledge of modern theory of microphysics—that is, quantum theory. A meaningful explanation, and one that certainly suffices for most engineering purposes, can be presented by using the concept of magnetic moment (introduced in Section 3.25).

3.26.5.1 Magnetization: A Result of Line-up of Atomic Magnetic Moments

In the model of the atom, the electrons perform orbital motions around the nucleus of the atom. Since a moving charge is a current, the negatively charged electrons orbiting the nucleus correspond to currents in the electron orbits. The net effect of all the orbital "currents" will be the existence of a *magnetic orbital moment* m_{orb} as shown in Figure 3.34a.

In addition, each electron is associated with a *magnetic spin moment* m_{spin}. This can be understood if the electron is pictured as a negatively charged, spinning sphere (Figure 3.34b). According to quantum theory, both of these moments can have only discrete or quantized values.

Consider an iron core inside a coil (Figure 3.35). In its unmagnetized state, the elementary atomic magnetic moments have a random orientation. If current is supplied to the coil, a magnetic flux is created in the core, and this flux subjects the elementary electron moments to torques that tend to line them up with the field created. Why this line-up takes place in iron but not in copper or other nonmagnetic materials is a result of the difference in atomic structure and binding or restraining forces.

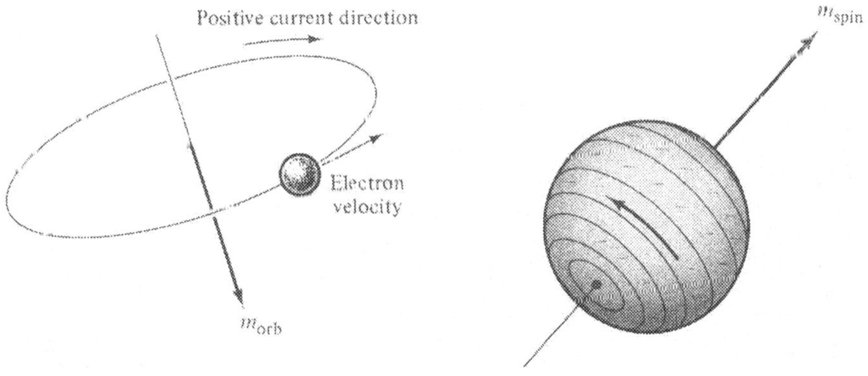

(a) (b)

Figure 3.34

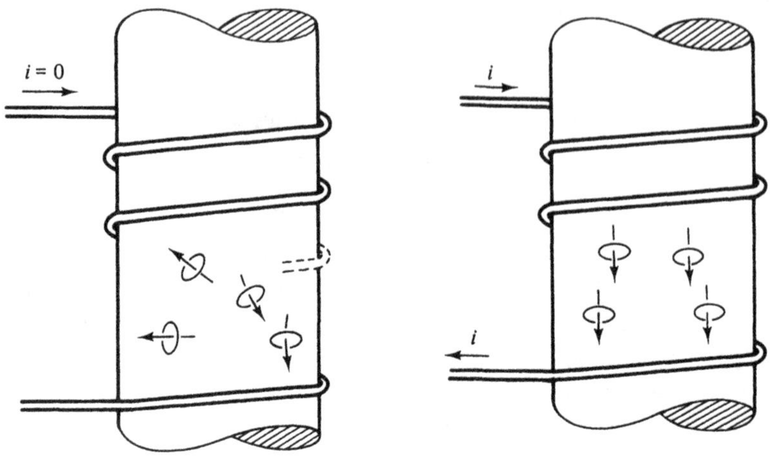

Figure 3.35

As the electron moments are being lined up, their mmf's are added to that of the coil, and the result is an amplification of the mmf. When the coil mmf reaches a level corresponding to point 3 in Figure 3.33, a huge mmf amplification takes place. This suggests that the line-up process takes place by the simultaneous "snapping" into position of a large number of elementary moments.

When all or most of the magnetic moments have been lined up, no additional mmf amplification can be achieved—we have now reached the saturation level corresponding to about point 4, as shown in Figure 3.33. When the coil current is reduced to zero, some of the magnetic moments lose their orientation, while others remain in their lined-up position. This accounts for the residual magnetism of the core.

3.26.5.2 "Bound" Currents

We can visualize each atomic magnetic moment as a small coil. As these coils are lined up under the influence of the magnetizing field, it becomes clear (Figure 3.36a) that the currents in the neighboring coil sides will neutralize each other. It is assumed that the magnetization (or *magnetic moment density*) is uniform throughout the core material. Only those coil currents located at the core surface (Figure 3.36a) remain unneutralized. The unneutralized currents add up to a surface sheet current referred to as the "bound" current (Figure 3.36b). These "frozen" currents in the core material itself cannot be measured. However, their field and force effects are as real as those that would be created by corresponding currents in an equivalent coil.

For example, consider the cylindrical iron sample shown in Figure 3.37. If it is magnetized, the core sample will possess bound surface currents that constitute a

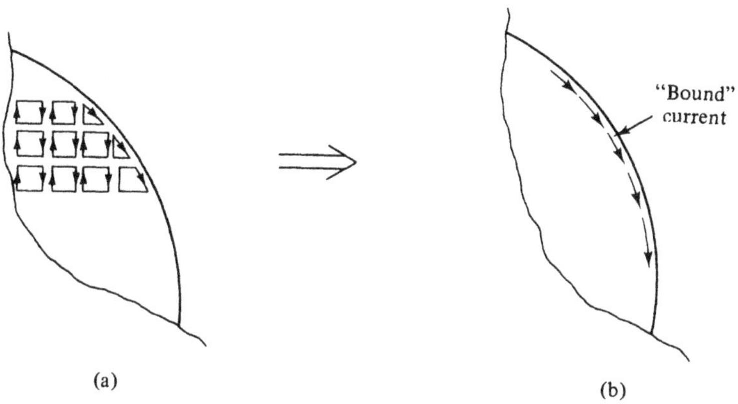

Figure 3.36

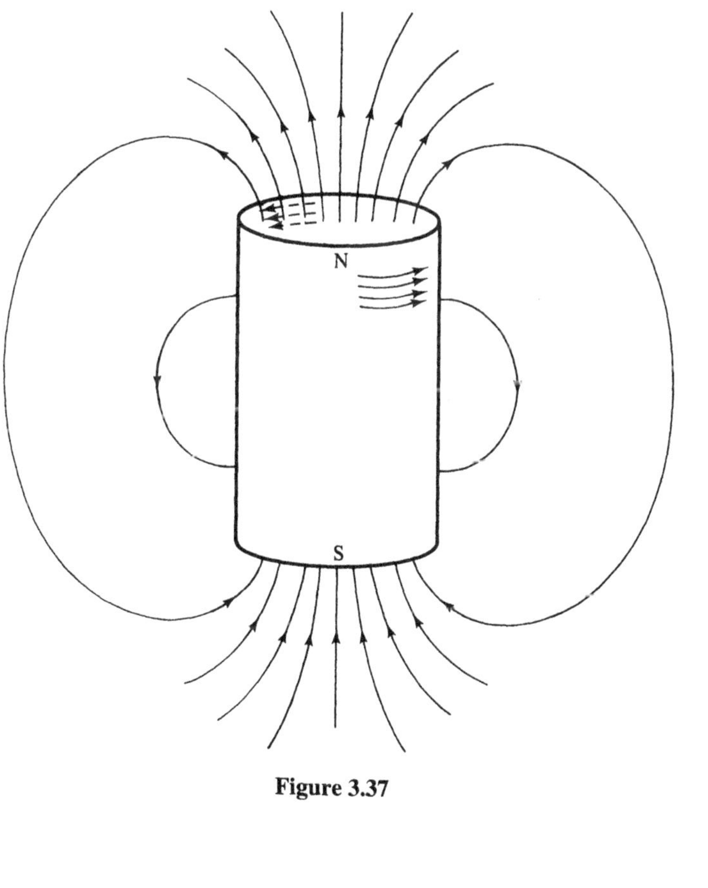

Figure 3.37

cylindrical shell. The magnetic flux distribution in and around the core will be similar to that associated with the cylindrical coil shown in Figure 3.18.

3.26.5.3 The Effects of Magnetized Material on Force and Torque

The forces and torques acting on magnetized cores have important practical applications because they are of considerable magnitude. The nature of these forces and torques can be ascertained by the use of the concepts of "bound" surface currents. We give two important examples:

1. Consider the lift magnet shown in Figure 3.38. As the magnet approaches the unmagnetized load object, the magnetic field from the magnet will magnetize the object (Figure 3.38b). The bound currents in the load object and the actual currents

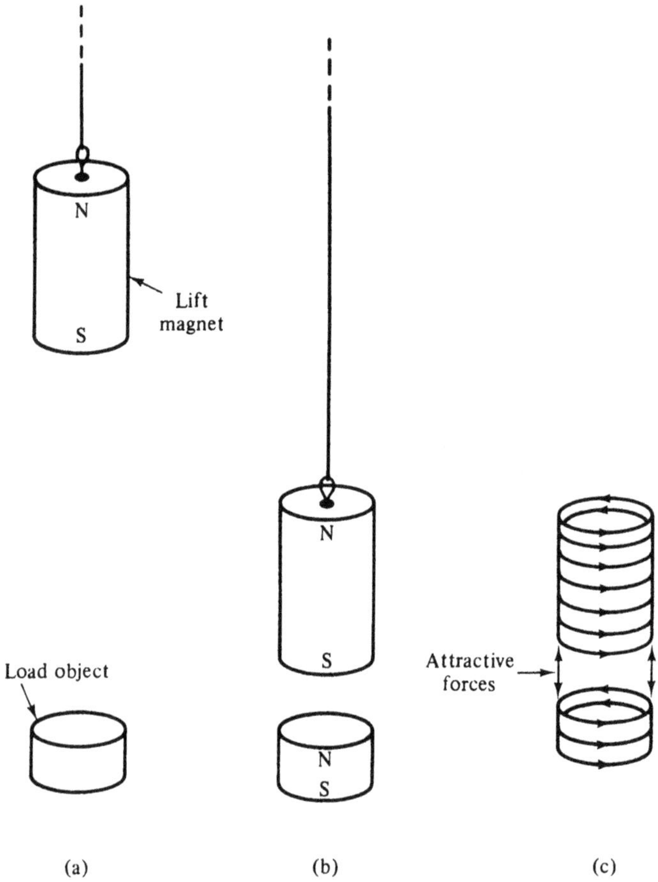

Figure 3.38

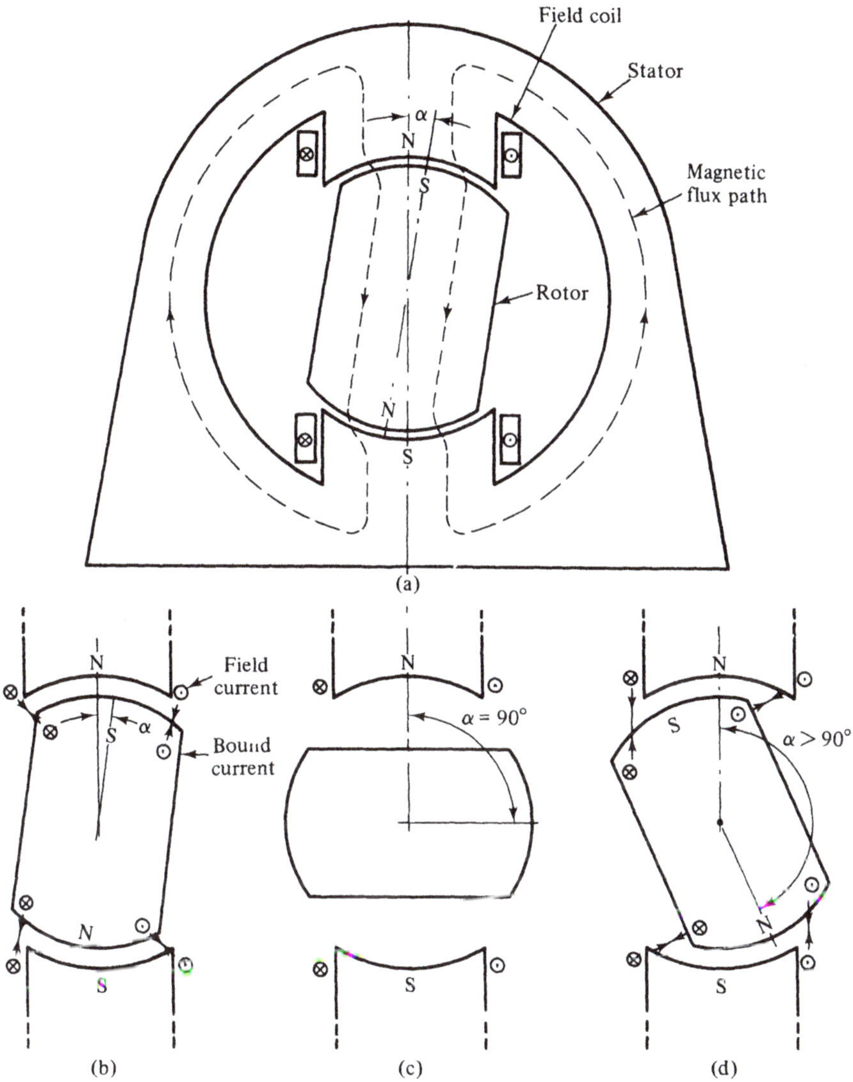

Figure 3.39

in the lift magnet coil are shown in Figure 3.38c. As we determined in Example 3.19, this current geometry results in an attraction between the magnet and the load.

2. Some types of electric machines operate on the basis of what is called "reluctance torque." Consider the arrangement shown in Figure 3.39. The dc current in the stator field coils sets up a flux along the magnetic paths indicated. The flux will magnetize the rotor, but its polarity (N or S) will depend on its angular position.

Three different rotor positions are shown. In each position we have indicated polarity by N or S and the corresponding bound current directions. In Figure 3.39b the attractive forces between the field currents and the bound currents (cf. Example 3.19) have been shown. As the rotor angle α is increased, these forces exert a CCW torque on the rotor (you may think of the forces as rubber bands).

The torque reaches a maximum when $\alpha = 45°$ and then decreases to zero as the rotor takes a horizontal position characterized by zero magnetization.

When the angle α exceeds 90° the polarity of the rotor is reversed and so does the torque. Figure 3.40 shows how the reluctance torque varies as a function of rotor position. Note that the torque completes a full cycle for a half-turn of the rotor. (In Chapters 4 and 8 we discuss the practical significance of this type of torque further.)

3.27 Summary

We have described in this chapter those electric and magnetic phenomena that form the basis of electric power engineering technology. Following a brief comparison with the concepts of mass, gravity, gravitational potential, and energy, we introduced the concepts of "static" electricity, electric field, electric potential, and energy. Electrostatic force and its effects as well as energy storage possibilities were discussed.

The latter part of the chapter dealt with electromagnetic phenomena. The importance of the electromagnetic force and the induction law to electric power engineering was discussed. A number of examples, all pertaining to practical apparatus, were given. The chapter ended with a brief summary of the most important aspects of ferromagnetism.

This chapter contains a wide spectrum of important material. The remainder of the book rests heavily on it.

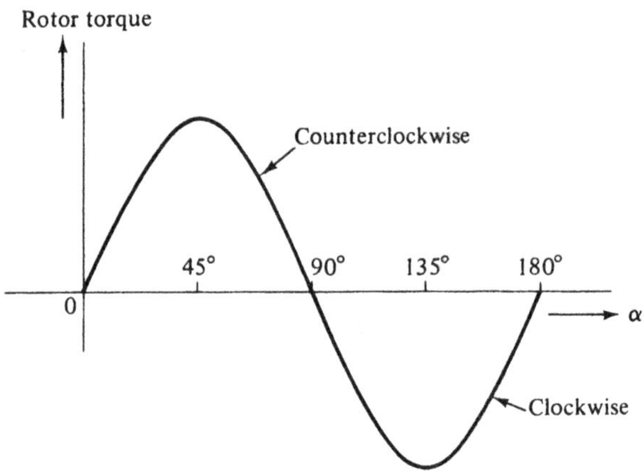

Figure 3.40

EXERCISES

3.1 A plate capacitor is charged from a battery of voltage v_0 and then disconnected from the battery. It is assumed that no charge subsequently leaks across the dielectric. The stored electric energy is

$$w_e = \frac{1}{2} C v_0^2 \qquad [\mathrm{J}]. \tag{3.96}$$

The plates are now moved apart so that the plate distance d is tripled. According to equation (3.19) this will decrease the capacitance to the new value $C/3$.

a) What will be the new capacitor voltage?

b) What will be the new stored electric energy?

c) You will find, if you obtained the correct answer to part b that the stored energy has increased. Where did the additional energy come from?

3.2 One of the two identical capacitors in Figure 3.41 is charged to V volts. After the switch S is closed and the system has reached a new steady state, each capacitor will have a voltage of $V/2$ volts across it. The stored energy before the closing of S is

$$w_e' = \tfrac{1}{2} C V^2 \qquad [\mathrm{J}]. \tag{3.97}$$

The stored energy after closing S will be

$$w_e'' = 2 \times \frac{1}{2} C \left(\frac{V}{2}\right)^2 = \frac{1}{4} C V^2 \qquad [\mathrm{J}]. \tag{3.98}$$

Half the initial energy has "disappeared." What is your explanation of this phenomenon?

3.3 Consider the previous problem. Put a resistor, R into the circuit before you close S.

a) Derive an expression for the current i as a function of time on the closing of S. (Compare it to Example 3.13.)

b) Determine and plot *both* capacitor voltages against time.

c) Compute the energy dissipated in R during the charge redistribution process.

d) In view of your findings in part c, suggest an explanation for the results obtained in Exercise 3.2!

3.4 Ten equal capacitors are connected in parallel and charged from a 1000-V battery. The capacitors are disconnected from the battery and then reconnected in series.

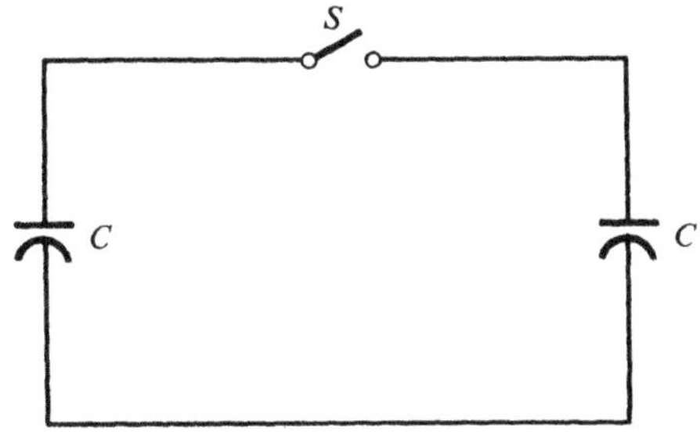

Figure 3.41

What will be the voltage across this capacitor pile? It is assumed that no charge leakage takes place. (This is how impulse test voltages are obtained in high-voltage laboratories.)

3.5 Find the capacitance of the layered capacitor in Figure 3.9. There are a total of 20 conductor foils, each of dimension 20 × 2000 cm, and 20 insulator sheets, each having a thickness of 0.1 mm. The foils are rolled into a cylindrical bundle. The insulator material is characterized by $\varepsilon = 5$.

3.6 A capacitor C is to be charged from a battery having an emf E. To limit the in-rush current a resistor R is placed in series with the capacitor.

a) Show that the charging current following the closing of the switch is of the form

$$i = \frac{E}{R} e^{-t/RC} \qquad [\text{A}]. \tag{3.99}$$

b) When the capacitor is fully charged its stored energy will be

$$w_e = \tfrac{1}{2} CE^2 \qquad [\text{J}]. \tag{3.100}$$

Show that exactly the same amount of energy is dissipated in the resistor during the charging process. (This means that the charging efficiency is only 50%.)

3.7 Three equal positive charges of 1 μC each are placed at the three corners of an equilateral triangle. The distance between the charges is 1 [m].

a) Find the magnitude and direction of the forces acting on each charge.

b) Compute the magnitude and direction of the electric field at the midpoint of the base of the triangle.

c) Show that the electric field at the centroid of the triangle of charges is zero.

[HINT: You know the voltage v and field intensity E emanating from one charge. The effects of several charges are obtained by superposition. Note, however, that E is a vectorial quantity. The potential v is a scalar.]

3.8 Eight charges of 1 μC each are placed so as to form the corners of a cube with 1-[m] sides.

a) Find the electric potential at the centroid of the cube.

b) Find the electric field strength at the centroid of the cube.

c) Compute the energy required to move a 1-μC charge from "infinity" to the centroid of the cube.

3.9 An ac current in a conductor can be expressed in the following form:

$$i = 100 \cdot \sin(314\,t) \qquad [\text{A}]. \tag{3.101}$$

a) What is the frequency?

b) The "free electron cloud" in a conductor carrying the current will evidently oscillate back and forth in the conductor covering a distance d. Find this distance under the following assumptions: (1) conductor cross-sectional area = 10 mm^2, (2) one free electron per atom, (3) 10^{20} atoms per mm^3.

3.10 A copper wire 1 km in length and 1 mm^2 cross-sectional area is connected across the terminals of a 12-V automobile battery. Neglecting the internal battery resistance, find the following:

a) Current.

b) Ohmic power dissipation in the wire.
c) Ohmic dissipation per meter of the wire. Do you think the wire will melt?

3.11 Repeat all parts of the previous problem assuming that the wire length is shortened to 1 [m]. Do you think the wire will melt?

3.12 An HVdc (high-voltage dc) transmission system transmits energy over a distance of 900 km. Each of the two conductors has a resistance of 0.014 Ω/km. The load voltage (v_l in Figure 3.13) is equal to 600 kV. The load power, p_l is 800 MW. Compute the
a) Current in each conductor,
b) Ohmic power loss in the transmission line,
c) Generator voltage, v_g,
d) Generator power, p_g,
e) Transmission efficiency.

3.13 Consider the coil in Example 3.15. If the current in the straight conductor is a saw-tooth-type ac current as shown in Figure 3.42, find the emf induced in the coil.

3.14 The two coils in Figure 3.27 are interconnected to form a single coil. Two cases are possible depending upon the polarity chosen.
Show that the self-inductance of the resulting coil is

$$L = L_1 + L_2 \pm 2M \qquad [\text{H}]. \tag{3.102}$$

Be careful to indicate to which case the plus and minus signs refer.

3.15 Twenty axial conductors, each of length 1 [m], are attached to the surface of a cylindrical rotor of diameter 1 [m] (the cylindrical rotor is made of an insulating material). Each conductor carries a current of magnitude 100 [A]. A magnetic field of flux density 1 [T] penetrates the cylindrical rotor surface radially. Will the rotor be subject to a torque? Find the magnitude of the torque in [N · m].

3.16 The N spokes in a wheel (Figure 3.43) each carry a current i [A]. The wheel is located in a magnetic field B perpendicular to the plane of the wheel. Show that the wheel will be subjected to a torque of magnitude:

$$T = \tfrac{1}{2}NBi(R^2 - r^2) \qquad [\text{N} \cdot \text{m}]. \tag{3.103}$$

3.17 Figure 3.44 shows the mechanism of an ammeter. The permanent magnet M produces a magnetic flux across the air gaps. We can assume that the flux density in the air gap is radial, uniform, and of magnitude B [T]. A small rectangular, N-turn coil is supported on its axis and it is free to rotate in the air gaps against the restraining

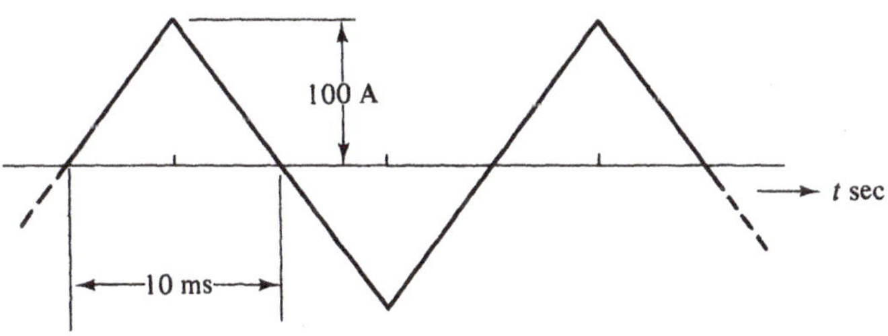

Figure 3.42

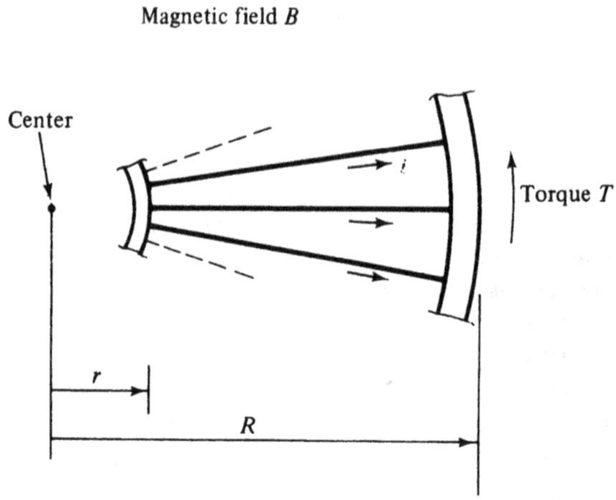

Figure 3.43

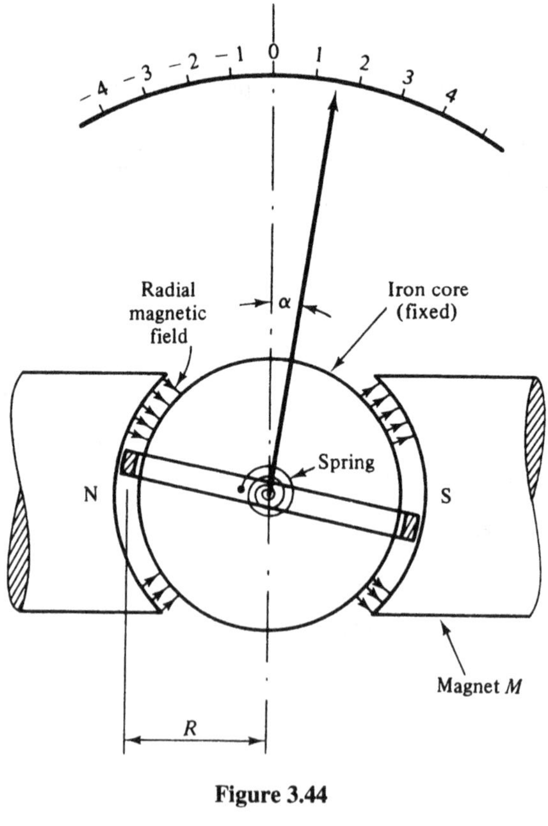

Figure 3.44

torque of a spring. The spring torque is proportional to the angle of rotation and is equal to T [N · m/degree].

Show that the coil will rotate through an angle α, which is proportional to current i. Find α for the following numerical data:

$$B = 0.3\ \text{T};$$

$$i = 0.1\ \text{mA};$$

$$N = 500\ \text{turns};$$

$$T = 10^{-5}\ \text{N}\cdot\text{m/degree}.$$

3 cm of each coil side is located in the magnetic flux. The coil radius $R = 2$ cm.

3.18 Find the self-inductance of the 100-turn coil placed on the toroidal core discussed in Example 3.22. Treat the two cases with and without the air gap.
[HINT: Find the total linked flux caused by a 1-[A] coil current.]

3.19 For the toroid shown in Figure 3.45, calculate the current required to set up a flux density of 1.5 [T] in the air-gap when

$$r = 50\ \text{cm}$$

$$N = 100$$

$$A = 20\ \text{cm}^2$$

$$x = 5\ \text{mm}$$

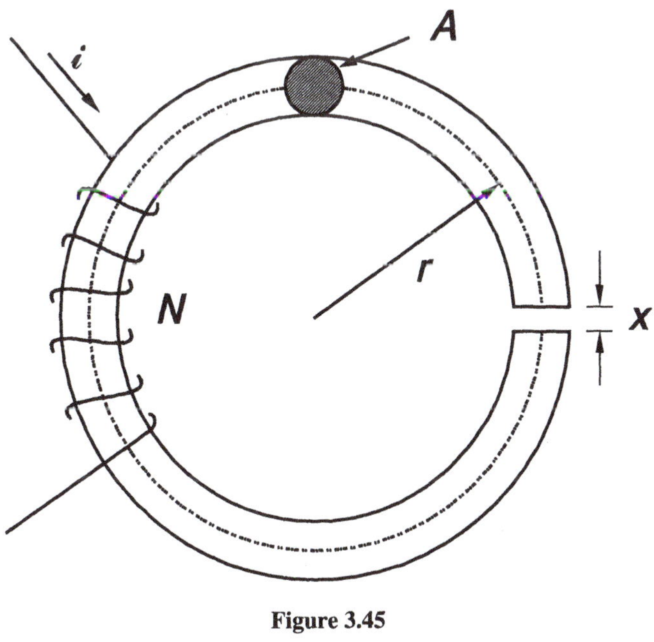

Figure 3.45

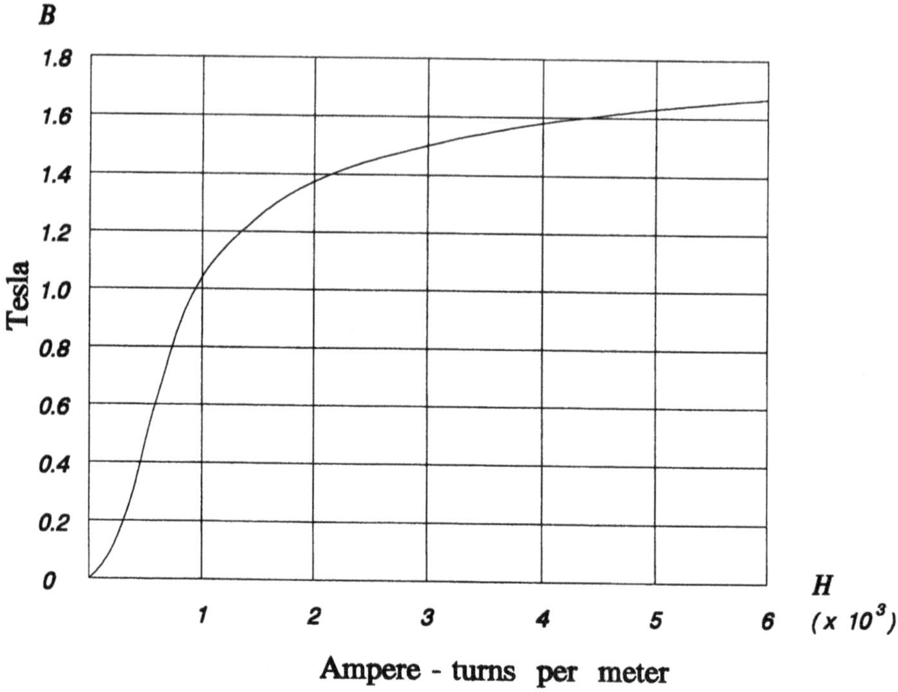

Figure 3.46

The B–H characteristics of the toroid material is given in Figure 3.46. State any assumptions you make.

3.20 Figure 3.47 shows a circular coil of radius r [m] with N turns rotating at n revolutions per minute about the axis X–X' in a magnetic field of uniform flux density B [T]. The inductance of the coil is L [H] and a load of resistance R [Ω] is connected across the terminals of the coil.

a) Derive an expression for the emf that would appear across the terminals of the coil on open circuit.

b) Derive an expression for the torque required to rotate the coil when the resistance R is connected to the coil, assuming that there are no losses except in R.

3.21 Figure 3.47 shows a circular coil of radius, r [m] in a magnetic field of flux density B [T] or [Wb/m^2]. A current i [A] flows in the coil, which has N turns.

a) Derive an expression for the force on *half* the coil when the plane of the coil is perpendicular to the direction of the flux.

b) Derive an expression for the torque on the coil when it is inclined at an angle α to the plane perpendicular to direction of the flux.

c) Indicate the direction of rotation if the coil were free to rotate about the axis X–X' when its plane is parallel to the magnetic field.

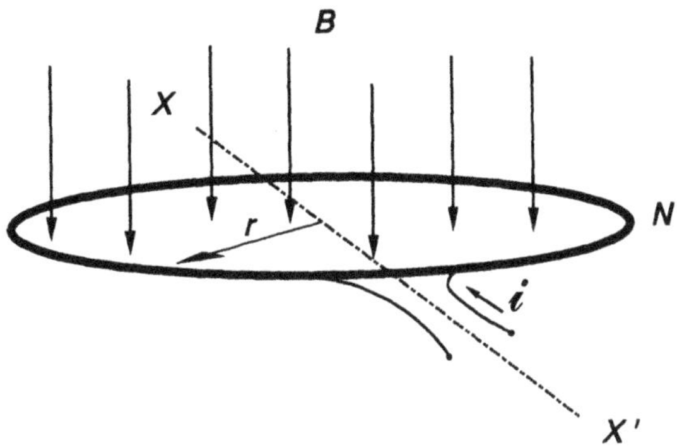

Figure 3.47

References

Broch, E., Lysne, D.K. *Hydro Power '92*. Proceedings of the Second International Conference on Hydropower, Lillehammer, Norway, June 1992. Rotterdam: Balkema, 1992.

International Energy Agency. *New Electricity 21: Designing a Sustainable Electric System for the 21st Century*. Paris: France, May, 1996.

West, M., White, P., Duckers, L., Loughridge B., Lockett, P., Peatfield, T., Hall, C. (Editors) *Alternative Energy Systems: Electrical Integration and Utilization*, Proceedings of the Conference in Coventry, England, Oxford: Pergamon Press, 1984.

Yamayee, Z.A., Bala, J.L. *Electromechanical Energy Devices and Power Systems*. New York: John Wiley & Sons, 1993.

4

Synchronous Machine

In the previous chapter, the physical laws that form the basis of electric energy technology were described. The remainder of the book will be devoted to the various components and systems, the design and operation of which are the concern of the electric power engineer.

It would be entirely beyond the scope of this book to attempt an inclusion of all types of machines that have been produced by electric power technology. It would also be unrealistic at this introductory level to develop mathematical models that describe various phenomena in power apparatus, such as the behavior of a three-phase synchronous generator subject to unbalanced fault currents. We shall feel content if we are able to convey an understanding of the operation of this important but rather complex machine under normal balanced operation.

4.1 Direct Current Versus Alternating Current

Before we begin the story of electric-generating technology, we must settle the question of "ac versus dc." Consider the simple electric transmission system in Figure 3.13. Let us assume that the load is of the simplest possible kind—a resistance R. For simplicity we shall disregard the line losses, that is, we neglect all line resistance.

We compare the power flow in the system in the two cases as follows.

CASE 1 ***Direct Current.*** The "generator" (e.g., an automobile battery) can now be represented by an emf e of constant magnitude. The line current will have the value

$$i = \frac{e}{R} \qquad [\mathrm{A}]. \tag{4.1}$$

Because e is constant, the current will also be constant. The power flow in the line will be

$$p = vi = ei = \frac{e^2}{R} = i^2 R \qquad [\text{W}], \tag{4.2}$$

and it too will be constant.

CASE 2 ***Single-Phase Alternating Current.*** The "generator" may consist of the rotating coil shown in Figure 3.22. The emf is now sinusoidal, that is, of the form

$$e(t) = e_{\max} \sin \omega t \qquad [\text{V}]. \tag{4.3}$$

The current will therefore also be sinusoidal:

$$i(t) = \frac{e(t)}{R} = \frac{e_{\max}}{R} \sin \omega t \qquad [\text{A}]. \tag{4.4}$$

For the instantaneous power flow in the line we obtain

$$p(t) = v(t)i(t) = e(t)i(t) = \frac{e_{\max}^2}{R} \sin^2 \omega t = \frac{e_{\max}^2}{2R}(1 - \cos 2\omega t) \qquad [\text{W}]. \tag{4.5}$$

We note that the power is pulsating, at the double radian frequency 2ω, between zero and the maximum value

$$p_{\max} = \frac{e_{\max}^2}{R} \qquad [\text{W}]. \tag{4.6}$$

The average power is equal to

$$p_{\text{ave}} = \frac{p_{\max}}{2} = \frac{e_{\max}^2}{2R} \qquad [\text{W}]. \tag{4.7}$$

The instantaneous, maximum, and average power in the circuit are shown in Figure 4.1a. A comparison of the expressions for power, (4.2) and (4.5), as well as the waveform of $p(t)$, immediately reveals the superiority of dc over single-phase ac. The dc power flow is smooth. The pulsating single-phase ac power would cause unacceptable vibration problems in both the generator and the load—certainly at power levels in the megawatt range. Fortunately, the pulsation of power (see Section 4.4) can be eliminated by the use of *multiphase* ac, particularly the important *three-phase* system. In this book, the three-phase system is discussed exclusively.

The three-phase ac offers the "smooth" features of dc power system plus the following additional advantages:

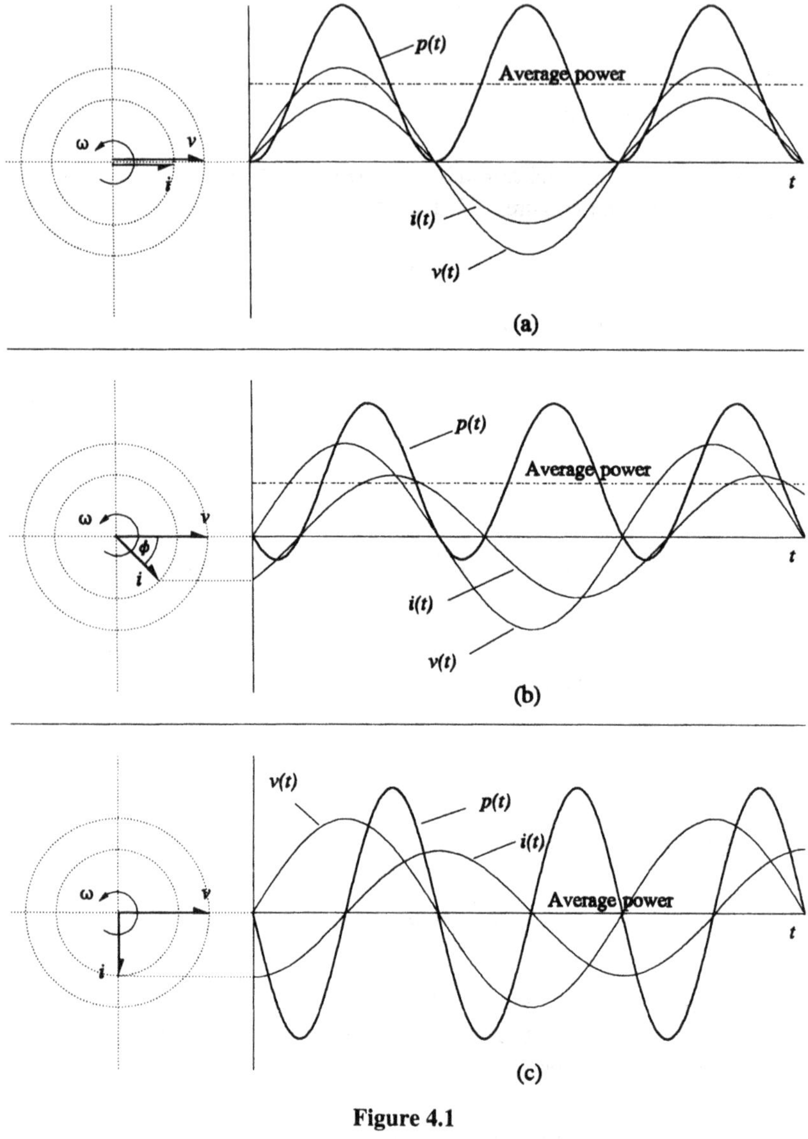

Figure 4.1

1. Easy generation
2. Easy transformation to and from high-voltage levels (which permits low-loss transmission)
3. Cheap and simple motor design

As a result, practically 100% of all bulk electric energy in the world is of the three-phase ac variety. In those relatively few instances where, for very special reasons (HVdc transmission,[1] dc motors, automobile batteries, etc.), dc is preferred over ac, one invariably obtains the dc power by rectification of ac power.

4.2 Power in Single-Phase Alternating Current

As ac power is of dominant concern throughout the remainder of this book, we shall review some of its important characteristics. This material is presented for the single-phase system, but it can be applied directly to the all-important three-phase system.

4.2.1 Real and Reactive Powers

Consider again the transmission system shown in Figure 3.13. Now remove the restriction made in Section 4.1 that the load be resistive. For a load, in general, the current and voltage will not be in phase; assume an angle ϕ between them. We then have

$$v(t) = v_{\max} \sin \omega t \qquad [\mathrm{V}]. \tag{4.8}$$

$$i(t) = i_{\max} \sin (\omega t - \phi) \qquad [\mathrm{A}]. \tag{4.9}$$

A positive ϕ value means that the current *lags* the voltage. This case (as shown in Figure 4.1b) is obtained when the load is *inductive*. A negative ϕ value implies that the current *leads* the voltage, which would be the case if the load were *capacitive*.

Figure 4.1b shows the instantaneous, maximum, and average values of the power flowing in the line. Note that the power pulsates at twice the radian frequency as before, but for a short period the power is negative, that is, energy is flowing from the load to the generator during this time. Figure 4.1c shows the instantaneous, maximum, and average values of the power flowing in the line when the angle ϕ is 90°. The power still pulsates at twice the radian frequency, but the energy supplied by the generator to the load during one-quarter of the cycle is returned during the second quarter. The load is described as *lossless*: a pure inductance or capacitance. Figure 4.1b represents a load that is made up of resistive and inductive components.

The reader no doubt is familiar with *phasor* representation of sinusoidal variables. (Appendix A summarizes the most important features of phasor analysis.) If

[1] High-voltage dc (HVdc) transmission is preferred in those cases where ac transmission would be either impractical or physically impossible. For example, in transmitting electric energy over large bodies of water, one is forced to use submarine cables. A cable is characterized by a very large shunt capacitance (because of the proximity between the conductor and the outer shield), and its shunt impedance, $1/\omega C$, is small. The result is that unacceptably large capacitive currents will flow. A dc cable has zero capacitive current. HVdc is also preferred when energy is transmitted over extremely long distances (>600 km) where ac can cause stability problems.

we represent the voltage and current with the phasors V and I, respectively, then we obtain the *phasor diagram* of Figure 4.1b. Note that the phase angle ϕ is defined by the relationship

$$\phi \equiv \angle V - \angle I. \tag{4.10}$$

For the instantaneous transmitted power we get

$$p(t) = v(t)\,i(t) = v_{\max} i_{\max} \sin \omega t \sin(\omega t - \phi) \qquad [\text{W}]. \tag{4.11}$$

If we now make use of the trigonometric identity

$$\sin \alpha \sin \beta = \frac{1}{2}\left[\cos(\alpha - \beta) - \cos(\alpha + \beta)\right]. \tag{4.12}$$

we can write the expression for power in the following form:

$$p(t) = \frac{v_{\max} i_{\max}}{2}\left[\cos \phi - \cos(2\omega t - \phi)\right] \qquad [\text{W}]. \tag{4.13}$$

We introduce the *effective*, or root-mean-square (*rms*), values of the voltage and current as

$$\begin{aligned} |V| &= \frac{1}{\sqrt{2}} v_{\max} \qquad [\text{V}]; \\ |I| &= \frac{1}{\sqrt{2}} i_{\max} \qquad [\text{A}]. \end{aligned} \tag{4.14}$$

and can then write expression (4.13) as follows:

$$p(t) = |V|\,|I| \cos \phi - |V|\,|I| \cos(2\omega t - \phi) \qquad [\text{W}]. \tag{4.15}$$

The line power evidently pulsates (Figure 4.1b) around an *average power* ($|V|\,|I|$ $\cos\phi$) at double radian frequency 2ω as expected.

From Figure 4.1b it is clear that the current I can be resolved into two components as follows:

1. $|I| \cos \phi$: in phase with the voltage
2. $|I| \sin \phi$: orthogonal to the voltage

The in-phase component of the current will follow the pattern shown in Figure 4.1a and produce *real* power.

$$P \equiv p_{\text{real}} = |V|\,|I| \cos \phi \qquad [\text{W}]. \tag{4.16}$$

The out-of-phase component of the current will follow the pattern shown in Figure 4.1c and produce *reactive* power

$$Q \equiv p_{\text{reactive}} = |V|\,|I| \sin\phi \qquad \text{[VAr] (volt-amperes reactive).} \qquad (4.17)$$

Clearly, from Figure 4.1c, the average value of the reactive power is zero.

These concepts are of such fundamental importance that we find it appropriate to say a few words about their meaning:

1. The real power P is defined as the average value of $p(t)$ and is therefore the useful power transmitted.
2. The reactive power Q is by definition equal to the peak value of the power component that travels back and forth on the line. Its average value is zero, and it is therefore not capable of useful work.

Both P and Q have the dimension *watts*. To emphasize the fact that the latter represents a nonactive or reactive power, it is measured in *volt-amperes reactive* [VAr]. Larger and more practical units are kilovar [kVAr] and megavar [MVAr], related to the basic unit as follows:

$$1\ \text{[MVAr]} = 10^3\ \text{[kVAr]} = 10^6\ \text{[VAr]}. \qquad (4.18)$$

4.2.2 Effects of Various Types of Load

In analyzing the real and reactive power for a specific type of load the electrical engineer makes use of concepts like *impedance*, *phasor*, and *complex algebra*. These topics belong to courses in electric circuit analysis. However, they have been summarized in Appendix A for those readers who wish to have a compact review.

Table 4.1 summarizes the real and reactive power for the most common types of load.

Note, in particular, that an inductive load draws reactive power from the source. In power lingo, an inductor is said to "consume" reactive power. A capacitive load, on the other hand, draws negative Q. Hence a capacitor is said to "generate" reactive power.

Example 4.1

Derive expressions for the real and reactive power in a circuit consisting of a resistance R in series with an inductance L when a voltage V (rms) is applied across it [Table 4.1(4)].

Solution: The impedance for the circuit is

$$Z = R + j\omega L \qquad [\Omega]. \qquad (4.1.1)$$

Thus,

$$|Z| = \sqrt{R^2 + (\omega L)^2} \qquad [\Omega] \qquad (4.1.2)$$

Table 4.1

	Load type	Phasor relation	Phase angle	Power absorbed by load P	Power absorbed by load Q
1	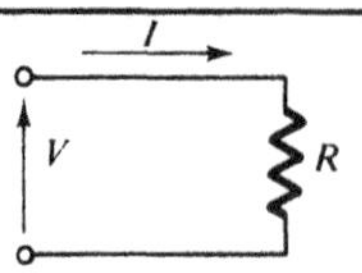	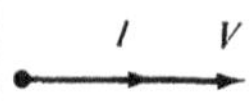	$\varphi = 0$	$\frac{\lvert V\rvert^2}{R}$	0
2	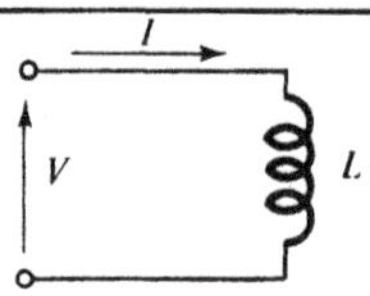	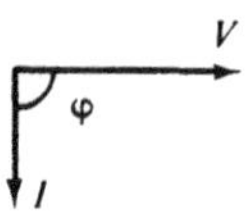	$\varphi = +90°$	0	$+\frac{\lvert V\rvert^2}{\omega L} = \omega L\lvert I\rvert^2$
3	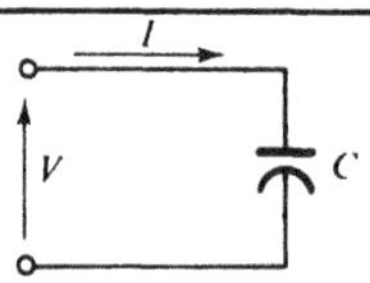	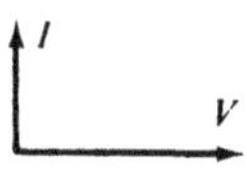	$\varphi = -90°$	0	$-\omega C\lvert V\rvert^2 = -\frac{1}{\omega C}\lvert I\rvert^2$
4	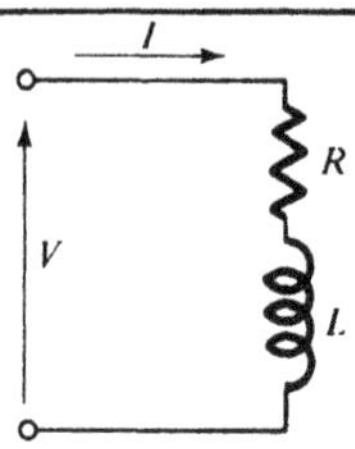	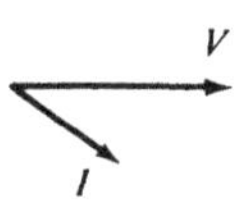	$\varphi = \tan^{-1}\frac{\omega L}{R}$	$\frac{R\lvert V\rvert^2}{R^2+(\omega L)^2} = R\lvert I\rvert^2$	$\frac{\omega L\lvert V\rvert^2}{R^2+(\omega L)^2} = \omega L\lvert I\rvert^2$
5	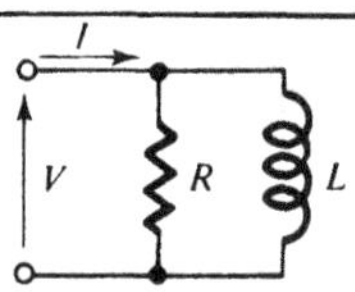		$\varphi = \tan^{-1}\frac{R}{\omega L}$	$\frac{\lvert V\rvert^2}{R}$	$\frac{\lvert V\rvert^2}{\omega L}$
6	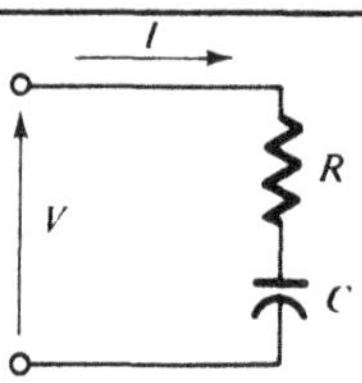	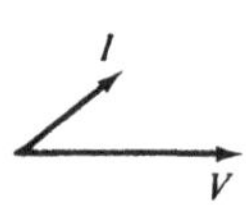	$\varphi = -\tan^{-1}\frac{1}{\omega CR}$	$\frac{R\lvert V\rvert^2}{R^2+\frac{1}{\omega^2C^2}} = R\lvert I\rvert^2$	$-\frac{\lvert V\rvert^2}{\omega C(R^2+\frac{1}{\omega^2C^2})} = -\frac{1}{\omega C}\lvert I\rvert^2$
7	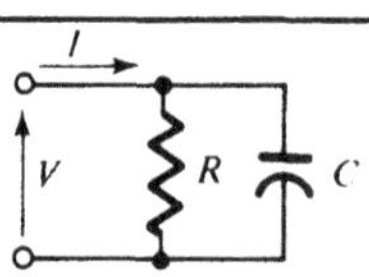		$\varphi = -\tan^{-1}\omega CR$	$\frac{\lvert V\rvert^2}{R}$	$-\omega C\lvert V\rvert^2$

and

$$\phi = \angle V - \angle I = \angle Z = \tan^{-1}\left(\frac{\omega L}{R}\right). \tag{4.1.3}$$

From which we get

$$\begin{aligned} \sin\phi &= \frac{\omega L}{\sqrt{R^2 + (\omega L)^2}}; \\ \cos\phi &= \frac{R}{\sqrt{R^2 + (\omega L)^2}}. \end{aligned} \tag{4.1.4}$$

If the rms value of V is $|V|$, then the rms of I is equal to

$$|I| = \frac{|V|}{\sqrt{R^2 + (\omega L)^2}} \qquad [\mathrm{A}]. \tag{4.1.5}$$

Substitution of these expressions for $|I|$, $\sin\phi$, and $\cos\phi$ into expressions (4.16) and (4.17) yields the tabulated values for P and Q.

Example 4.2

In this example we demonstrate the physical relationship between reactive power and the energy stored in the magnetic field of the coil used in the previous example.

Solution: The instantaneous current in the RL circuit is of the form

$$i = \sqrt{2}\frac{|V|}{\sqrt{R^2 + (\omega L)^2}} \sin(\omega t - \phi) \qquad [\mathrm{A}]. \tag{4.2.1}$$

According to equation (3.76), we can write, for the energy stored in the magnetic field,

$$W_{\mathrm{mag}} = \frac{1}{2}Li^2 = \frac{1}{2}L\left[\sqrt{2}\frac{|V|}{\sqrt{R^2 + (\omega L)^2}} \sin(\omega t - \phi)\right]^2 \qquad [\mathrm{J}]. \tag{4.2.2}$$

The rate of change of w_{mag} is

$$\begin{aligned} \frac{dw_{\mathrm{mag}}}{dt} &= 2\omega L\frac{|V|^2}{R^2 + (\omega L)^2} \sin(\omega t - \phi)\cos(\omega t - \phi) \\ &= \omega L\frac{|V|^2}{R^2 + (\omega L)^2} \sin 2(\omega t - \phi) \qquad [\mathrm{W}]. \\ &= Q \sin 2(\omega t - \phi) \end{aligned} \tag{4.2.3}$$

The last step follows from Table 4.1. We conclude:

The *rate* of change of the energy stored in the magnetic field varies harmonically at a frequency of 2ω. It has zero average value and a peak value equal to Q.

Note that the rate of change of energy (power) is *periodic* at twice the supply frequency. The reactive power flows into and is stored in the magnetic field of the coil during one-quarter of the cycle, and it is returned to the source during the second quarter of the cycle, and so on.

An analysis of capacitive circuits reveals a similar relationship between the reactive power and the stored energy in the electric field of the capacitor.

4.2.3 The Concept of Complex Power

The voltage and current phasors in Figure 4.1 can be described in the form

$$V = |V|e^{j\angle V} \qquad [\mathrm{V}];$$
$$I = |I|e^{j\angle I} \qquad [\mathrm{A}]. \qquad (4.19)$$

We introduce the *conjugate* current defined by

$$I^* \equiv |I|e^{-j\angle I} \qquad [\mathrm{A}]. \qquad (4.20)$$

Then we form the product

$$S = VI^* \qquad [\mathrm{VA}]. \qquad (4.21)$$

This product, called *complex power,* has a very useful property, which we shall discover by the following analysis.

Substitution of I^* into equation (4.21) gives

$$S = VI^* = |V|e^{j\angle V}|I|\,e^{-j\angle I} = |V|\,|I|\,e^{j(\angle V - \angle I)} \qquad [\mathrm{VA}]. \qquad (4.22)$$

In view of equation (4.10) we get

$$S = |V|\,|I|e^{j\phi} = |V|\,|I|(\cos\phi + j\sin\phi) = P + jQ \qquad [\mathrm{VA}]. \qquad (4.23)$$

The last step follows directly from the definition of P and Q. In words:

The real and reactive power can be obtained as the real and imaginary parts of S.

The magnitude $|S|$ of the complex power is referred to as *apparent power.* It can be expressed in any one of several ways:

$$|S| = |VI^*| = |V||I| = \sqrt{P^2 + Q^2} \qquad [\mathrm{VA}]. \qquad (4.24)$$

The unit of $|S|$ is obviously volt-amperes, but often we prefer the larger units kilovolt-amperes or megavolt-amperes. The practical significance of apparent power is as a rating unit for generators and transformers (see Section 5.5.3).

Example 4.3

Find the real and reactive powers consumed by the circuit shown in Figure 4.2. The voltage V has the rms value 100 V.

Solution: We first find the impedance Z_p of the parallel branch:

$$Z_p = \frac{(6 - j3)(5 + j8)}{11 + j5} = 5.199 + j0.637 \qquad [\Omega]. \tag{4.3.1}$$

The total impedance Z_{tot} of the circuit is

$$Z_{tot} = 10 + Z_p = 15.199 + j0.637 \qquad [\Omega]. \tag{4.3.2}$$

For the current, and its conjugate, we get

$$I = \frac{V}{Z_{tot}} = \frac{100}{15.199 + j0.637} = 6.574\angle - 2.400° \qquad [A]. \tag{4.3.3}$$

Therefore,

$$I^* = 6.574\underline{/+2.400} \qquad [A]. \tag{4.3.4}$$

The complex power

$$S = VI^* = 100 \cdot 6.574e^{j2.400} = 656.8 + j27.5 \qquad [VA]. \tag{4.3.5}$$

Thus the circuit consumes

$$P_{tot} = 656.8 \qquad [W]. \tag{4.3.6}$$

$$Q_{tot} = 27.5 \qquad [VAr]. \tag{4.3.7}$$

4.3 The Single-Phase Alternating-Current Generator

Although we concluded earlier that the single-phase ac generator is of very limited[2] practical significance, it serves as a natural takeoff point in our presentation of the three-phase generator. We analyzed the induction of an ac emf in a coil rotating in a uniform magnetic field (Section 3.19.1). This simple system is a prototype of all ac generators. A basic difference is, however, that in a practical ac generator the coils are generally stationary and the magnetic field rotates.

4.3.1 Alternating-Current Generator Design

Figure 4.3 shows the two main components of a *synchronous* ac generator, the *rotor* and the *stator*. The rotor consists of an even number (four in this case) of

[2] Small, single-phase units, where the vibration problem is controllable, find limited use in cases when single-phase ac power is required in relatively small quantities.

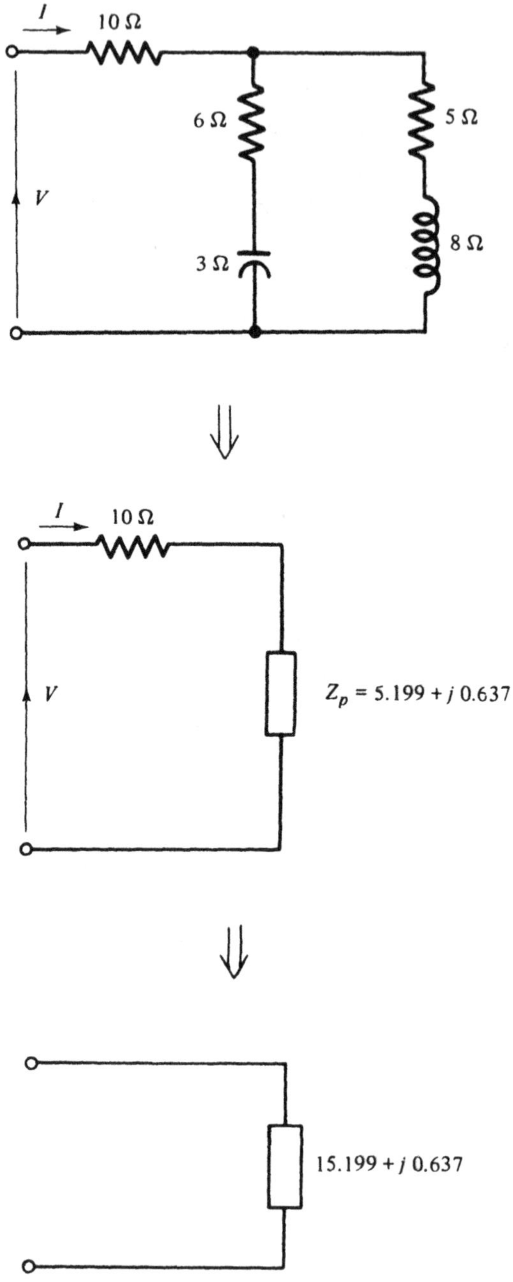

Figure 4.2

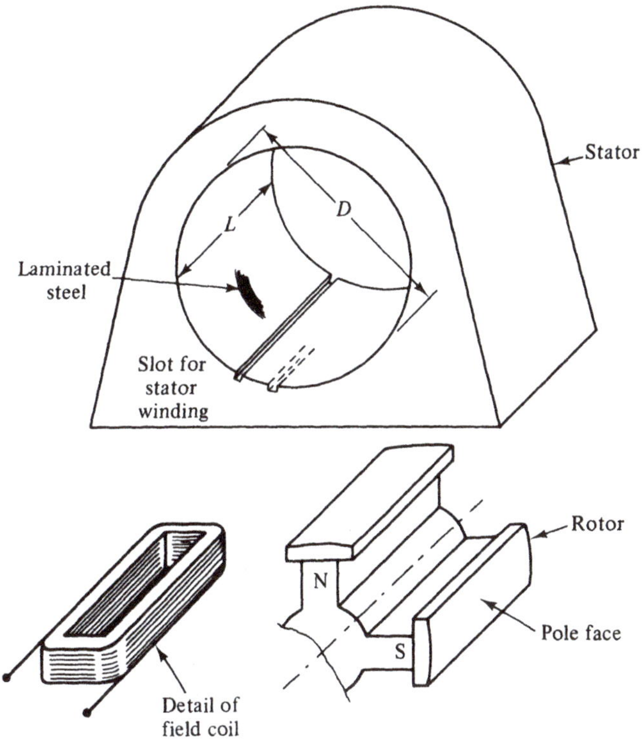

Figure 4.3

poles of alternating polarity. On each pole is placed a field coil, the detail of which is shown in Figure 4.3. The field coils are connected together to form a *field winding* (shown schematically in Figure 4.4). An *exciter*[3] feeds dc current into the field winding, and the resulting mmf creates the magnetic flux in the paths indicated in Figure 4.4. Note that the rotor can be replaced by a suitable permanent magnet. However, a permanent magnet is not used in a practical synchronous machine because heat and vibration will, in due course, destroy the flux. Note also that the flux paths cross the *air gaps* (the airspace between rotor and stator) twice.

The stator or *armature winding*, in which the emf's are generated, are placed in equidistant *slots* on the stator surface (only one slot is shown in Figure 4.3). The stator winding consists of *coils* placed so that the coil sides are one pole division

[3] The exciter may be a regular dc generator (Chapter 7) driven by the same prime mover that drives the synchronous generator. In this case the dc current is fed into the rotor field windings via brushes and slip-rings. In a "brushless" exciter the dc current is obtained from a separate ac winding placed on a separate rotor, connected directly to the main rotor. The ac voltage is rectified in a rectifier circuit placed on the rotor.

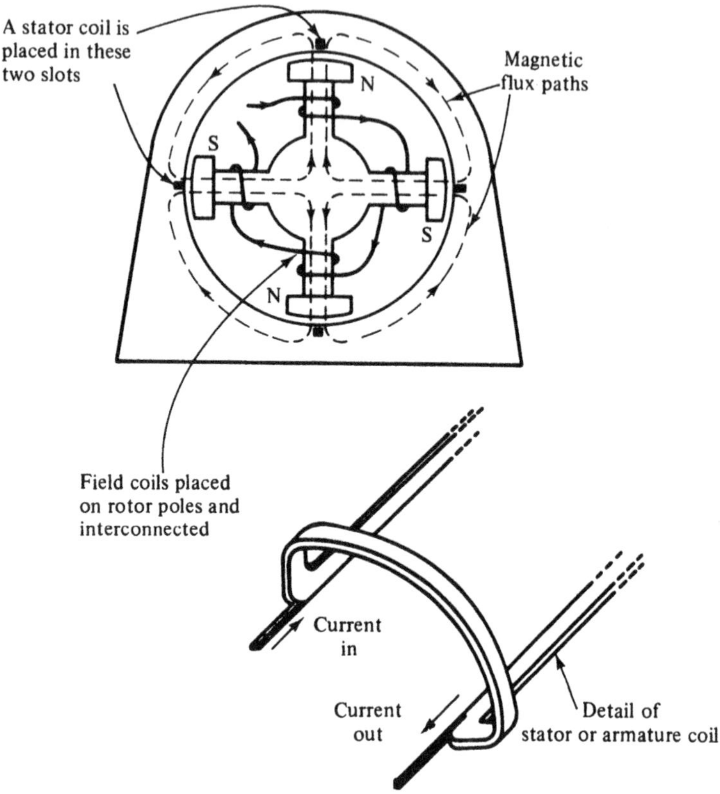

Figure 4.4

(90° in Figure 4.4) apart. Figure 4.4 also shows details of the stator coils. As the rotor spins, the flux sweeps by the armature winding, causing the stator iron to experience a changing flux. If the stator iron were solid, induced currents would flow in it, resulting in core losses and hence elevated temperatures. To prevent this, the stator core is made of *laminated* iron sheets individually insulated from each other (see Section 4.8).

4.3.2 Frequency, Poles, and Speed

The dotted flux path shown in Figure 4.4 tells us that, for the rotor position indicated in the figure, the total magnetic flux linked to one of the armature coils is zero. If the rotor turns through 45°, the linked coil flux will reach a maximum. An added rotation of 45° reduces the flux to zero, and after an additional 45° the coil flux will reach a maximum of opposite polarity, and so on. We have a periodic flux change, and, according to Faraday's law, we can expect a periodic emf to be induced in the armature coil.

We conclude from this that a full cycle of emf will be obtained when the rotor of the 4-pole generator has turned 180 *mechanical* degrees. A full cycle of emf represents 360 *electrical* degrees for the voltage wave (cf. Figures 3.22 and 4.1) and by extending this finding to a p-pole generator (p must always be an *even* integer), we discover the following important relation between mechanical rotor angles α_{mech} and electrical angles α_{elec}:

$$\alpha_{\text{elec}} = \frac{p}{2}\alpha_{\text{mech}}. \tag{4.25}$$

Because $p/2$ emf cycles will be generated for one complete rotor revolution and because the rotor completes $n/60$ turns per second, we obtain the following relation between electrical frequency f Hz and mechanical rotor speed n rpm:

$$f = \frac{p}{2} \cdot \frac{n}{60} = \frac{pn}{120} \qquad \text{[Hz]}. \tag{4.26}$$

Example 4.4

How fast must a 6-pole generator run if operated in a 60-Hz network?

Solution: Equation (4.26) gives

$$n = \frac{120f}{p} = \frac{120 \cdot 60}{6} = 1200 \text{ [rpm]} \tag{4.4.1}$$

4.3.3 Saliency and Nonsaliency

The 4-pole generator shown in Figures 4.3 and 4.4 is of *salient rotor* design. This is a typical rotor design in cases where low-speed prime movers (e.g., hydroturbines) are used.[4] In the United States most electric generators are driven by *steam turbines*. Expanding steam has high velocity, and consequently steam-driven generators run at high speed. The two most common designs have 2-pole and 4-pole rotors running at 3600 and 1800 rpm, respectively. (In countries where the frequency is 50 Hz the corresponding speeds are 3000 and 1500 rpm.)

The rotor coils in a salient rotor machine could not endure the centrifugal stresses for such high speeds; therefore, steam turbine-driven generators are designed with *nonsalient* rotors, as shown in Figure 4.5. Here the dc rotor field winding, like the ac stator armature winding, is placed in slots. This design evidently permits a better mechanical support and hence higher speed of operation.

[4] The speed of a hydroturbine is determined by the speed of the falling water. From our discussions in Chapter 2, it is clear that the hydroturbine speed must decrease (and the number of poles thus increase) with decreased water head.

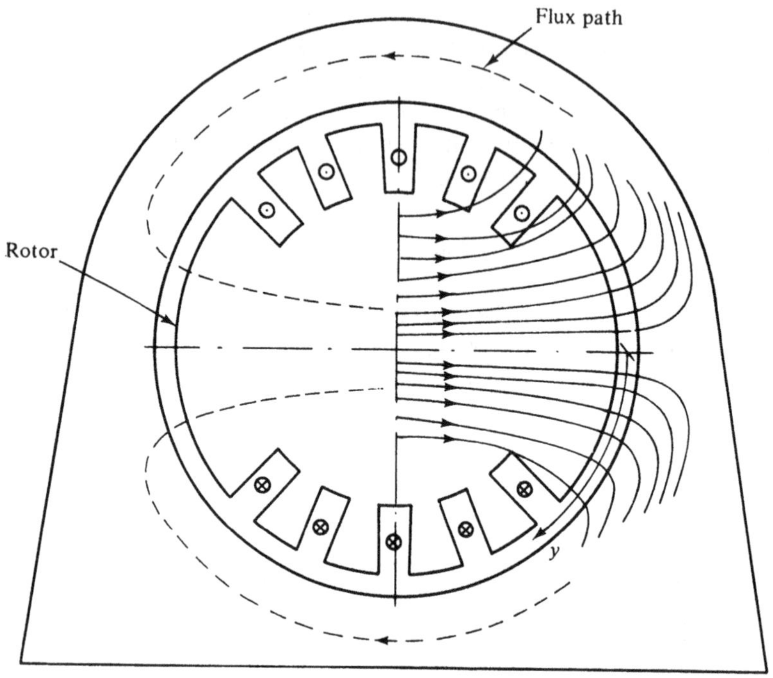

Figure 4.5

4.3.4 The Air-Gap Flux in Terms of Rotor Coordinates

For a salient-pole rotor (Figure 3.31a) the pole face is shaped so that the width of the air gap reaches a minimum at the center of the pole. Consequently, the magnetic flux is a maximum at this point, and it decreases to zero at the midpoint between the poles and reaches a negative maximum at the center of the adjacent pole. By properly changing the length of the air gap as we move away from the pole center, we can make the flux density vary harmonically[5] along the periphery. Figure 4.6 shows the flux density distribution of an 8-pole machine.

The flux density at any point in the air gap is given by

$$B = B_{\max} \cos \beta y \qquad [\mathrm{T}], \tag{4.27}$$

where y is a coordinate fixed with respect to the rotor (Figure 4.7). (The coordinate y is "curved" around the periphery of the air gap. In all the graphs that follow we have "straightened out" the axes for easier drawing.)

[5] By "harmonic function" we mean one that can be expressed as a sine or cosine of an independent variable.

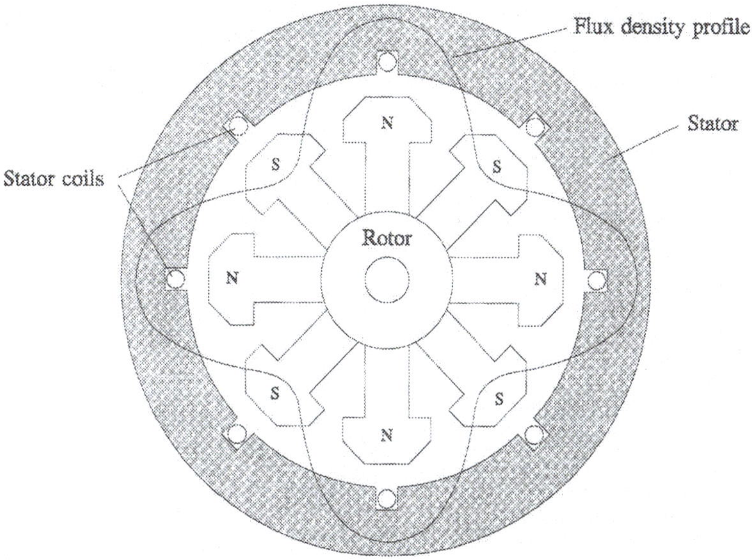

Figure 4.6

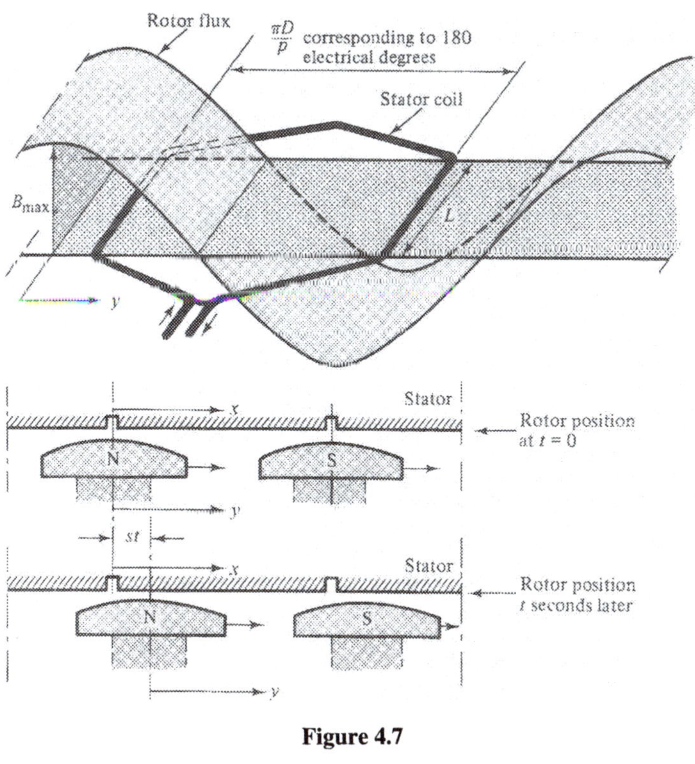

Figure 4.7

The factor β is determined from the fact that the angle βy must be equal to $(p/2) \cdot 2\pi$ radians for $y = \pi D$, D being the air gap diameter shown in Figure 4.3. Thus,

$$\frac{p}{2} \cdot 2\pi = \beta \pi D. \tag{4.28}$$

This equation yields

$$\beta = \frac{p}{D}. \tag{4.29}$$

Equation (4.27) can be written as

$$B = B_{\max} \cos\left(\frac{py}{D}\right) \qquad [\mathrm{T}]. \tag{4.30}$$

In the case of a nonsalient rotor the harmonic relationship in equation (4.30) is ensured by the proper shape and distribution of the rotor slots. (Refer to comment in Section 3.26.2.)

4.3.5 The Traveling Flux Wave

The reason for designing the rotor so as to achieve a harmonic flux distribution along its periphery will be clear as we compute the induced emf in the stator winding. Preparatory to doing this we shall write the flux in terms of a coordinate x, fixed with respect to the stator (Figure 4.7). The reason for doing this is the need to express the flux in terms of coordinates fixed with respect to the stator winding. From Figure 4.7 we note the relationship

$$x = y + st \qquad [\mathrm{m}], \tag{4.31}$$

where s, the tangential speed of the rotor, is equal to

$$s = 2\pi \cdot \frac{n}{60} \cdot \frac{D}{2} = \frac{n\pi D}{60} \qquad [\mathrm{m/s}]. \tag{4.32}$$

By substitution into equation (4.30), we get

$$B = B_{\max} \cos\left(\frac{px}{D} - \frac{pn\pi}{60} t\right) \qquad [\mathrm{T}], \tag{4.33}$$

or, in shorter form,

$$B = B_{\max} \cos(\beta x - \omega t) \qquad [\mathrm{T}], \tag{4.34}$$

where

$$\omega \equiv \frac{pn\pi}{60} = 2\pi f \qquad \text{[rad/s]}. \tag{4.35}$$

[The last step follows from equation (4.26).]

Equation (4.34) is the mathematical expression for a traveling wave. If we plot B versus x for two different moments t_1 and t_2, we obtain the plots, shown in Figure 4.8, of a harmonic wave traveling from left to right with speed s [m/s].

4.3.6 The Induced Electromotive Force in the Stator

We turn our attention to an N-turn coil of the type shown in Figures 4.4 and 4.7. Note that the coil spans $360/p$ mechanical degrees in order to link the total flux emanating from one pole. The distance in meters between the coil sides is $\pi D/p$, measured along the periphery.

The differential flux penetrating the differential window of width dx (see Figure 4.8) is

$$d\Phi = BL\,dx \qquad \text{[Wb]}, \tag{4.36}$$

where L is the axial length of the rotor (see Figure 4.3). By integrating over the total coil span we get the total flux, Φ, passing through the coil as

$$\Phi = \int_{x=0}^{\frac{\pi D}{p}} BL\,dx = \int_{x=0}^{\frac{\pi D}{p}} LB_{\max}\cos(\beta x - \omega t)\,dx \qquad \text{[Wb]}. \tag{4.37}$$

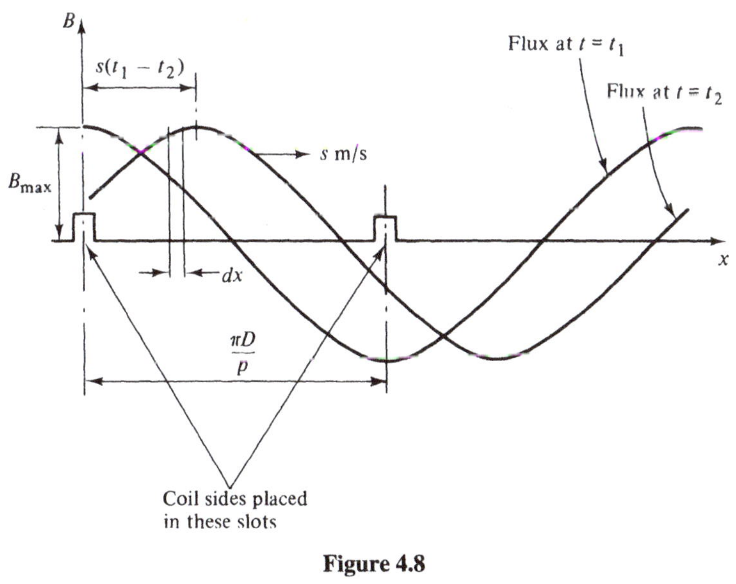

Figure 4.8

Integration yields

$$\Phi = 2\frac{B_{max}LD}{p}\sin \omega t \qquad [\text{Wb}]. \tag{4.38}$$

In order to find the total induced emf we make the following observations:

1. The derivative $d\Phi/dt$ will give the emf in one turn of the coil placed in the two slots, as shown in Figure 4.8.
2. In reality we have N conductors per slot (or N turns per coil).
3. Also, we have similar coils placed in similar slots located under all $(p/2)$ pole pairs (as shown in Figure 4.4). Not only are the emf's induced in those coils identical in magnitude but they are also of equal phase.

If e_1 represents the total emf induced in the coils in the p equidistant slots, and we assume all conductors to be connected in series, as shown in Figure 4.9, we have (according to Faraday's law)

$$e_1 = N\cdot\frac{p}{2}\cdot\frac{d\phi}{dt} = \omega LDNB_{max}\cos\omega t \qquad [\text{V}], \tag{4.39}$$

or, in view of equation (4.35),

$$e_1 = \frac{\pi}{60}pnNLDB_{max}\cos\omega t \qquad [\text{V}]. \tag{4.40}$$

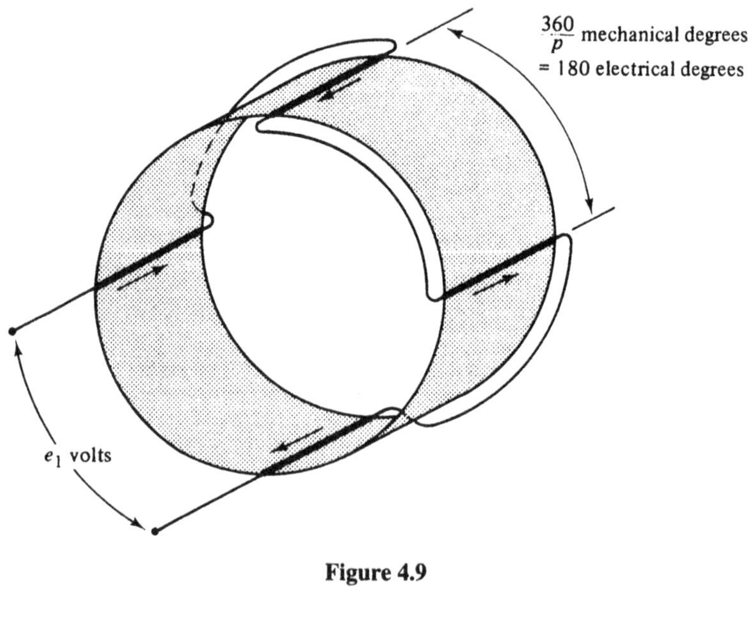

Figure 4.9

The rms value of the generated emf is equal to

$$|E_1| = \frac{\pi}{60\sqrt{2}} pnNLDB_{\max} \qquad [\mathrm{V}]. \tag{4.41}$$

We make the following two important observations:

1. Because the rotor was designed to achieve harmonic flux variation with respect to the space coordinates y or x, the coil flux, and thus the emf, will be harmonic with respect to time t.[6]
2. The emf reaches its peak value at the instant when the pole centers of the N and S poles are opposite to the slots in which the coil sides are placed.

4.3.7 Distribution Effects

Equation (4.41) gives the rms value of the induced voltage in the stator, assuming that the winding consists of only one coil per pole pair. Of course, in a practical design, one would utilize the total stator surface by arranging the stator winding in many slots distributed around the periphery, as shown in Figure 4.10. Assume that

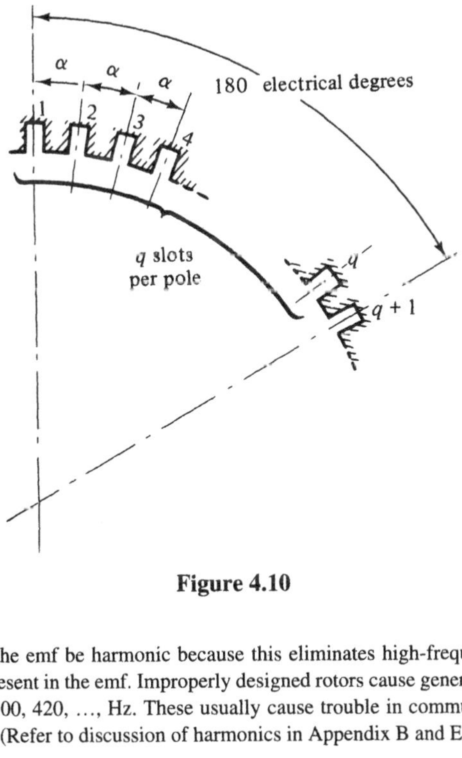

Figure 4.10

[6] It is important that the emf be harmonic because this eliminates high-frequency components that would otherwise be present in the emf. Improperly designed rotors cause generation of components of the frequencies 180, 300, 420, ..., Hz. These usually cause trouble in communication networks and also add to the losses. (Refer to discussion of harmonics in Appendix B and Exercise 4.14.)

the slots are placed α electrical degrees apart and that there are a total of q slots per pole. What would be the total emf obtained by connecting all these coils in series?

Clearly the emf induced in slot 2 will lag the emf induced in slot 1 by α degrees. The emf induced in slot 3 will lag that in slot 1 by 2α degrees in phase, and so on. By showing the emf's as phasors (Appendix A) designated $E_1, E_2, \ldots, E_q$, respectively, one can obtain the phasor diagram in Figure 4.11. The total stator emf, E, is the complex sum of phasors E_1, E_2, and so on:

$$E = E_1 + E_2 + \cdots + E_q \qquad [\text{V}]. \tag{4.42}$$

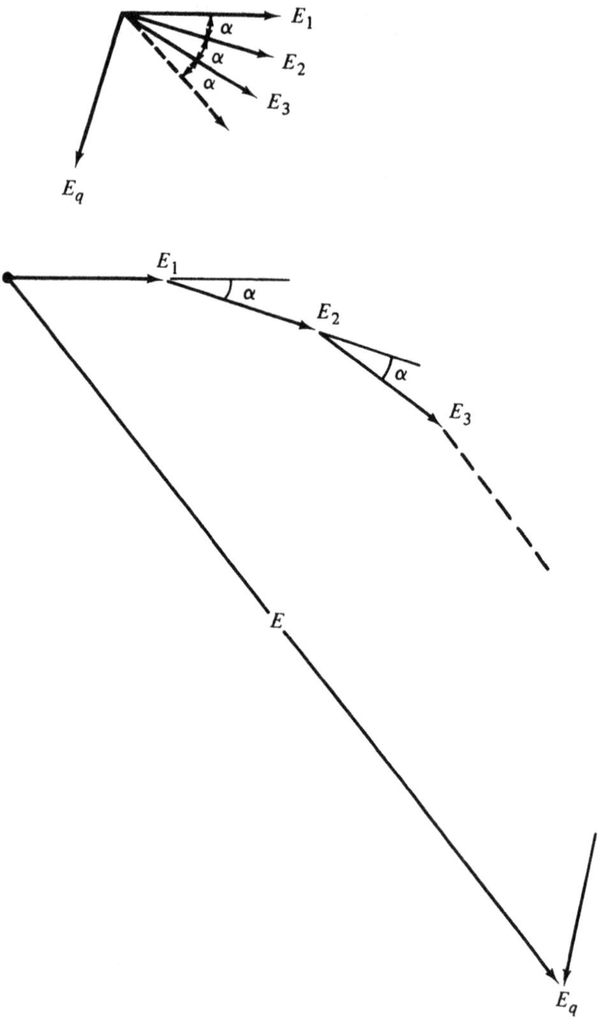

Figure 4.11

But from Figure 4.11 we have

$$\begin{aligned} E_2 &= E_1 e^{-j\alpha} \\ E_3 &= E_1 e^{-j2\alpha} \\ &\vdots\ \vdots \\ E_q &= E_1 e^{-j(q-1)\alpha} \qquad [\text{V}]. \end{aligned} \tag{4.43}$$

By substitution into (4.42) we obtain

$$\begin{aligned} E &= E_1 + E_1 e^{-j\alpha} + \cdots + E_1 e^{-j(q-1)\alpha} \\ &= E_1(1 + e^{-j\alpha} + \cdots + e^{-j(q-1)\alpha}) \qquad [\text{V}]. \end{aligned} \tag{4.44}$$

The expression within the parentheses is a geometric series, the sum of which can be written in closed form, as

$$1 + e^{-j\alpha} + \cdots + e^{-j(q-1)\alpha} = \frac{1 - e^{-jq\alpha}}{1 - e^{-j\alpha}}. \tag{4.45}$$

We then have for the phasor,

$$E = E_1 \frac{1 - e^{-jq\alpha}}{1 - e^{-j\alpha}} \qquad [\text{V}], \tag{4.46}$$

and for the rms value of $|E|$, we have

$$|E| = |E_1| \left| \frac{1 - e^{-jq\alpha}}{1 - e^{-j\alpha}} \right| = |E_1| \frac{|1 - \cos q\alpha + j \sin q\alpha|}{|1 - \cos \alpha + j \sin \alpha|} \qquad [\text{V}]; \tag{4.47}$$

$$|E| = |E_1| \frac{\sqrt{(1 - \cos q\alpha)^2 + \sin^2 q\alpha}}{\sqrt{(1 - \cos \alpha)^2 + \sin^2 \alpha}} = |E| \frac{\sin(q\alpha/2)}{\sin(\alpha/2)} \qquad [\text{V}]. \tag{4.48}$$

Example 4.5

A 2-pole, 3600-rpm turbogenerator has the dimensions $L = 2$ m, $D = 0.7$ m. The flux wave has a peak value of density $B_{max} = 1.5$ [T], and the number of turns $N = 4$.

(a) Find the induced emf in the coil.

(b) If the stator winding is placed in 18 equidistant slots, compute the total stator emf, assuming all coils are connected in series.

Solution:

(a) Equation (4.41) yields

$$|E_1| = \frac{\pi}{60\sqrt{2}} \cdot 2 \cdot 3600 \cdot 4 \cdot 2 \cdot 0.7 \cdot 1.5 = 2239 \qquad [\text{V}] \tag{4.5.1}$$

(b) Since this is a 2-pole generator, we have

$$q = \frac{18}{2} = 9 \qquad \text{[slots/pole]}. \tag{4.5.2}$$

Because electrical and mechanical degrees are identical for $p = 2$, we have

$$\alpha = \frac{360^\circ}{18} = 20^\circ \tag{4.5.3}$$

Hence

$$\frac{\sin(q\alpha/2)}{\sin(\alpha/2)} = \frac{\sin 90^\circ}{\sin 10^\circ} = 5.759. \tag{4.5.4}$$

The emf per coil is

$$|E_1| = 2239 \qquad \text{[V]}. \tag{4.5.5}$$

From equation (4.48) we have for the total stator emf

$$|E| = 2239 \cdot 5.759 = 12{,}894 \qquad \text{[V]}. \tag{4.5.6}$$

Note:

(i) We have used in **(b)** nine times more copper wire in the stator than in **(a)**.
(ii) The machine emf has increased by a factor of 5.8.
(iii) The amount of iron in the stator and rotor remain unchanged.

4.4 The Three-Phase Generator

We have pointed out that the one great drawback of single-phase ac power is its pulsating character. The power pulsations can be eliminated by designing the stator windings of the synchronous machine to form a three-phase system.[7]

A three-phase electric generator (Figure 4.12a) supplies, at the *phase terminals a, b,* and *c*, three-phase voltages V_a V_b, V_c, all of equal rms value, $|V|$, but of different phase angles.[8] V_b lags V_a by 120° and V_c lags V_b by the same angle; we talk about the generator having the *phase sequence abca....* The phase voltages are all measured relative to a fourth terminal, the *neutral, n,* which is usually grounded.

[7] The three-phase system has certain cost advantages over other multiphase systems.

[8] We prefer the letter symbol E for emf and V for terminal voltage. Sometimes the two are equal, sometimes not. For example, should the generator in Example 4.5 be open-circuited (i.e., unloaded) then the terminal voltage would be equal to the emf:

$$V = E$$

If the generator is loaded, a certain stator current I would exist, and this current would cause a voltage drop IZ across the winding impedance, Z. We would now have

$$V = E - IZ.$$

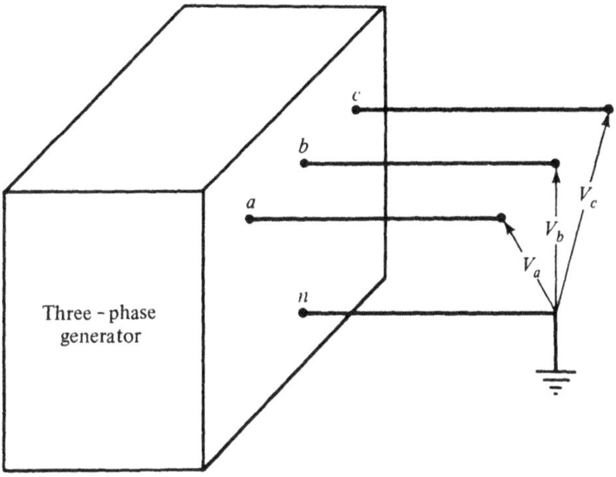

(a)

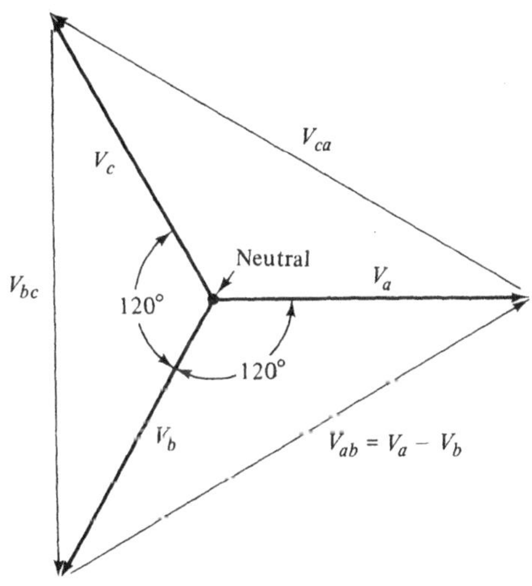

(b)

Figure 4.12

The three voltages are shown in the phasor diagram in Figure 4.12b; they are said to form a *symmetrical three-phase set.*

If V_a is chosen as the reference voltage, we can express the three-phase voltages as

$$\begin{aligned} V_a &= |V|, \\ V_b &= |V| e^{-j120^\circ}, \\ V_c &= |V| e^{-j240^\circ} \qquad [\mathrm{V}]. \end{aligned} \tag{4.49}$$

Alternately, the voltages and corresponding currents can be expressed as

Phase a:

$$v_a = \sqrt{2}\,|V_{ph}| \sin \omega t$$
$$i_a = \sqrt{2}\,|I_{ph}| \sin (\omega t - \phi). \tag{4.50}$$

Phase b:

$$v_b = \sqrt{2}\,|V_{ph}| \sin\left(\omega t - \frac{2}{3}\pi\right)$$
$$i_b = \sqrt{2}\,|I_{ph}| \sin\left(\omega t - \frac{2}{3}\pi - \phi\right). \tag{4.51}$$

Phase c:

$$v_c = \sqrt{2}\,|V_{ph}| \sin\left(\omega t - \frac{4}{3}\pi\right)$$
$$i_c = \sqrt{2}\,|I_{ph}| \sin\left(\omega t - \frac{4}{3}\pi - \phi\right). \tag{4.52}$$

The instantaneous three-phase power,

$$p_{3\Phi} = v_a i_a + v_b i_b + v_c i_c \qquad [\mathrm{W}]. \tag{4.53}$$

$$\begin{aligned} p_{3\Phi} &= 2|V_{ph}|\,|I_{ph}| \sin \omega t \sin(\omega t - \phi) \\ &+ 2|V_{ph}|\,|I_{ph}| \sin\left(\omega t - \frac{2\pi}{3}\right) \sin\left(\omega t - \frac{2\pi}{3} - \phi\right) \\ &+ 2|V_{ph}|\,|I_{ph}| \sin\left(\omega t - \frac{4\pi}{3}\right) \sin\left(\omega t - \frac{4\pi}{3} - \phi\right) \qquad [\mathrm{W}]. \end{aligned} \tag{4.54}$$

Using

$$2 \sin \alpha \sin \beta = \cos (\alpha - \beta) - \cos (\alpha + \beta), \tag{4.55}$$

$$\begin{aligned} p_{3\Phi} &= |V_{ph}|\,|I_{ph}| [\cos \phi - \cos (2\omega t - \phi)] \\ &+ |V_{ph}|\,|I_{ph}| \left[\cos \phi - \cos\left(2\omega t - \frac{4\pi}{3} - \phi\right)\right] \\ &+ |V_{ph}|\,|I_{ph}| \left[\cos \phi - \cos\left(2\omega t - \frac{8\pi}{3} - \phi\right)\right] \qquad [\mathrm{W}]. \end{aligned} \tag{4.56}$$

Now,

$$\cos \theta + \cos\left(\theta - \frac{4\pi}{3}\right) + \cos\left(\theta - \frac{8\pi}{3}\right) = 0, \tag{4.57}$$

where

$$\theta = 2\omega t - \phi. \tag{4.58}$$

Therefore,

$$p_{3\Phi} = 3\left|V_{\text{ph}}\right| \left|I_{\text{ph}}\right| \cos\phi \qquad \text{[W]}. \tag{4.59}$$

The sum of the three individual pulsating phase powers is a constant, nonpulsating, total power of magnitude three times the real power in a single phase.

As the instantaneous power in a three-phase balanced system is a constant with respect to time, the machine torque will be a constant, and hence the machine will run with a minimum of vibration and noise. Figure 4.13 shows the voltages and currents for the three phases as well as the three-phase power plotted against time. Without loss of generality, in Figure 4.13 it has been assumed that, in each of the three phases, the voltage is in phase with the current.

4.4.1 Three-Phase Winding Design

We proceed to show how a set of three-phase voltages can be generated. Consider the two-pole synchronous generator discussed in Example 4.5. We had earlier

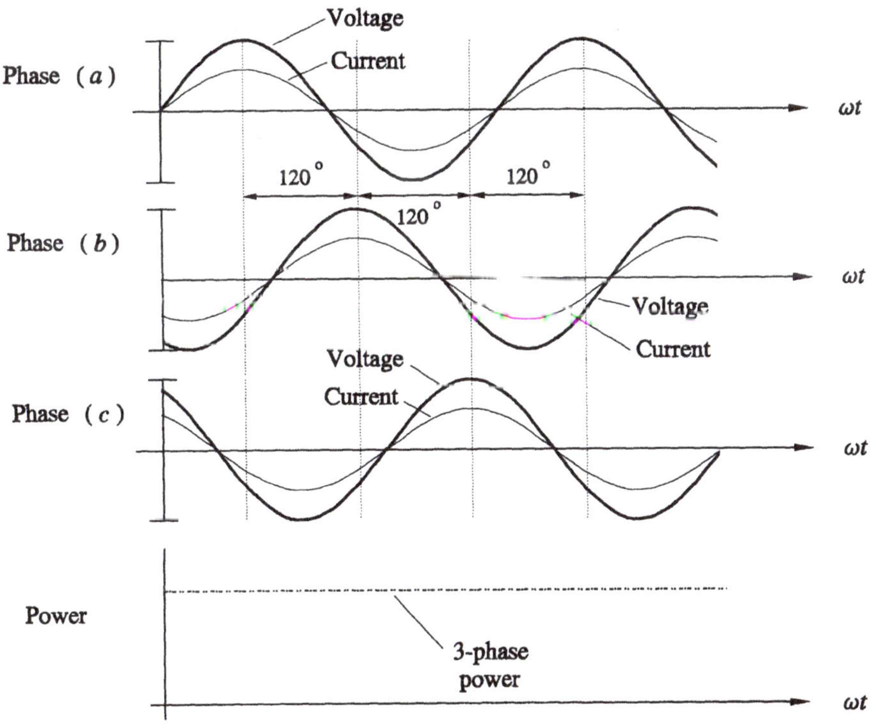

Figure 4.13

connected the total number of conductors placed in its 18 slots in series and obtained a single winding, which produced a single-phase emf of 12,894 V rms. We can convert this machine into a three-phase generator by connecting the first three coils in series to form a winding (a–a'), referred to as phase a (Figure 4.14b). Since phase b has to lag phase a by 120°, phase b starts in slot 7 and it is made up of the next three coils with terminals (b–b'). Similarly phase c starts in slot 13 and incorporates the last three coils with terminals (c–c'). The three ends of the windings a', b', and c' are joined together to form the *neutral node* (Figure 4.14f) and hence to give a *symmetrical three-phase* winding. The neutral is normally grounded.

Consider the emf E_a generated in phase a. We determined in Example 4.5 that the emf generated in each coil is 2239 V. The total phase-to-neutral emf is obtained by vectorial addition of the three individual coil emf's. The electrical phase angle between the coil emf's being 20°. By arbitrarily designating the emf of the coil whose sides are placed in slots 2 and 11 (Figure 4.14b) as the reference phasor we obtain the phasor diagram shown in Figure 4.14c. The total emf, E_a of the phase-a winding is obtained from equation (4.56). Note that q in this equation, which in the single-phase case represented the number of slots per pole, now must mean "number of slots per pole and phase." Therefore, we have

$$q = \frac{18}{2 \times 3} = 3. \tag{4.60}$$

Equation (4.48) gives

$$|E_a| = |E_1| \frac{\sin(3 \times 20/2)}{\sin(20/2)} = 2239 \frac{0.5000}{0.1736} = 6447 \qquad [\mathrm{V}]. \tag{4.61}$$

Note also that the phase angle of E_a is equal to that of the reference emf. E_b must lag E_a by 120° and similarly, E_c must lag E_b by 120°.

In summary, we have obtained the following symmetrical three-phase emf set:

$$\begin{aligned} E_a &= 6447, \\ E_b &= 6447e^{-j120^\circ}, \\ E_c &= 6447e^{-j240^\circ} \qquad [\mathrm{V}]. \end{aligned} \tag{4.62}$$

Should this generator be operated on open-circuit (no load) the voltages that can be measured between its three-phase terminals a, b, and c and the neutral are those shown in the phasor diagram in Figure 4.14f. It is easy to verify that for the symmetrical three-phase emf set we have the important relationship:

$$E_a + E_b + E_c = 0. \tag{4.63}$$

When the individual phase voltages have the relative time phase relationships shown in Figure 4.14f, one talks about the generator having the phase sequence

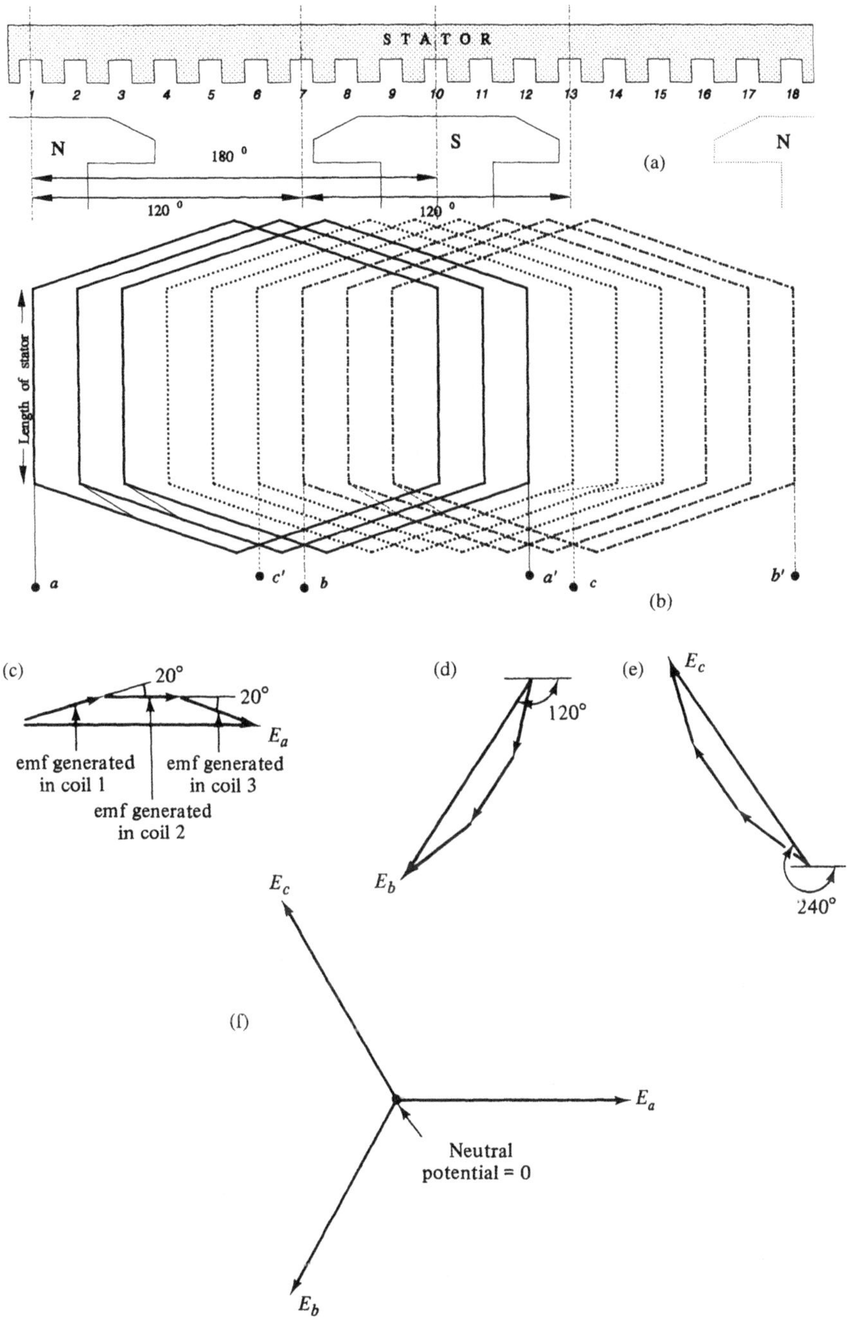
STATOR
1 2 3 4 5 6 7 8 9 10 11 12 13 14 15 16 17 18
N
S
N
180°
120°
120°
(a)
Length of stator
a
c'
b
a'
c
b'
(b)
(c)
20°
20°
E_a
emf generated in coil 1
emf generated in coil 2
emf generated in coil 3
(d)
120°
E_b
(e)
E_c
240°
(f)
E_c
E_a
E_b
Neutral potential = 0

Figure 4.14

abca ⋯. Should the rotor turn in the opposite direction, the induced emf's would have the reversed phase sequence *acba* ⋯. Note that the phase sequence can be changed from one to the other by interchanging any two of the terminals without reversing the direction of rotation.

4.4.2 Phase and Line Voltages

As was pointed out earlier, if we load the generator in a *balanced* manner (see Section 4.5) the voltages at the terminals *a*, *b*, and *c* will be different from the emf's, but they will still constitute a symmetrical three-phase set.

The voltages V_a, V_b, and V_c (Figure 4.12b) measured between the phase terminals and ground (or neutral) are referred to as *phase* or *phase-to-ground* voltages. The voltages that can be measured between terminals *a*, *b*, and *c* are referred to as *line* or *phase-to-phase* voltages. (Whenever one refers to the voltage of a three-phase system one invariably implies the *line* voltage.) The three line voltages V_{ab}, V_{bc}, V_{ca} are obtained from the phase voltages as follows:

$$\begin{aligned} V_{ab} &= V_a - V_b, \\ V_{bc} &= V_b - V_c, \\ V_{ca} &= V_c - V_a \qquad \text{[V]}. \end{aligned} \tag{4.64}$$

These voltage phasors are shown in Figure 4.12b. Using equation (4.49), we get

$$V_{ab} = V_a - V_b = V_a - V_a e^{-j120°} = \sqrt{3} V_a e^{j30°} = \sqrt{3}|V|e^{j30°} \qquad \text{[V]}. \tag{4.65}$$

Similarly,

$$V_{bc} = \sqrt{3}|V|e^{-j90°} \qquad \text{[V]} \tag{4.66}$$

and

$$V_{ca} = \sqrt{3}|V|e^{-j210°} \qquad \text{[V]}. \tag{4.67}$$

Note that the rms value of the line voltages are $\sqrt{3}$ times the rms value of the phase voltages. This can be derived graphically from the phasor diagram shown in Figure 4.12. The reader should confirm this. Note also that the line voltages form a symmetrical three-phase set.

Example 4.6

Show how by disconnecting the three-phase winding developed above and by proper reconnection as a single-phase winding we obtain anew the single-phase generator of Example 4.5(b).

Solution: We start with the three-phase winding in Figure 4.15, having its neutral grounded. (The winding is graphically displayed in a Y so as to give a direct symbolic display of the phase relationship of the three emf's.)

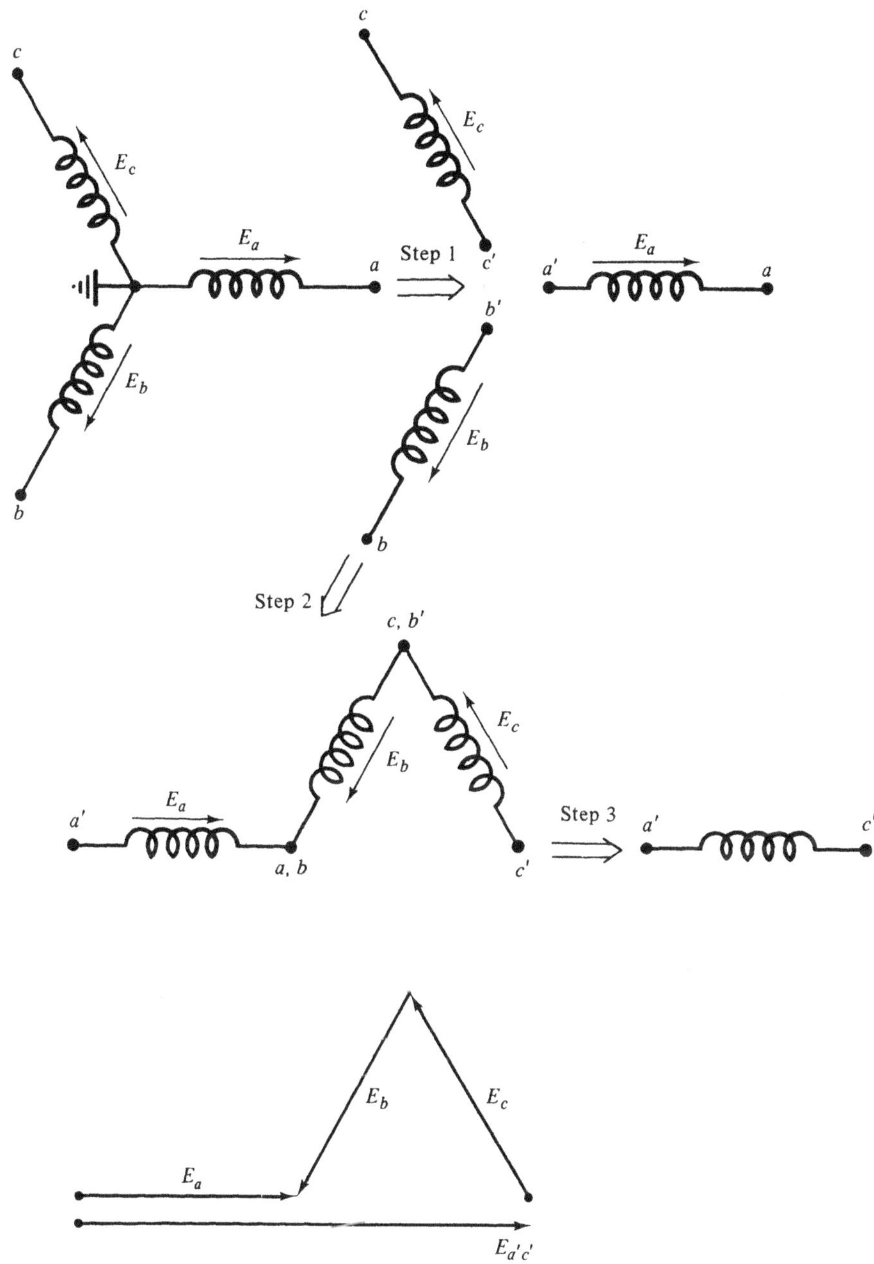

Figure 4.15

In three easy-to-follow steps we first "dissolve" the neutral, then reconnect the three-phase windings to finally obtain a single-phase winding with terminals marked a'–c'. The emf $E_{a'c'}$ which we can measure across these new terminals is clearly

$$E_{a'c'} = E_a - E_b - E_c \qquad [\text{V}]. \tag{4.6.1}$$

But from equation (4.63) we have

$$E_b + E_c = -E_a \qquad [\text{V}], \tag{4.6.2}$$

and by substitution into equation (4.6.1) we get

$$E_{a'c'} = 2E_a \qquad [\text{V}]. \tag{4.6.3}$$

(An equally simple deduction can be made graphically as shown in Figure 4.15.)

The numerical rms value of $E_{a'c'}$ is

$$|E_{a'c'}| = 2E_a = 2 \cdot 6447 = 12{,}894 \ [\text{V}], \tag{4.6.4}$$

which agrees with the earlier result in Example 4.5b.

4.5 Balanced Three-Phase Loading

Assume that the three-phase generator discussed in the previous section is used to supply a load. Loading can be achieved in several ways:

1. A single load element, for example an electric heater, is connected either between one-phase terminal and neutral, or between two-phase terminals.
2. Three identical load elements are connected between the phase terminals and ground to form a *Y-connected* three-phase load. (It is demonstrated in Chapter 8 that a symmetrical three-phase motor is equivalent to a Y-connected load.)
3. Three identical load elements are connected between the phase terminals to form a Δ-*connected* three-phase load.
4. The generator is *synchronized* onto an existing electrical power network and then made to supply a share of the power to a city, for example.

In the first case the generator would be loaded *unsymmetrically* or *unbalanced*. This would defeat the real purpose of three-phase power and will not be discussed further.[9] The three remaining loading methods all result in a *balanced* or *symmetrical* load. We treat them separately.

[9] Single-phase loading is very common in domestic distribution systems. However, the power company distributes the customers in certain areas equally between the three phases, so as to achieve balanced *overall* loading (also see Chapter 6).

4.5.1 Balanced Loading Between Phase Terminals and Ground (Y-Connected Load)

The Y-connected load is shown in Figure 4.16a. The three *phase currents* are related to the phase voltages by the following equations:

$$
\begin{aligned}
I_a &= \frac{V_a}{Z} \quad [\mathrm{A}],\\
I_b &= \frac{V_b}{Z} \quad [\mathrm{A}],\\
I_c &= \frac{V_c}{Z} \quad [\mathrm{A}].
\end{aligned}
\tag{4.68}
$$

If the impedance Z is written as

$$Z = |Z| e^{j\phi} \quad [\Omega], \tag{4.69}$$

then from equation (4.49) the phase currents can be written as

$$
\begin{aligned}
I_a &= \frac{|V|}{|Z|} e^{-j\phi} \quad [\mathrm{A}],\\
I_b &= \frac{|V|}{|Z|} e^{-j(120^\circ + \phi)} \quad [\mathrm{A}],\\
I_c &= \frac{|V|}{|Z|} e^{-j(240^\circ + \phi)} \quad [\mathrm{A}].
\end{aligned}
\tag{4.70}
$$

The current and voltage phasors are shown in Figure 4.16b. We conclude that the currents constitute a symmetrical three-phase set.

By vectorial addition, we can readily confirm that

$$I_a + I_b + I_c = 0 \quad [\mathrm{A}]. \tag{4.71}$$

This means that the current I_n in the neutral lead is zero; in other words, a return conductor is superfluous.

Example 4.7

A 60-Hz, 220-V, three-phase generator is loaded by three equal impedances,

$$Z = 1.21 + j0.31 \ [\Omega] \tag{4.7.1}$$

connected between each phase and ground. Find the phase currents.

Solution: The voltage rating of three-phase generators is always given in line value. For the phase voltages we have

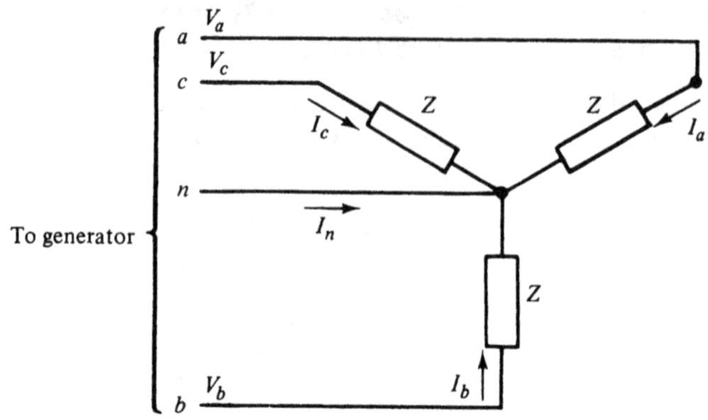

(a)

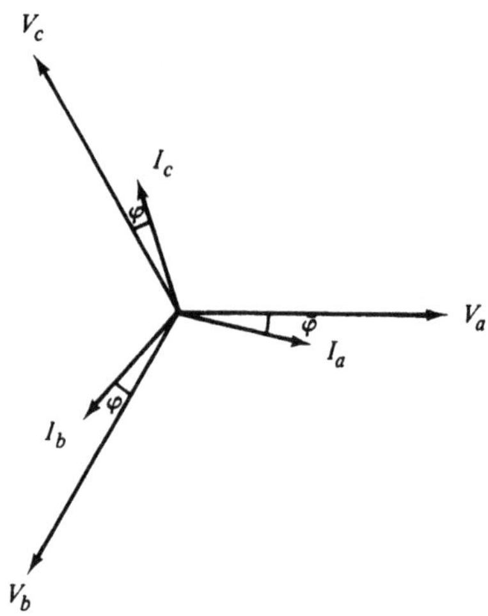

(b)

Figure 4.16

$$V_a = \frac{220}{\sqrt{3}} = 127.0V; \tag{4.7.2}$$

$$V_b = 127.0e^{-j120^\circ} \qquad [\text{V}]; \tag{4.7.3}$$

$$V_c = 127.0e^{-j240^\circ} \qquad [\text{V}]. \tag{4.7.4}$$

As the impedance is

$$Z = 1.249\ e^{j14.4^\circ} \qquad [\Omega], \tag{4.7.5}$$

we obtain from equation (4.70)

$$I_a = 101.7e^{-j14.4°},$$
$$I_b = 101.7e^{-j134.4°}, \qquad (4.7.6)$$
$$I_c = 101.7e^{-j254.4°} \qquad [A].$$

Note, for a Y-connected load, the phase current is the same as the line current. The phase voltage is $1/\sqrt{3}$ times the line voltage.

4.5.2 Balanced Loading Between Phase Terminals (Δ-Connected Load)

The three equal impedances, Z, are now connected across the three line voltages as shown in Figure 4.17a. For the three Δ-*phase currents*, we have

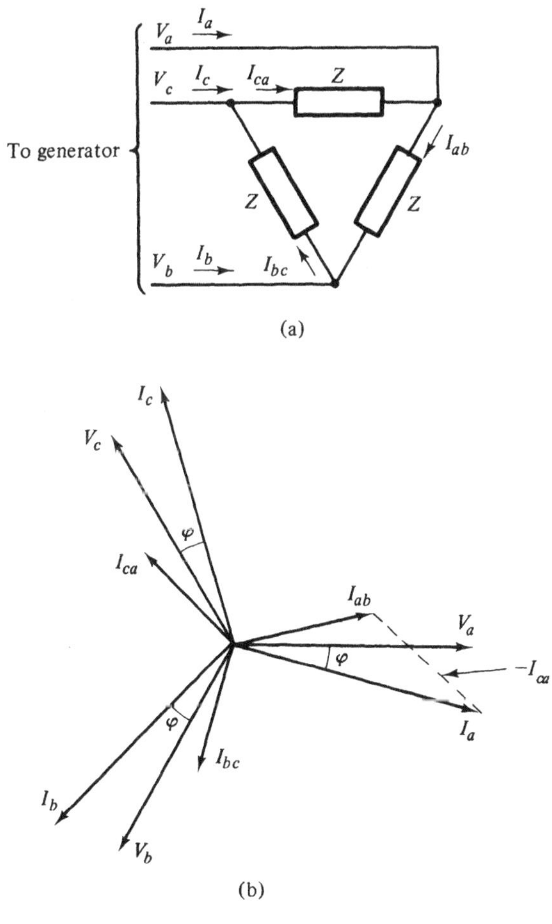

Figure 4.17

$$I_{ab} = \frac{V_{ab}}{Z} \quad [\mathrm{A}],$$

$$I_{bc} = \frac{V_{bc}}{Z} \quad [\mathrm{A}], \tag{4.72}$$

$$I_{ca} = \frac{V_{ca}}{Z} \quad [\mathrm{A}].$$

By using the expression for the line voltages, (4.65), (4.66), and (4.67), we solve for the Δ-phase currents:

$$I_{ab} = \sqrt{3}\frac{|V|}{|Z|} e^{j(30^\circ - \phi)},$$

$$I_{bc} = \sqrt{3}\frac{|V|}{|Z|} e^{j(-90^\circ - \phi)}, \tag{4.73}$$

$$I_{ca} = \sqrt{3}\frac{|V|}{|Z|} e^{j(-210^\circ - \phi)} \quad [\mathrm{A}].$$

With a knowledge of the Δ-phase currents we can, finally, compute the *line currents*:

$$I_a = I_{ab} - I_{ca} = \sqrt{3}\frac{|V|}{|Z|}\left(e^{j(30^\circ - \phi)} - e^{j(-210^\circ - \phi)}\right) \quad [\mathrm{A}]. \tag{4.74}$$

Therefore,

$$I_a = 3\frac{|V|}{|Z|} e^{-j\phi} \quad [\mathrm{A}]. \tag{4.75}$$

Similarly,

$$I_b = 3\frac{|V|}{|Z|} e^{-j(120^\circ + \phi)} \quad [\mathrm{A}]; \tag{4.76}$$

$$I_c = 3\frac{|V|}{|Z|} e^{-j(240^\circ + \phi)} \quad [\mathrm{A}]. \tag{4.77}$$

Figure 4.17b shows the relationship between the phase voltages, the Δ-phase currents and the line currents.

We conclude that

1. The Δ-phase currents have the same phase relation to their respective line voltages as in the Y-connected case, but their rms values are three times larger (assuming, of course, the same load impedance).
2. The rms value of the line current is $\sqrt{3}$ times the rms value of the Δ-phase current.

Example 4.8

Reconnect the three impedances in Example 4.7 into a Δ-load. Find all the currents.

Solution: The line currents will have an rms value of

$$3 \cdot 101.7 = 305.1 \qquad [\text{A}]. \tag{4.8.1}$$

The rms value of the Δ-phase currents will be

$$\frac{305.1}{\sqrt{3}} = 176.2 \qquad [\text{A}]. \tag{4.8.2}$$

Example 4.9

The terminal voltage of a three-phase generator measured phase-to-phase (line) is equal to 13.2 kV. It is symmetrically loaded and delivers an rms current of 1.230 kA per phase at a phase angle of $\phi = 18.3°$ lagging (meaning the current lags the voltage as shown in Figure 4.16). Compute the power delivered by the machine.

Solution: We first compute the rms value, $|V|$, of the phase voltage.

$$|V| = \frac{13.2}{\sqrt{3}} = 7.621 \qquad [\text{kV/phase}]. \tag{4.9.1}$$

Equations (4.16) and (4.17) then yield

$$P = 7.621 \cdot 1.230 \cdot \cos 18.3° = 8.900 \qquad [\text{MW/phase}], \tag{4.9.2}$$

$$Q = 7.621 \cdot 1.230 \cdot \sin 18.3° = 2.943 \qquad [\text{MVAr/phase}]. \tag{4.9.3}$$

The power in the phases a, b, and c will pulsate as shown in Figure 4.1, with

$$\begin{aligned} p_a &= 8.900(1 - \cos 2\omega t) - 2.943 \sin 2\omega t, \\ p_b &= 8.900(1 - \cos 2[\omega t - 120°]) - 2.943 \sin 2(\omega t - 120°), \\ p_c &= 8.900(1 - \cos 2[\omega t - 240°]) - 2.943 \sin 2(\omega t - 240°). \end{aligned} \tag{4.9.4}$$

The total three-phase power will be

$$p_{3\phi} = 3 \cdot 8.900 = 26.700 \qquad [\text{MW}]. \tag{4.9.5}$$

These various single-phase powers are plotted in Figure 4.18.

***Important Note*:** The fact that the three-phase power is constant tempts us to believe that the reactive power in a three-phase system is zero (as in a dc circuit). However, the reactive power is present in each phase as given in equation (4.9.4). The reactive power per phase is 2.943 MVAr. In power engineering lingo one

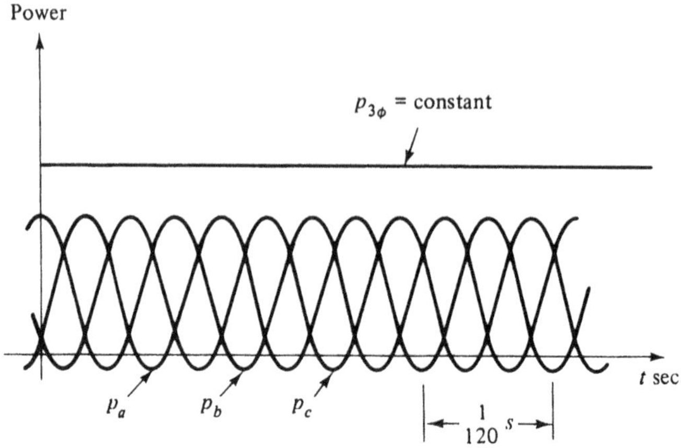

Figure 4.18

would say that the reactive power produced by the generator is 8.829 ($3 \cdot 2.943$) MVAr three-phase. (This is natural because the real three-phase power was found to be three times the per-phase value.)

Example 4.10

Find the power delivered to the load in Examples 4.7 and 4.8.

Solution: In Example 4.7 we had

$$|V| = 127.0 \qquad [\text{V}], \tag{4.10.1}$$

$$|I| = 101.7 \qquad [\text{A}], \tag{4.10.2}$$

$$\cos\phi = 0.969; \tag{4.10.3}$$

$$\therefore p_{3\phi} = 3 \cdot 127.0 \cdot 101.7 \cdot 0.969 = 37.5 \qquad [\text{kW}]. \tag{4.10.4}$$

In the Δ-connected case (Example 4.8) the current is tripled. Thus the power will also triple, that is,

$$p_{3\phi} = 3 \cdot 37.5 = 112.5 \qquad [\text{kW}]. \tag{4.10.5}$$

4.5.3 The Generator Operating as Part of a Power Grid

This is the most important form of loading. Most three-phase generators operate in this mode. We shall treat this case separately in Section 4.7. Here, it will suffice to say that even in this case the phase currents will constitute a symmetrical three-phase set.

4.6 Torque Mechanism in a Three-Phase Generator

When a loaded three-phase synchronous generator delivers a constant electric power $p_{3\phi}$, it is clear that the power originates from the prime mover that drives the generator. Let us now consider the important mechanism whereby this mechanical power is transformed into electric power. The mechanical power p_{mech} delivered by the prime mover must be equal to the electric power $p_{3\phi}$ plus the losses p_{loss} sustained in the process of transformation. These losses are, in general, relatively small, hence we can write:

$$p_{\text{mech}} = p_{3\phi} + p_{\text{loss}} \approx p_{3\phi} \qquad [\text{W}]. \tag{4.78}$$

The mechanical power, according to equation (2.24), is associated with a mechanical torque, delivered by the prime mover:

$$T_{\text{mech}} = \frac{p_{\text{mech}}}{\omega_{\text{mech}}} \qquad [\text{N} \cdot \text{m}], \tag{4.79}$$

where ω_{mech} is the synchronous angular velocity of the rotor measured in [rad/s]. Both the velocity and mechanical power are constant and, consequently, so is the torque. The direction of the torque is such as to maintain the speed, that is, T_{mech} tends to accelerate the rotor in the ω direction.

The fact that the rotor does not experience an acceleration is because of the existence of a counteracting torque T_{em} of equal magnitude but opposite in direction. The *electromechanical torque* is created by the interaction between the rotor-bound flux and the stator-bound current. The presence of T_{em} acting on the rotor necessitates a *reaction torque* of equal magnitude, but in the opposite direction, acting on the stator. This reaction torque evidently tends to tilt the stator in the direction of rotation. The stator is prevented from rotating by the common mechanical support (concrete floor) of both the prime mover and the generator. There is, of course, also a reaction torque acting on the static portion of the prime mover. This torque is equal in magnitude to T_{mech} but is in the opposite direction, that is, it tends to tilt the prime mover housing in the direction opposite to ω_{mech}.

The various torques and their directions as well as the direction of rotation of the rotor and power flow are shown in Figure 4.19.

Example 4.11

A 4-pole synchronous generator delivers an electric power of 1000 MW. Determine the magnitude of T_{mech} and T_{em}

Solution: The speed of a 4-pole generator (assuming 60 Hz) is 1800 rpm. Thus,

$$\omega_{\text{mech}} = \frac{1800}{60} \cdot 2\pi = 188.5 \qquad [\text{rad/s}] \tag{4.11.1}$$

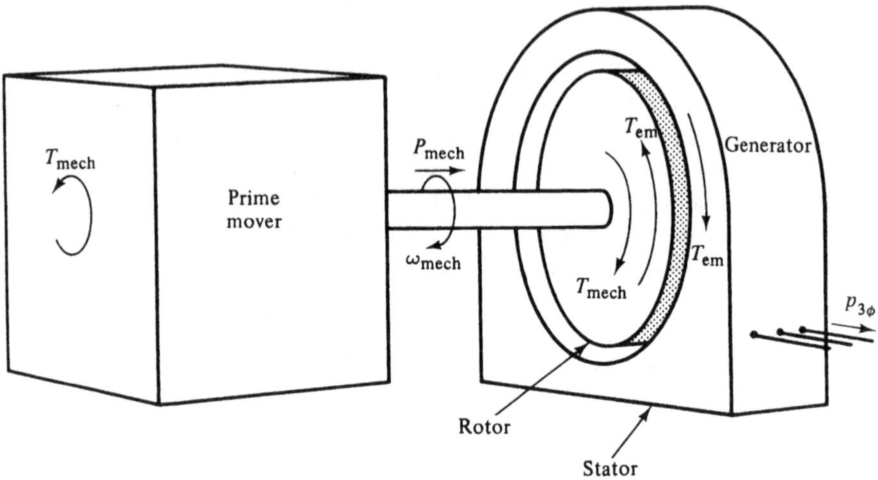

Figure 4.19

Equation (4.79) then yields

$$T_{em} = T_{mech} = \frac{1000 \times 10^6}{188.5} = 5.31 \times 10^6 \qquad [\text{N} \cdot \text{m}] \qquad (4.11.2)$$

or 541 ton-meters.

4.6.1 The Stator "Current Wave"

The electromechanical torque T_{em} acting across the air gap of the generator is evidently the key factor in the energy transformation process. How is it formed? What characteristics does it have? All rotating electric energy converters, motors and generators alike, differ in essence only in the formation of their electromechanical torque. In our study of these devices we are, therefore, most interested in these torques. Let us investigate the features of the torque of a symmetrically loaded three-phase synchronous machine.

The torque, as already mentioned, arises from the interaction between the rotor flux and the stator current. We know the former to be a flux with a sinusoidally distributed density, fixed with respect to the rotor (Figure 4.6). As the shaft rotates, the flux rotates with it, and it is then a constant-amplitude traveling flux wave. The voltage induced in the stator coils must be sinusoidal, and, assuming a linear load, the currents must also be sinusoidal. It follows that the currents must combine to form a constant-amplitude traveling wave moving along the periphery of the stator with a speed equal to that of the rotor flux wave.

Of course, the stator currents exist only in the stator slots. However, if we view the stator slot currents, in a "macrosense," as being "smeared" over the total surface, then the stator current can be considered a *surface* or *sheet* current.

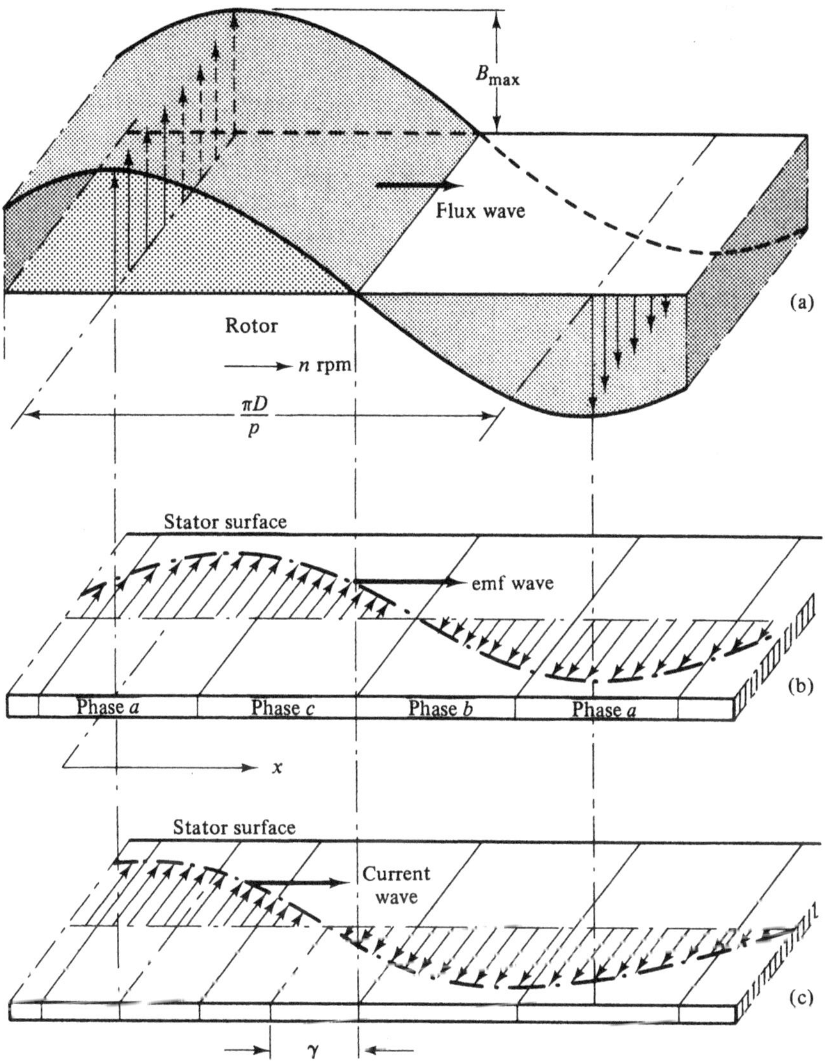

Figure 4.20

4.6.2 Torque and Power

Figure 4.20 summarizes the system of waves existing in a loaded synchronous machine. (All high-frequency harmonics have been left out.) We have the magnetic flux wave bound to the rotor and traveling with constant velocity and constant amplitude B_{max}. It is described by equation (4.34) and shown in Figure 4.20a. As the flux wave sweeps by the stator conductors it induces an emf wave that follows the B wave. The crest of the emf wave coincides with that of the B wave (Figure 4.20b).

Then, there is the current wave, bound to the stator (Figure 4.20c). We shall assume here that, in general, the current in phase a lags the emf in phase a by the angle γ. This means, that the current wave trails the emf (or flux) wave by γ radians. Because the current wave has the same speed as the flux wave and an equal number of maxima, we can, by analogy with equation (4.34), express it in the form:

$$A = A_{\max} \cos\left(\frac{px}{D} - \omega t - \gamma\right) \qquad \text{[A/m]}. \tag{4.80}$$

(Note that the physical dimension of A is amperes per (tangential) meter, that is, it is a *surface current density*.)

If we consider an elemental strip of the stator surface (Figure 4.21) of axial length L and tangential width dx, it is clear that the elemental current flowing in the elemental strip will interact with the flux to produce a tangential force, df. According to equation (3.56) the force on the strip will be

$$df = LB\,di = LBA\,dx \qquad \text{[N]}. \tag{4.81}$$

Therefore,

$$df = LB_{\max} \cos\left(\frac{px}{D} - \omega t\right) A_{\max} \cos\left(\frac{px}{D} - \omega t - \gamma\right) dx \qquad \text{[N]}. \tag{4.82}$$

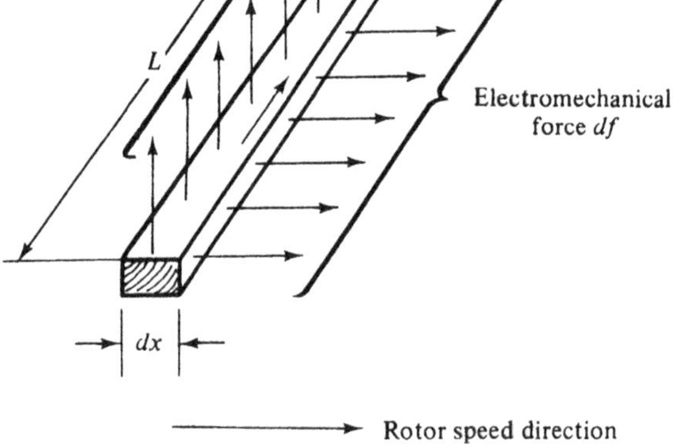

Figure 4.21

Integration from $x = 0$ to $x = \pi D$ yields the total force on the stator:

$$f = LB_{\max}A_{\max}\int_{x=0}^{\pi D}\cos\left(\frac{px}{D} - \omega t\right)\cos\left(\frac{px}{D} - \omega t - \gamma\right)dx \qquad [\mathrm{N}]. \tag{4.83}$$

The integration gives

$$f = \frac{1}{2}\,\pi LD^2 B_{\max}A_{\max}\cos\gamma \qquad [\mathrm{N}]. \tag{4.84}$$

Multiplication by D/2 gives the electromechanical torque:

$$T_{\mathrm{em}} = \frac{1}{4}\,\pi LD^2 B_{\max}A_{\max}\cos\gamma \qquad [\mathrm{N}\cdot\mathrm{m}]. \tag{4.85}$$

Further multiplication by $\omega_{\mathrm{mech}} = (n\pi)/30$ yields the corresponding electrical power:

$$p_{3\Phi} = \frac{1}{120}\,\pi^2 nLD^2 B_{\max}A_{\max}\cos\gamma \qquad [\mathrm{W}]. \tag{4.86}$$

By introducing the rotor volume

$$V_{\mathrm{rot}} = \frac{\pi}{4}LD^2 \qquad [\mathrm{m}^3], \tag{4.87}$$

we can write expressions for the torque and power as

$$T_{\mathrm{em}} = V_{\mathrm{rot}}B_{\max}A_{\max}\cos\gamma \qquad [\mathrm{N}\cdot\mathrm{m}] \tag{4.88}$$

and

$$p_{3\Phi} = \frac{\pi}{30}\,nV_{\mathrm{rot}}B_{\max}A_{\max}\cos\gamma \qquad [\mathrm{W}], \tag{4.89}$$

respectively.

4.6.3 Some Practical Observations

We can draw some important conclusions from the equations just derived.

- The electromechanical force, f, and thus the torque, T_{em}, acting on the stator have a direction that agrees with that depicted in Figure 4.19.
- As the angle γ is a constant, both the torque and power must be constants. This reinforces our earlier findings regarding three-phase power.
- The power and torque increase in direct proportion to the volume of the machine.
- The power increases in direct proportion to speed. (This means, for example, that, everything else being equal, one can obtain 10 times more "kilowatts per kilogram"

from a 3600 rpm steam turbine than from a slow-running 360 rpm hydrogenerator. (This is true for all types of motors as well.)

- Power and torque increase in direct proportion to the magnetic flux density used. Magnetic saturation sets the limit to it. If we wish to "squeeze" more flux out of the rotor, then we must use a disproportionate rotor field current, and this will cause heating problems in the rotor winding, as well as additional mechanical stresses on the rotor and its bearings.
- Power and torque increase in direct proportion to the current density in the stator surface. Ohmic losses and resulting temperature elevation in the windings set practical limits. In the last two decades we have seen dramatic increases in current densities due to new and better insulating materials that can withstand higher temperatures, but the most important development has been in improved methods of *forced cooling*. Typically, water or hydrogen is pumped at high velocity through hollow stator conductors, thus removing the ohmic heat and allowing higher current densities. By *supercooling* of the windings, the resistance, and thus the ohmic losses, disappear altogether. One should then be able to use limitless current densities. This technology is still in the experimental stages (see Chapter 1).

4.6.4 Power and the Angle γ

An interesting aspect of equations (4.88) and (4.89) is the dependence of the generated power on the angle γ. As this angle is permitted to vary throughout the range

$$-\pi < \gamma < +\pi, \tag{4.90}$$

the power (or torque) will vary, as shown in Figure 4.22. In the range

$$-\frac{\pi}{2} < \gamma < +\frac{\pi}{2}, \tag{4.91}$$

the force on the stator has the direction shown in Figure 4.21; the reaction force on the rotor is in the opposite direction. This agrees with the torque directions

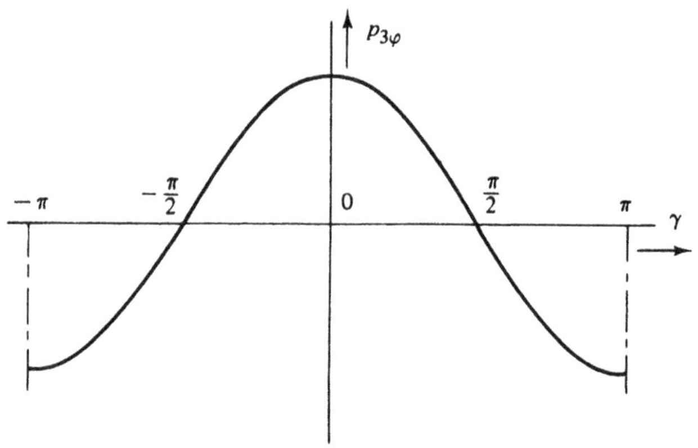

Figure 4.22

shown in Figure 4.19, and the machine is now operating as a *generator*. When γ exceeds $\pi/2$ (or is less than $-\pi/2$) the force and torque change polarity and the machine is now a *motor*. For $\gamma = \pi/2$ or $-\pi/2$ the torque and power are both equal to zero, although the stator windings may carry full-load currents. The machine is now operating as a *synchronous condenser* or a *synchronous inductor* (see Example 4.12).

Clearly, the angle γ determines whether the synchronous machine is operating as a generator or motor and whether it is producing a large or small power output. The phase angle γ between the flux and current waves becomes the most important factor in controlling the flow of power from or to the machine. How this can be accomplished is our next topic.

4.7 The Synchronous Machine as Part of a Power Grid

The idea of the power grid is to connect many synchronous machines in parallel so that if the local generating station were to go out of service, power can still be supplied from the other stations on the grid. Figure 4.23 shows a typical power grid. The geographic area covered by a grid can be of the order of hundreds of thousands of square kilometers with hundreds of generating stations involved. Figure 4.24 shows how the generating station(s) and the load(s) are connected

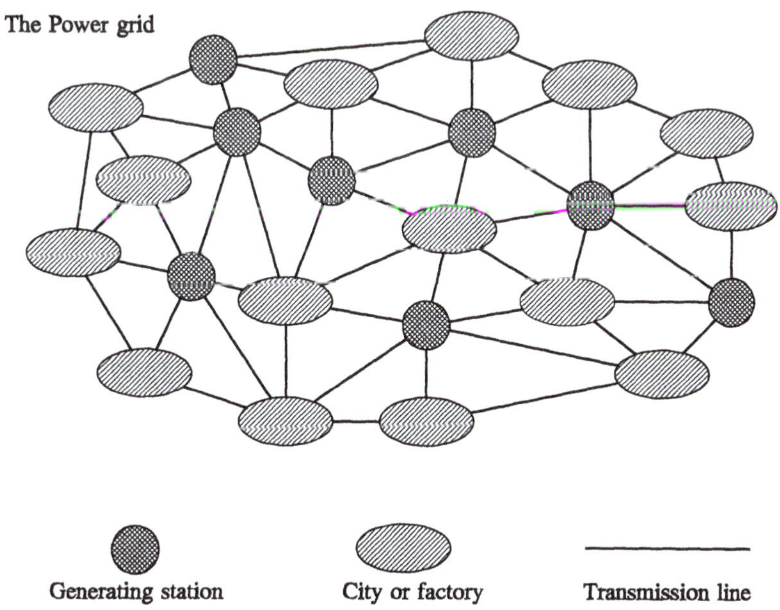

Figure 4.23

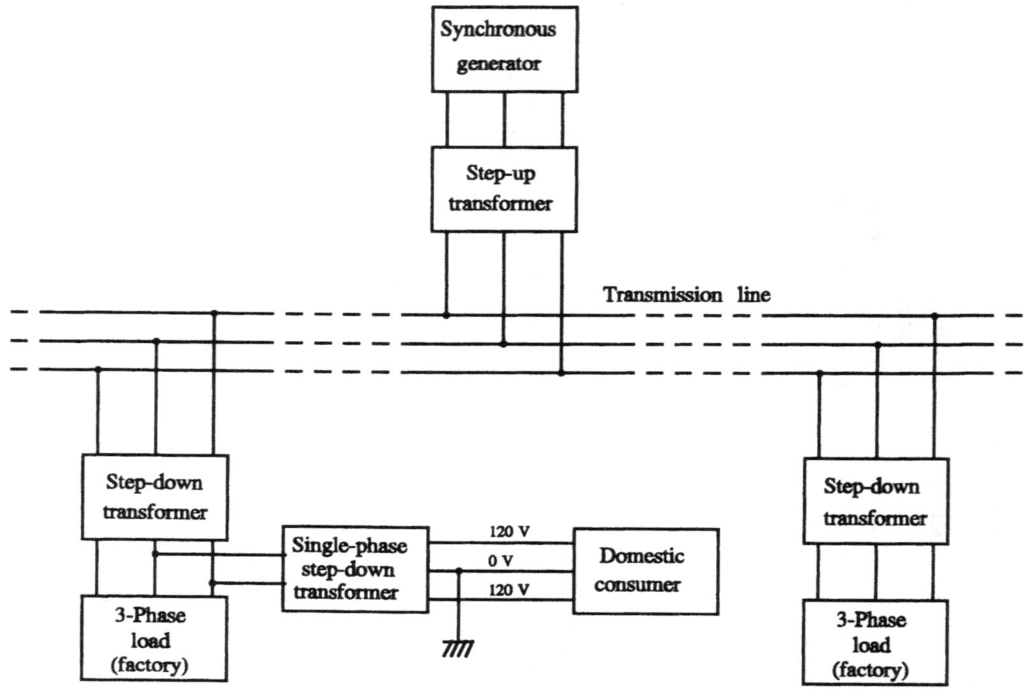
Synchronous generator
Step-up transformer
Transmission line
Step-down transformer
3-Phase load (factory)
Single-phase step-down transformer
120 V
0 V
120 V
Domestic consumer
Step-down transformer
3-Phase load (factory)

Figure 4.24

through transformers. The role played by the transformer in the power grid is discussed in Chapter 5.

The total power capability of the grid is so large that even a 1000-MVA machine forms a very small fraction of the total. The system is described as an *infinite source*. It is assumed that it is capable of supplying infinite current at constant voltage (ideal voltage source) and constant frequency (60.00 ± 0.05 Hz is the standard in North America).

4.7.1 Synchronization of the Machine to the Grid

Before a synchronous machine can be connected to the grid, the following conditions must exist:

1. The frequency of the grid and machine emf must be the same (the prime mover is used to turn the machine at the correct rpm: 3600 rpm for a 2-pole; 1800 for a 4-pole, etc.).
2. The phase sequence of the grid and machine must be the same (the phase sequence is either *abca* ··· or *acba* ···; it can be changed by interchanging any two of the three terminals).
3. The magnitude of the machine emf must be equal to that of the grid voltage (the machine emf is adjusted by varying the rotor field current).
4. The phase difference between the machine emf and the grid voltage must be zero (the prime mover is used to adjust the speed very gently until the instant when all three voltage phasors are coincident).

If and only if all four conditions are satisfied can the machine be connected safely to the grid.

The circuit for synchronizing the machine is as shown in Figure 4.25. When the switches *a*, *b*, and *c* are open and the machine is stationary, the three lamps blink 120 times a second due to the grid frequency of 60 Hz. The blinking is not visible to the human eye. The prime mover is used to apply torque to the synchronous machine, and, as it accelerates, the lamps can be observed to blink (more slowly) at a rate equal to twice the difference in frequency between the grid and the machine emf.

When all four conditions have been met, all the lamps go off simultaneously—the synchronous machine is applying identical voltages to the left terminals of the three lamps as the grid is applying to the right terminals; zero volts appear across each lamp. The ganged switch can be closed. The machine is then said to be *synchronized* to the grid. To understand the phenomenon of synchronization, it is necessary to examine the nature of the magnetic field that is created when the stator of the synchronous machine is connected to the grid. Figure 4.26 shows a simplified stator winding of a 2-pole machine.

When $\omega t = 0$, the current i_a in coil $(a\text{–}a^-)$ is at its peak positive value, and, by using the right-hand rule, it would create a flux $\Phi_a = \Phi_{\text{peak}}$ perpendicular to the

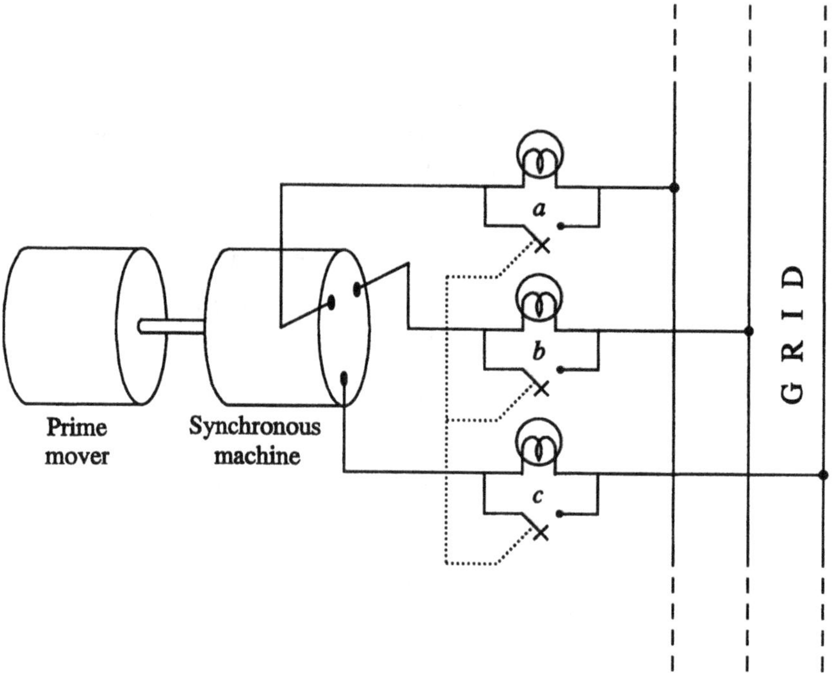

Figure 4.25

plane of coil (a–a^-) as shown. At the same instant, the current i_b is negative and equal to half the magnitude of i_a and it creates a magnetic field Φ_b, which is half the value of Φ_{peak} with an orientation perpendicular to the coil (b–b^-), as shown. Similarly, the current i_c creates a flux Φ_c that is half the value of Φ_{peak} and oriented at right angles to the plane of the coil (c–c^-), as shown. The sum of the three fluxes is

$$\Phi_T = \frac{3}{2}\Phi_{\text{peak}} \qquad [\text{Wb}], \tag{4.92}$$

and its orientation is perpendicular to the plane of the coil (a–a^-).

At the instant when $\omega t = \pi/3$, i_a and i_b are positive and half the value of the peak current. The flux $\Phi_a = \Phi_b = \frac{1}{2}\Phi_{\text{peak}}$, and their orientation is as shown in the diagram. The current i_c is at its negative peak value and produces a flux Φ_{peak} with an orientation as shown. The sum of the three fluxes Φ_T is again $3/2\Phi_{\text{peak}}$ and its orientation is perpendicular to the plane of the coil (c–c^-).

Following the procedure given above, it can be seen that when $\omega t = \frac{2}{3}\pi$, the sum of the fluxes is Φ_T, and it is perpendicular to the coil (b–b^-), as shown.

It is clear from the above discussion that the stator windings of the synchronous machine, when connected to the grid, produces a rotating flux of constant

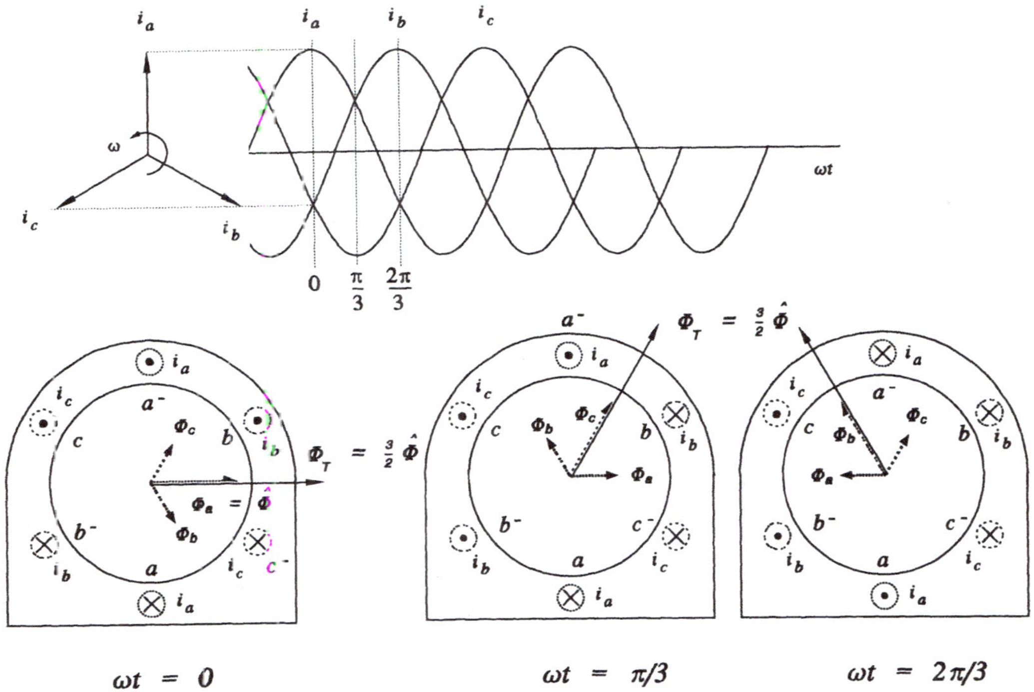

i_a
i_b
i_c
ω
ωt
0
$\frac{\pi}{3}$
$\frac{2\pi}{3}$
$\Phi_T = \frac{3}{2}\hat{\Phi}$
$\Phi_a = \hat{\Phi}$
Φ_a
Φ_b
Φ_c
a
a^-
b
b^-
c
c^-
$\omega t = 0$
$\omega t = \pi/3$
$\omega t = 2\pi/3$

Figure 4.26

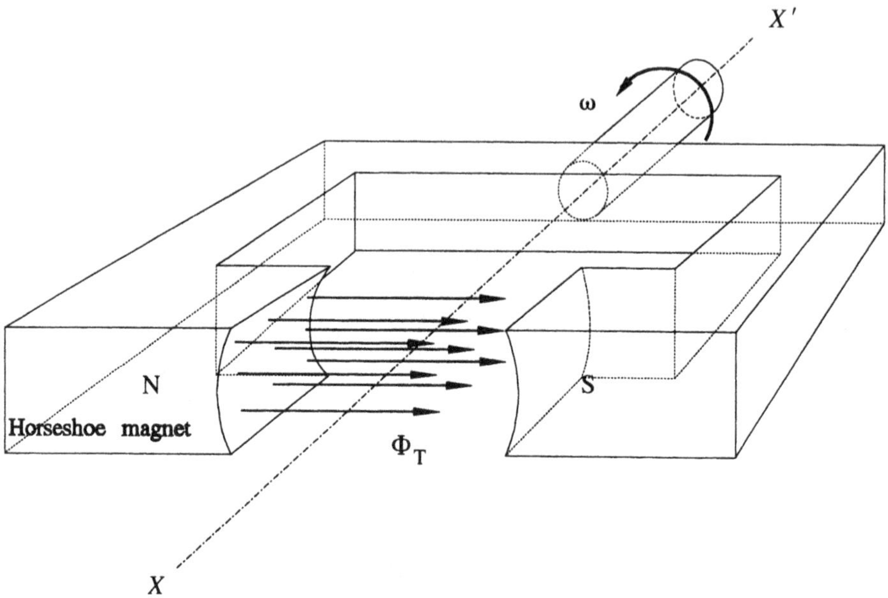

Figure 4.27

magnitude Φ_T. The action of the stator is exactly the same as that of the horseshoe magnet, shown in Figure 4.27; when it is rotated mechanically about the axis X–X' at a constant angular velocity (synchronous speed), it produces a rotating flux of constant magnitude.

The rotor of a 2-pole synchronous machine, when supplied with dc, may be represented by a simple bar magnet. With the rotor in place, an exact analog of the synchronous machine is as shown in Figure 4.28. With no load on the shaft of the bar magnet, the magnetic fields will line up, and the bar magnet will rotate at synchronous speed. As the load on the shaft is increased, the bar magnet will respond by increasing the misalignment between the fields of the two magnets without slowing down. Figure 4.29 shows a sketch of the distorted field.

4.7.2 Synchronous Machine Control

After synchronization, changes can be made to the machine parameters, depending on its intended function.

4.7.2.1 Effects of Prime Mover Torque

As we increase the prime mover torque, the generator will tend to accelerate away from the grid. However, it is locked to the grid, and, rather than pulling away, the

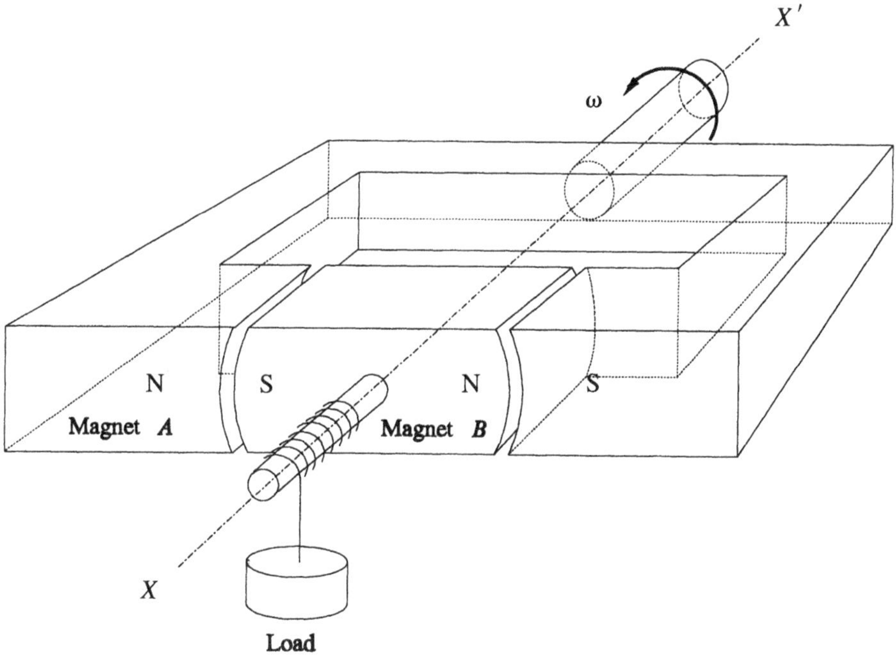

Figure 4.28

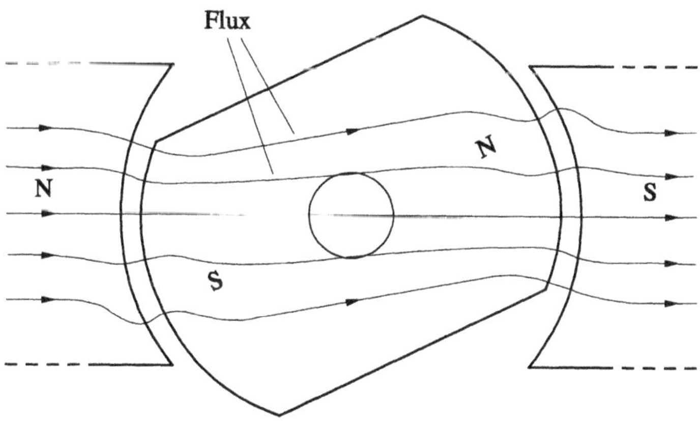

Figure 4.29

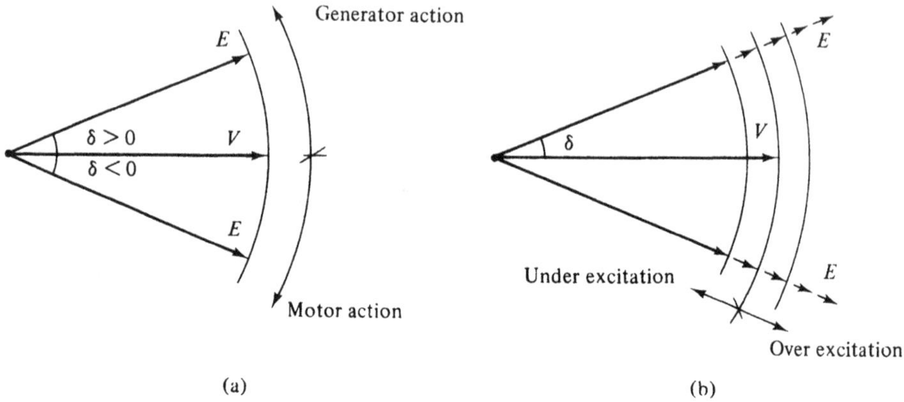

Figure 4.30

rotor will advance its phase relative to the grid voltage by some angle. As the emf wave follows the rotor (Figure 4.20) the effect will be an advancement in the phase of E. This is shown in Figure 4.30a. The *power angle,* δ, is a measure of the real power, $p_{3\phi}$, delivered by the machine. It is positive when E leads V.

If we apply a negative torque by letting the machine pull a mechanical load, thus acting as a motor, then E would lag V and the power angle would become negative. The grid will then be supplying the power for pulling the load attached to the machine.

4.7.2.2 Effects of Field-Current Control

Adjustment of the field current to the rotor cannot affect the real power or torque delivered by the prime mover. Thus no change can take place in the real power delivered by the generator, and no direct change of the power angle δ can occur. (An indirect change will occur, which is explained below.)

An increase in field current results in an increase in the magnitude of B_{max} and hence that of $|E|$, thus making $|E| > |V|$. We refer to this case as *overexcitation.* A decrease in field current will decrease $|E|$, thus making $|E| < |V|$. This is *underexcitation.*

4.7.2.3 Summary

A change in *torque* moves the tip of E tangentially along concentric circles. A change in *field current* moves the tip of E radially between the circles. These are shown in Figure 4.30.

4.7.3 Phasor Diagram

As the machine emf E changes both in phase and magnitude and the grid voltage V stays constant, we clearly obtain a difference voltage:

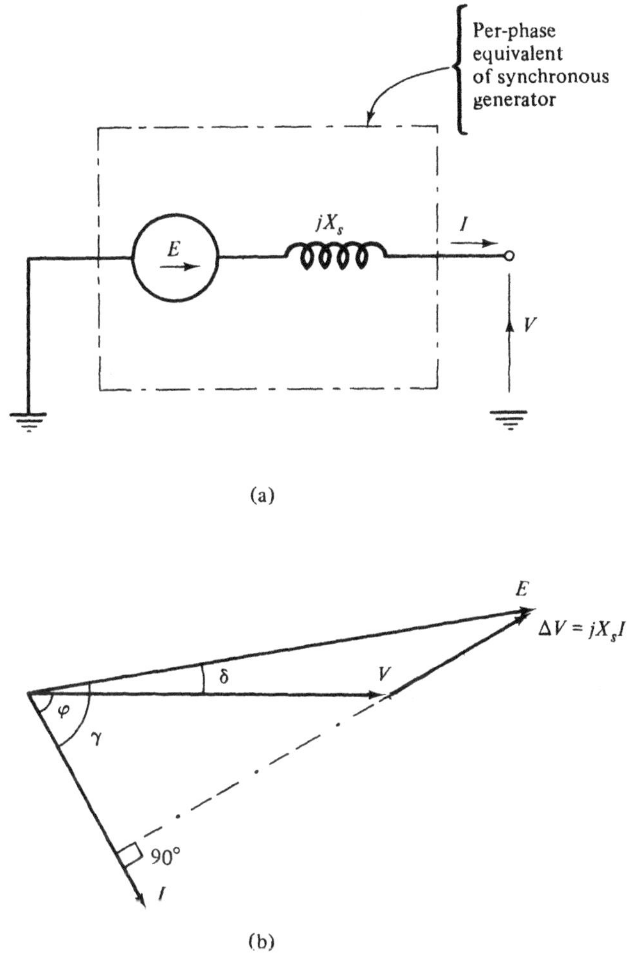

Figure 4.31

$$\Delta V = E - V \qquad \text{[V/phase]}. \tag{4.93}$$

This will give rise to a current I (as shown in Figure 4.31a) of value

$$I = \frac{\Delta V}{Z_s} \qquad \text{[A/phase]}, \tag{4.94}$$

where Z_s represents the *synchronous impedance* of the stator winding.[10] Z_s has a real and a reactive part and can be written as

[10] We have neglected any impedance between the generator terminals and the system bus. Therefore, V is the terminal voltage of the machine.

$$Z_s = R_s + j\omega L_s \qquad [\text{V/phase}]. \tag{4.95}$$

R_s is the resistance and L_s is the inductance, both measured per phase. The inductance, which is a measure of the coupled magnetic flux per ampere, is relatively large. This is due to the fact that the magnetic flux path, except for the small air gap, is iron. For a typical machine we have

$$\omega L_s \gg R_s, \tag{4.96}$$

and with good approximation we can set

$$Z_s \approx j\omega L_s \equiv jX_s \qquad [\text{V/phase}], \tag{4.97}$$

X_s being referred to as the *synchronous reactance*. The generator may then be viewed as an emf E in series with the synchronous reactance X_s, which corresponds to the *equivalent circuit* shown in Figure 4.31a.

As the three phases are symmetrical, the voltage differences, ΔV, will constitute a symmetrical three-phase set. Consequently, the same will apply to the currents. We are now able to draw the phasor diagram of Figure 4.31b. Note that, in view of equation (4.97) the current will lag ΔV by 90°. Note also that we have identified the angles γ and ϕ earlier defined in Figures 4.20 and 4.16, respectively. In view of the three-phase symmetry we need to show only one phase.

Example 4.12

Following synchronization a generator is subjected to field current control. Consider these two cases:

CASE A. The field current is raised so that $|E|$ is increased by 20% (overexcitation).

CASE B. The field current is lowered so that $|E|$ is decreased by 20% (underexcitation).

Draw phasor diagrams and compute currents and powers, assuming the following machine data: rated terminal voltage = 15.0 kV, synchronous reactance = 11.0 Ω/phase. Assume zero prime mover torque.

Solution: We construct the two phasor diagrams in Figure 4.32. Note that because the torque is zero, $\delta = 0$ in both cases. Note also that $\gamma = \phi = \pm 90°$, confirming zero real power both from equations (4.59) and (4.89).

We first compute the phase voltage:

$$|V| = \frac{15.0}{\sqrt{3}} = 8.660 \qquad [\text{kV/phase}]. \tag{4.12.1}$$

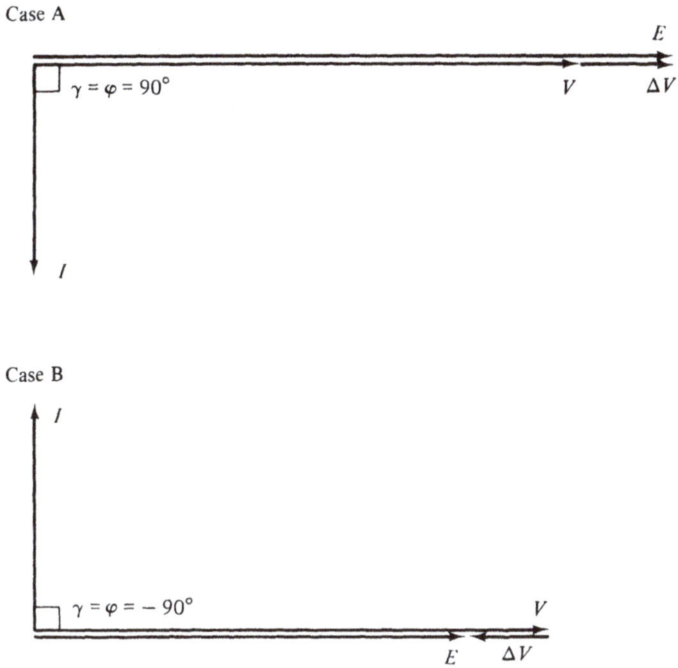

Figure 4.32

For both cases we have for the current

$$|I| = \frac{|\Delta V|}{X_s} = \frac{0.2 \cdot 8.660}{11.0} = 0.1575 \qquad \text{[kA/phase]}. \qquad (4.12.2)$$

Real powers are zero in both cases, but for the reactive powers we have

CASE A

$$\phi = +90°, \qquad (4.12.3)$$

$$Q = 8.660 \cdot 0.1575 \cdot \sin(+90°) = 1.364 \qquad \text{[MVAr/phase]} \quad (4.12.4)$$

(or 4.092 MVAr three-phase).

CASE B

$$\phi = -90°; \qquad (4.12.5)$$

$$Q = 8.660 \cdot 0.1575 \cdot \sin(-90°) = -1.364 \qquad \text{[MVAr/phase]} \quad (4.12.6)$$

(or −4.092 MVAr three-phase).

Summary

The *overexcited* machine delivers 4.09 MVAr to the network. From the network point of view, it acts like a capacitive load. (Operated in this manner the synchronous machine is referred to as a *synchronous condenser.*)

The *underexcited* machine draws 4.09 MVAr from the network. From the network point of view, it acts like an inductive load.

The example shows that by adjusting the field current we can make the synchronous machine either generate or consume reactive power. This feature finds important uses in power systems operation, as is demonstrated in Chapter 6.

4.7.4 Practical Expressions for Power

A disadvantage of equations (4.88) and (4.89) is that they explain the formation of torque and power of the machine in terms of internal physical variables, A_{max}, B_{max}, and γ. From a practical and operational viewpoint it would be better to express the power and torque in terms of externally measurable variables. We can readily obtain such expressions from the phasor diagram in Figure 4.31.

By projecting E on the reference vector V (and noting that ΔV leads V by the angle $90° - \phi$) we get

$$|E| \cos\delta = |V| + X_s |I| \sin\phi \qquad \text{[V/phase]}. \tag{4.98}$$

Similarly, by projecting E orthogonally to V, we obtain

$$|E| \sin\delta = X_s |I| \cos\phi \qquad \text{[V/phase]}. \tag{4.99}$$

We obtain from these equations:

$$\begin{aligned} |I| \cos\phi &= \frac{|E|}{X_s} \sin\delta \qquad \text{[A/phase]}; \\ |I| \sin\phi &= \frac{|E| \cos\delta - |V|}{X_s} \qquad \text{[A/phase]}. \end{aligned} \tag{4.100}$$

From equations (4.16) and (4.17) we get

$$P = \frac{|E||V|}{X_s} \sin\delta \qquad \text{[W/phase]} \tag{4.101}$$

and

$$Q = \frac{|V||E| \cos\delta - |V|^2}{X_s} \qquad \text{[VAr/phase]}. \tag{4.102}$$

Example 4.13

Consider the generator in the previous example. It is overexcited so that $|E|$ is 20% in excess of $|V|$. The prime mover torque is set at a value such that the machine delivers 12 MW (three-phase) to the network. Find δ, Q, and $|I|$.

Solution:

$$|E| = 1.20 \cdot |V| = 10.39 \qquad \text{[kV/phase]}, \tag{4.13.1}$$

Substituting into equation (4.101) we get

$$P = \frac{p_{3\phi}}{3} = \frac{12}{3} = 4 = \frac{10.39 \times 8.660}{11.0} \sin\delta \qquad \text{[W]}, \tag{4.13.2}$$

which gives

$$\therefore \delta = 29.3^{\circ}. \tag{4.13.3}$$

From equation (4.102) we have

$$Q = \frac{8.660 \times 10.39 \cos 29.3 - 8.660^2}{11.0} = 0.32 \qquad \text{[MVAr/phase]} \tag{4.13.4}$$

(or 0.96 MVAr three-phase), delivered to the network.

For the current, we obtain from equation (4.24):

$$|I| = \frac{\sqrt{4.00^2 + 0.32^2}}{8.660} = 0.463 \qquad \text{[kA/phase]}. \tag{4.13.5}$$

4.7.5 Pullout Power

In order to understand the phenomenon of *pullout*, we return to the analog of the synchronous machine shown in Figure 4.28. As a reminder, the effect of the currents in the stator coils is to produce a rotating magnetic field—represented by the horseshoe magnet driven at a constant speed. The rotor is represented by the permanent bar magnet.

It is evident that when the "machine" is operating in the "motor mode," increasing the torque required to drive the load will eventually force the bar magnet (rotor) to "break away" from the field of the horseshoe magnet (stator). In doing so, the bar magnet will slow down relative to the horseshoe magnet. After skipping one pole pair, it will attempt to latch onto the field of the horseshoe magnet but it cannot develop the necessary torque to keep up with the horseshoe magnet, much less to accelerate up to its speed. It will continue to skip pole pairs. The real synchronous machine will respond to excessive mechanical load by losing synchronism with the grid in a similar fashion. The removal of the load cannot make it recover synchronism; the machine has to be taken off the line and resynchronized.

A similar set of events will take place when the prime mover supplies excessive torque to the synchronous machine. In that case, the rotor flux (bar magnet) will "break away" from the rotating flux due to the stator currents (horseshoe magnet) and in doing so it will speed up relative to the rotating flux. Again, it will continue to skip pole pairs until it is taken off the line.

The behavior of the synchronous machine is similar. If the torque is increased slowly, the power angle δ grows, and eventually the rotor will "skip poles." Let us see when this will happen by considering the expression for power (4.101). If the field current is kept constant, then $|E|$ will be fixed. We had earlier pointed out that $|V|$ is essentially a constant. Under these assumptions P will reach a maximum when δ is equal to 90°. A negative maximum occurs for $\delta = -90°$. In Figure 4.33 we have plotted P versus δ.

From Figure 4.33, it can be seen that if we increase the torque further, that is, an increase in δ, this will result in a decrease in the electric power. We lose synchronism at this point. The maximum power, the pullout power is

$$P_{\text{po}} = \frac{|E||V|}{X_s} \qquad \text{[MW/phase].} \tag{4.103}$$

"Losing synchronism" implies that the rotor-bound B wave and the stator-bound A wave lose the "grip" between them. The rotor will now "slip" or "skip pole pairs" and will continue to do so until the machine is taken off the line. Excessive current will flow in the stator at the moments when the E and V phasors are 180° out of phase with each other. Dangerous heating and excessively large torque may

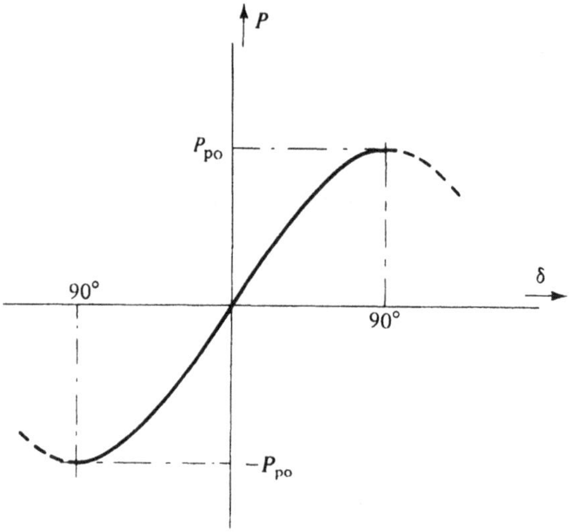

Figure 4.33

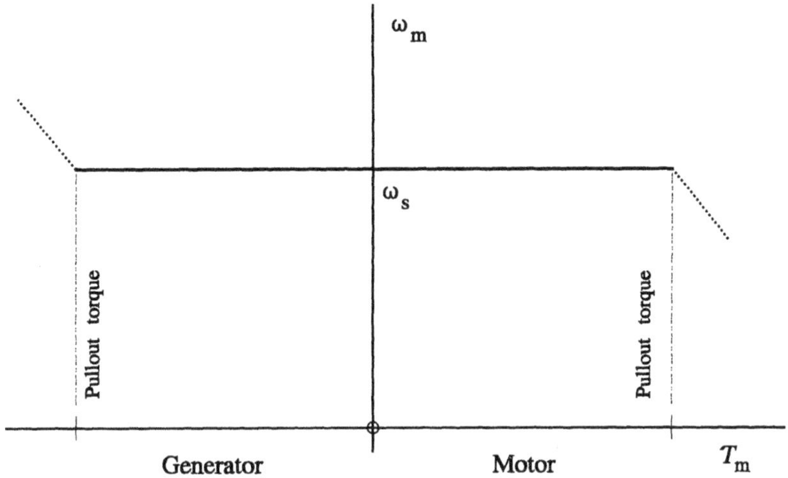

Figure 4.34

be produced. The torque-speed characteristics of the synchronous machine is shown in Figure 4.34.

Example 4.14

Consider the synchronous machine of Example 4.12. Let us assume that it is excited so that $|E| = |V|$. Find the real and reactive power and the current at the point of pullout. Draw a phasor diagram for the system.

Solution: As the torque is increased, the tip of E follows the circle shown in Figure 4.35. Pullout occurs when $\delta = 90°$. The current phasor has, at this point, advanced V by 45°, that is, $\phi = 45°$.

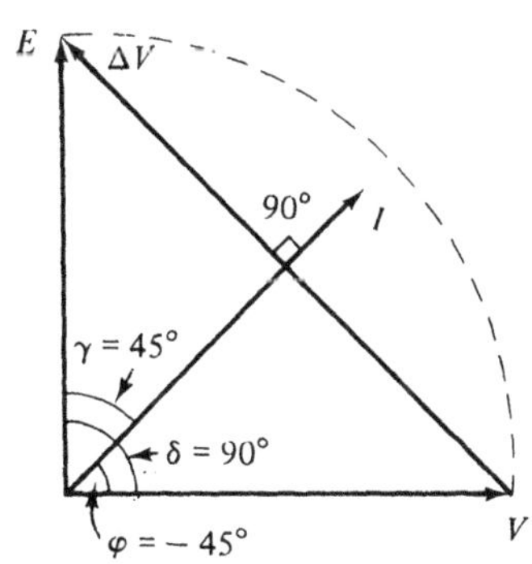

Figure 4.35

We have

$$P_{po} = \frac{8.660 \times 8.660}{11.0} = 6.818 \qquad \text{[MW/phase]} \tag{4.14.1}$$

or 20.5 MW total three-phase.

The current $|I|$ is obtained from the phasor diagram. We note that

$$X_s \cdot |I| = |\Delta V| = \sqrt{2}|V| \qquad \text{[V]}. \tag{4.14.2}$$

Therefore,

$$|I| = \frac{\sqrt{2} \times 8.660}{11.0} = 1.11 \qquad \text{[kA/phase]}. \tag{4.14.3}$$

Because $\phi = -45°$,

$$\sin\phi = -\cos\phi = -0.707. \tag{4.14.4}$$

Thus, $Q = -P = -20.5$ MVAr (three-phase). This means that at the point of pullout the generator was absorbing 20.5 MVAr (three-phase) from the grid. Note that under this condition, the angle γ is equal to 45°. (The reader should explain to his or her own satisfaction why the pullout power does not correspond to the maximum power in Figure 4.22.)

4.8 Summary and Some Final Observations

In this chapter we have tried to present a fairly complicated machine—the synchronous three-phase generator—as simply as possible. In order not to break up the continuity of our story we have occasionally taken "shortcuts," avoided unnecessary complications, or skipped important practical considerations altogether.

In summary, here are our most important findings:

1. The three-phase synchronous generator is the electric power engineer's "workhorse." Worldwide, well in excess of 99% of all bulk electric energy is generated by this type of machine.
2. Although single-phase ac power pulsates at twice the system frequency, balanced three-phase ac power is constant with respect to time.
3. The link between the mechanical power obtained from the prime mover and the electrical power delivered from the terminals of the generator is the electromechanical air-gap torque. This torque arises from an interaction between the rotor flux and the stator current.

In order to cover the fundamentals of the synchronous machine, we have had to leave out some important but nonessential topics. We summarize in the briefest of

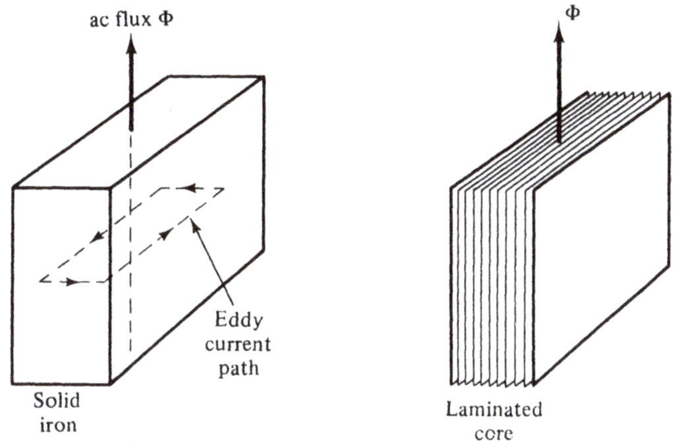

Figure 4.36

terms some points that need to be made in order to complete the basic presentation of the synchronous machine:

- Every part of the stator (but not the rotor) experiences an ac magnetic flux. According to Faraday's law, emf's are generated around the flux paths (Figure 4.36), and, if solid iron cores were used, *eddy currents* would flow perpendicular to the flux. This would cause intolerable heating effects and considerable losses. Laminated cores are used to break up the current paths and hence reduce the eddy current losses. Adding silicon to the core material gives an alloy with high resistivity, thus further reducing these losses. The ac flux also causes *hysteresis losses* in the iron. These losses are due to the reorientation of the magnetic moments that must take place 60 times per second. (One may think of an *internal magnetic friction* that must be overcome to turn the moments around.)

  Eddy current plus hysteresis losses are collectively referred to as *iron* or *core losses*.

- We assumed that the impedance of the stator winding could be represented by the reactance X_s This is not quite correct. To understand this statement, consider the two positions of the current wave shown in Figure 4.37. When the current wave is in position *A*, the wave will give rise to a flux wave (flux *A*) that will be lined up with the pole (*direct* or *d* axis).

  When the current wave is in position *B*, the flux will have its center between the poles ("quadrature" or *q* axis). As the flux will encounter a higher magnetic reluctance in the *q* direction than in the *d* direction, the stator reactance will vary with the position of the current wave between a maximum value X_d (direct-axis value) and a minimum value X_q (quadrature-axis value).

  When one takes this *saliency effect* into account (and does not make use of an average value as we have done in the text) the power formulas have to be modified. The modification results in a second-order correction term.

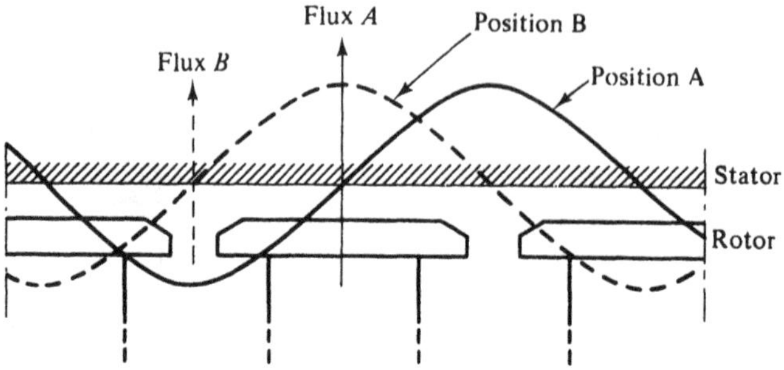

Figure 4.37

- In practice, the stator winding is designed somewhat differently from what is described in the text. For example, certain advantages (elimination of harmonics) can be gained by some overlap of the different phases. Also the "coil pitch" is not always 180° as discussed in the text.
- What happens to the higher harmonics present in the stator current wave? An analysis of their combined effect is more complex than that of the fundamental wave and is beyond the scope of this book. Generally, it can be said that their effect on power and torque is small. Not only is their amplitude much smaller than that of the fundamental wave but their effect can be minimized and even nullified by proper design of the winding (see previous point).
- To simplify the analysis we assumed that the generator was connected to an *infinite* network bus, the voltage of which did not change as we varied the torque and the emf of the generator.

  In reality, changing the torque and the emf of a single synchronous machine, affects the frequency and voltage of the grid. In general, the larger the synchronous machine is compared to the total generating capacity of the grid, the greater is its effect on the grid frequency and voltage. An increase in the torque will not only tend to accelerate the network—it *will* accelerate it. The frequency of the grid will, in fact, increase. Actually (see Chapter 6), by varying the prime-mover torque of individual generators we can, in effect, control the frequency of the power system (*load frequency control*).

  Similarly, changes of the generator field current (that is, changes to the magnitude of the emf) will be felt not only in the terminal voltage but, actually, as a change in the total voltage profile of the system.
- In practice, synchronous machines are provided with a *damper* (or *amortisseur*) winding on the rotor. The winding consists of short-circuited copper bars, resembling the cage windings in induction motors (see Chapter 8).

  Under normal operating conditions, the damper winding carries no currents and therefore has no influence on the torque. However, if there is a system disturbance, which subjects the rotor to transient position changes relative to the stator current wave (that is, changes in γ), emf's and currents will be induced in the damper winding. According to Lenz's law, these currents will have such directions as to counteract the position changes. This is tantamount to saying that the currents will have a damping effect on the transient rotor behavior. (Without this damping, the response of the rotor could be oscillatory.)

EXERCISES

4.1 A single-phase generator delivers at its terminals a voltage of 600 V rms and a current of 30 A rms. The real power delivered from its terminals is $P = 11.5$ kW. Find the reactive power Q. Note that there are two solutions.

4.2 Derive the expressions of item 6 in Table 4.1.

4.3 The 2-pole rotor of the generator discussed in Example 4.5 is replaced by a 4-pole rotor. This rotor is now run at 1800 rpm. The stator winding remains unchanged. Describe the terminal voltages you would expect to measure. Assume the same B_{max} as before. [HINT: Before you dive headlong into formulas, consider the physical layout of the machine first.]

4.4 Consider a 3-phase generator the terminal voltage of which is 8300 V rms line-to-line. Connect three equal 50-Ω resistors between each phase terminal and ground. Assume that the terminal voltage remains unchanged. (Normally, the current would cause a voltage drop across the synchronous reactance, and the terminal voltage would drop. In this case, we assume that a voltage regulator (see Chapter 6) eliminates this drop by increasing the field current.)

a) Use Table 4.1 to compute the real power dissipated in each resistor.

b) Compute the total three-phase power.

c) How many kilowatt-hours of energy are dissipated in the "load" over a period of 8 hours?

4.5 A 2-pole synchronous generator generates 60 Hz. Its stator has 36 equidistant slots. Each slot has two conductors. Each coil spans exactly 180°. The magnetic flux is sinusoidally distributed in the peripheral direction. The total flux leaving one pole is 2.5 Wb. Compute the rms value of the generated emf if the stator winding is connected as

a) a single-phase winding with all coils in series;

b) a three-phase winding.

4.6 Equation (3.47) gives the emf induced in a conductor that "cuts" a perpendicular magnetic field B at speed s [m/s]. Use this formula to derive equation (4.41).

4.7 A synchronous generator of salient pole design is driven by a slow-running hydroturbine at the rated speed, $n = 150$ rpm. The generator is generating 60 Hz. There are 576 equidistant stator slots with two conductors per slot. The air-gap dimensions are $D = 6.30$ m and $L = 1.11$ m. The maximum flux density is $B_{max} = 1.3$ [T].

a) Compute the stator emf (rms) if all conductors are connected in such a manner as to produce maximum terminal single-phase stator voltage.

b) Assume the winding to be arranged as a balanced three-phase winding. Find the emf generated per phase in this winding.

4.8 There are three impedances, each of which consists of a 10-Ω resistor, a 40-mH inductor, and a 300-μF capacitor, connected in series. These impedances are connected in Δ across the phase terminals of a three-phase generator that delivers a line voltage of 1 kV, 60 Hz. Find:

(a) currents in the delta (rms);
(b) line currents (rms);
(c) power delivered, three-phase, real, and reactive.

4.9 A three-phase synchronous generator operates onto a grid bus of voltage 12 kV (line value). The synchronous reactance is 5 Ω/phase. The magnitude of the generator emf is equal to the magnitude of the bus voltage. The machine delivers 18 MW to the grid. Find:
(a) the power angle δ;
(b) phase current, magnitude, and phase (relative to V);
(c) magnitude and direction of the reactive power.

4.10 Consider the generator in the previous example. The prime mover torque is kept constant at a value corresponding to 18 MW real power output. The magnitude of E is now lowered by a decrease in the field current. By how many percent can $|E|$ be decreased before the machine steps out of synchronism?

4.11 If you have done Exercise 4.10 correctly, you found that the machine absorbs reactive power from the network. Now keep the prime mover torque constant and increase $|E|$. As you do so explain by means of the power formulas why
a) the power angle δ decreases,
b) the absorbed reactive power decreases.
c) By how many percent must you increase $|E|$ in order for the reactive power absorption to reach zero (you now operate the generator with unity power factor)?

4.12 Consider again the three-phase generator in Exercise 4.9. It is delivering 10 MW and 5 MVAr (3-phase values) to the 12-kV grid bus. Find:
(a) the power angle δ;
(b) the phase angle ϕ;
(c) the magnitude of the emf, $|E|$.

4.13 In the previous exercise keep the torque constant corresponding to 10 MW real power. Is it possible by decreasing $|E|$ to reverse the reactive power flow from + 5 MVAr to − 5 MVAr? If it is possible, what will be the corresponding value of $|E|$?

4.14 In the three-phase generator discussed in Section 4.4.1, the rms value of the induced phase emf was 6447 V, based on a peak flux value of $B_{max} = 1.5$ [T]. According to equation (4.30), the flux wave can be expressed as

$$B = 1.50 \cos\left(\frac{py}{D}\right) \quad [\mathrm{T}]. \tag{4.104}$$

By increasing the rotor field current by 30%, one tries to increase the peak flux value by 30% to 1.95 [T], thereby also increasing the emf by 30% to 8381 V. Due to magnetic *saturation* the flux density increases only to 1.65 [T], and, furthermore, it is flattened out as indicated in Figure 4.38. A "harmonic analysis" (see Appendix B) of this wave reveals that it contains a considerable third harmonic component. In fact, we find that the equation for the flux wave can be expressed as

$$B = 1.80 \cos\left(\frac{py}{D}\right) - 0.15 \cos\left(3\,\frac{py}{D}\right) \quad [\mathrm{T}]. \tag{4.105}$$

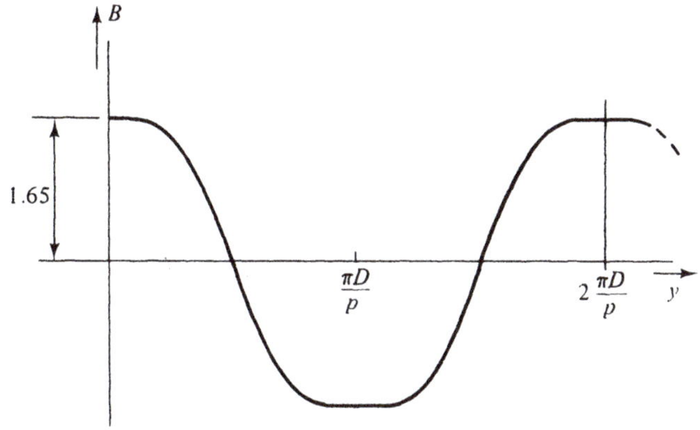

Figure 4.38

a) Show that the induced emf in each phase now contains a 180-Hz component. Find the rms value of this component.

b) Show that the 180-Hz emf components in the three-phase windings are in phase.

c) As a consequence of the finding in part b, show that no 180-Hz component appears in the line voltages.

d) What will be the rms value of the 60-Hz emf component in each phase?

[HINT: First, find the flux linked by one coil by using equation (4.37). Then take the distribution effect into account, noting that the α value in equation (4.48) refers to 60 Hz.]

References

Del Toro, V. *Electric Machines and Power Systems*. Englewood Cliffs, NJ: Prentice-Hall, 1985.

Engelmann, R.H. and Middendorf, W.H., *Handbook of Electric Motors*, Marcel Dekker, New York: 1995.

Sen, P.C. *Principles of Electric Machines and Power Electronics*. New York: John Wiley & Sons, 1989.

Slemon, G.R. *Electric Machines and Drives*. Reading, MA: Addison-Wesley, 1992.

5

The Power Transformer

5.1 Why Transformers?

The purpose of the power transformer is to raise the voltage from the level at which it is generated to a much higher value for transmission. This is done to minimize power loss and large voltage drops. Consider a nuclear-powered generator with a rated capacity of 1000 MW—a power rating well within current practices. Let us assume that we wish to transmit this block of power over a distance of 20 km.

The magnitude of the generator voltage is limited in practice (see Chapter 4) by the number of conductors that can be placed physically in the stator slots of the synchronous machine. We must remember that the conductors must have a minimum cross-sectional area in order to carry the required stator current. If we assume that the generator is rated at 20 kV (line-to-line). The magnitude of the current $|I|$ is computed from equation (4.16) as

$$|I| = \frac{P}{|V| \cos \phi} \qquad \text{[A per phase]}. \tag{5.1}$$

We have

$$P = \frac{1000}{3} = 333.3 \qquad \text{[MW per phase]}; \tag{5.2}$$

$$|V| = \frac{20}{\sqrt{3}} = 11.55 \quad \text{[kV per phase]}. \tag{5.3}$$

Thus,

$$|I| = \frac{333.3}{11.55} = 28.86 \text{ [kA per phase]}. \tag{5.4}$$

(We have assumed that the generator delivers the power at unity power factor, that is, $\cos \phi = 1$. Other values of $\cos \phi$ would give higher values of current.) If

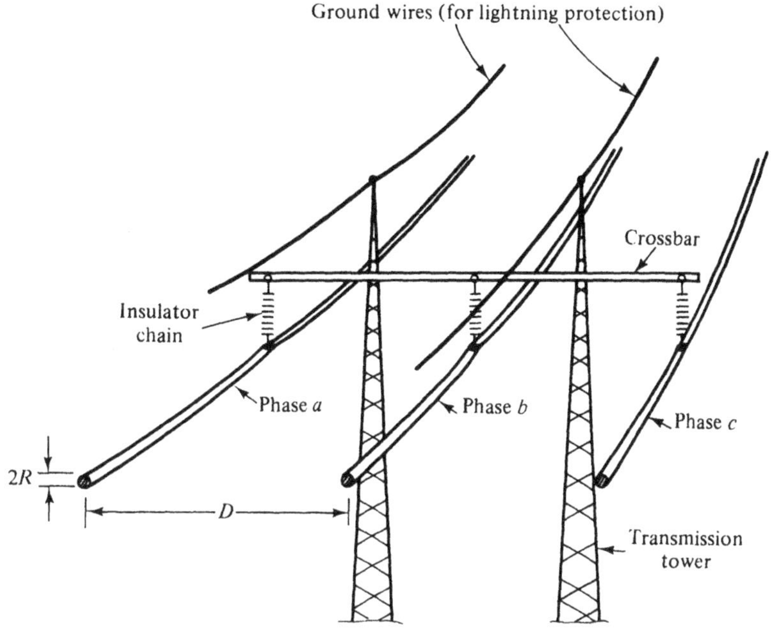

Figure 5.1

we assume that the current is to be carried in three identical overhead bare copper conductors (shown in Figure 5.1) each having a radius R of 25 mm. A smaller conductor would probably not accommodate the current (see below). The distance D between the conductors is 5 m, since a smaller distance would probably not be tolerated in view of insulation constraints.

Copper has a resistivity of $1.75 \cdot 10^{-8}\ \Omega \cdot$ m. According to equation (3.24) the resistance of the 20 km line would be

$$R = 1.75 \cdot 10^{-8} \frac{20 \cdot 10^3}{\pi \cdot 25^2 \cdot 10^{-6}} = 0.178 \qquad [\Omega \text{ per phase}]. \tag{5.5}$$

We can now compute the ohmic power loss:

$$P_\Omega = R \cdot |I|^2 = 0.178 \cdot 28.86^2 = 148 \qquad [\text{MW per phase}], \tag{5.6}$$

or

$$\frac{148 \times 10^3}{20 \times 10^3} = 7.4 \qquad [\text{kW/m}]. \tag{5.7}$$

(A higher power loss would probably melt the conductor. This is the reason we did not choose a smaller diameter conductor.) The power loss is

$$\frac{P_{\Omega}}{P_{\text{trans}}} = \frac{148}{333.3} \times 100 = 44.4\% \tag{5.8}$$

of the generator output. In other words, almost half the generated power is lost during transmission.

Even more disturbing results are obtained if we compute the voltage drop along the 20-km line caused by the current. For an overhead transmission line the series reactance is typically much larger than the resistance. Without proof[1] we give the following formula for the series reactance of the line:

$$X = \omega \frac{\mu_o}{2\pi}\left[\frac{1}{4} + \ln\left(\frac{\sqrt[3]{2}D}{R}\right)\right] \qquad [\Omega/\text{m}]. \tag{5.9}$$

The reactance of the 20-km line (computed for 60 Hz) is

$$X = 377\,\frac{4\pi\cdot 10^{-7}}{2\pi}\left(\frac{1}{4} + \ln\frac{\sqrt[3]{2}\cdot 5}{25\cdot 10^{-3}}\right)\cdot 20\cdot 10^{3} = 8.72 \qquad [\Omega\text{ per phase}]. \tag{5.10}$$

Therefore, the reactive voltage drop will be

$$X\cdot|I| = 8.72\cdot 28.86 = 251.7 \qquad [\text{kV per phase}]. \tag{5.11}$$

This result is, of course, absurd; the voltage drop cannot exceed the generator voltage, which was 11.55 kV per phase. We conclude that it is physically impossible to transmit 1000 MW over the 20-km line, if the voltage is 20 kV.[2]

If we choose instead a generator voltage of 200 kV, we get the following:

1. The phase current will be only 2.89 kA instead of 28.9 kA.
2. The power loss will be only 1.48 MW per phase instead of 148 MW, that is, only 0.44% of the generated power.
3. The voltage drop across the series reactance will be 25.2 kV per phase which is only 21.8% of the generator voltage (115.5 kV per phase).

Both power losses and voltage drops are within acceptable and physically realizable limits.

The example tells us vividly that power transmission, even over short distances, is possible only if we can work with voltage levels far exceeding those that can be generated directly in a normal synchronous machine. Power transformers transform the generator voltage to levels at which transmission becomes feasible, even over distances as long as 1000 km.

[1] For a discussion of transmission line parameters see Elgerd, 1971.

[2] In Chapter 6 we find that this line can transmit, at most, about 50 MW at a voltage of 20 kV.

5.2 The Single-Phase Transformer: Basic Design

The power transformer is always designed for single frequency operation (60 Hz for North America; 50 Hz in Europe). It comes in either single-phase or three-phase units. Sizes range from a few kilovolt-amperes for small distribution transformers to more than 1000 MVA for large transmission transformers. Compared to the synchronous generator, the power transformer is a relatively simple machine, due mainly to the fact that it has no moving parts.

In the simplest form, a power transformer (Figure 5.2) consists of two windings on an iron core. We refer to the windings as HV and LV (high and low voltage, respectively). The designations *primary* and *secondary* are also commonly used. Often additional windings (*tertiary*, etc.) are added.

5.3 The Concept of an "Ideal" Transformer

In developing a mathematical and electrical model of the power transformer, it is advantageous to start by making a number of assumptions that idealize the device.

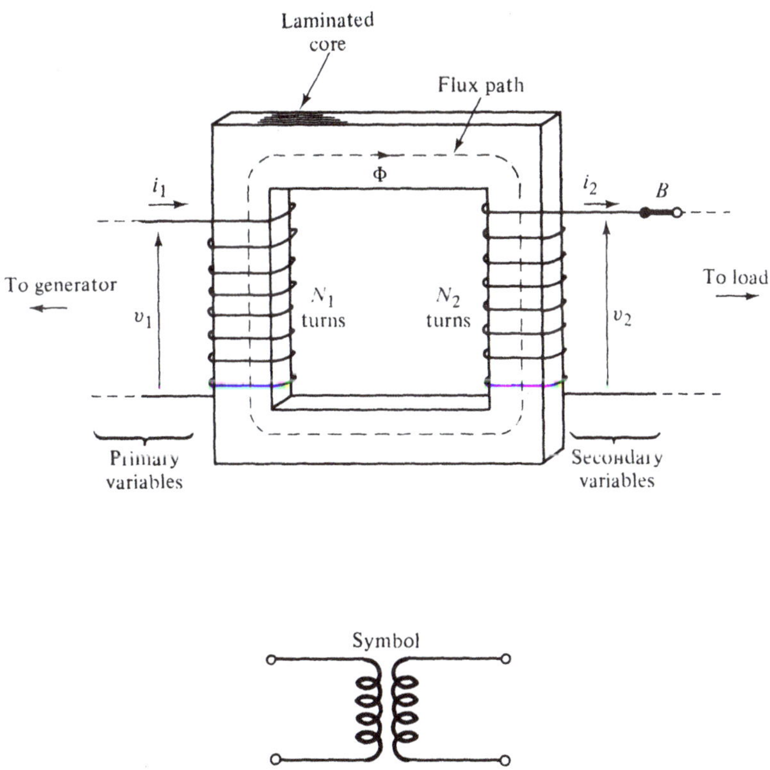

Figure 5.2

A simplified model can then be constructed. This model is called the "ideal" transformer (IT). After we have gained some insight into the characteristics and operation of the "ideal" transformer, we can introduce new elements into the model to account for the nonideal behavior of the practical device by dropping the assumptions on which the IT model is based. The following assumptions are appropriate:

1. The transformer windings have zero resistance. This means in effect that we neglect both the ohmic power losses and resistive voltage drops that occur in the actual device.
2. The transformer core is made of iron whose permeability is infinite. This assumption implies two things:
   a) It takes zero mmf to create the magnetic flux in the core.
   b) All flux is confined to the core (flux takes the path of least reluctance).
3. The transformer has zero core losses. In other words, there is neither hysteresis nor eddy-current loss.

5.3.1 The Ideal Transformer on No-Load

Consider the two-winding transformer shown in Figure 5.2. The N_1-turn primary winding is energized from a single-phase generator having a terminal voltage v_1. The N_2-turn secondary winding feeds a load (not shown). For the present, we consider the transformer operating at *no-load*, that is, the load circuit breaker B is open, and hence the secondary current is zero.

In view of the IT assumption 1, we can write, using Faraday's law,

$$v_1 = N_1 \frac{d\Phi}{dt} \qquad [\text{V}], \tag{5.12}$$

$$v_2 = N_2 \frac{d\Phi}{dt} \qquad [\text{V}], \tag{5.13}$$

where Φ is the core flux identified in Figure 5.2.

5.3.1.1 Voltage Relationship

Elimination of $d\Phi/dt$ from (5.12) and (5.13) yields

$$\frac{v_1}{v_2} = \frac{N_1}{N_2} \equiv a. \tag{5.14}$$

In words: The ratio of the primary to the secondary voltage is equal to the ratio of the turns in the primary to that in the secondary, a. The term a is referred to as the *transformer turns ratio*.

Note that equation (5.14) holds for arbitrary voltage waveforms (except dc, of course). If the voltages are sinusoidal, then equation (5.14) applies to the voltage phasors V_1 and V_2 as well, and we have

$$\frac{V_1}{V_2} = \frac{N_1}{N_2} = a. \tag{5.15}$$

5.3.1.2 Magnetic Flux in a Sinusoidally Excited Transformer

Assume that the generator in Figure 5.2 delivers a sinusoidal voltage of $|V_1|$ volts (rms). We can then express the primary voltage as

$$v_1 = \sqrt{2}|V_1| \sin \omega t \qquad \text{[V]}. \tag{5.16}$$

From equation (5.12) we get, for the derivative of the flux,

$$\frac{d\Phi}{dt} = \frac{\sqrt{2}|V_1|}{N_1} \sin \omega t \qquad \text{[Wb/s]}. \tag{5.17}$$

Integration of expression (5.17) gives

$$\Phi = -\frac{\sqrt{2}|V_1|}{\omega N_1} \cos \omega t = \frac{\sqrt{2}|V_1|}{\omega N_1} \sin\left(\omega t - \frac{\pi}{2}\right) \qquad \text{[Wb]}. \tag{5.18}$$

Note that the flux Φ lags V_1 (and V_2) by 90°.

The IT voltages and core flux are shown as phasors in the phasor diagram shown in Figure 5.3. The peak flux Φ_{max} is obtained from equation (5.18) as

$$\Phi_{max} = \frac{\sqrt{2}|V_1|}{\omega N_1} \qquad \text{[Wb]}. \tag{5.19}$$

5.3.1.3 Voltage per Turn

In terms of frequency f, effective[3] cross-sectional area of the core A, and maximum flux density B_{max}, equation (5.19) can be written as

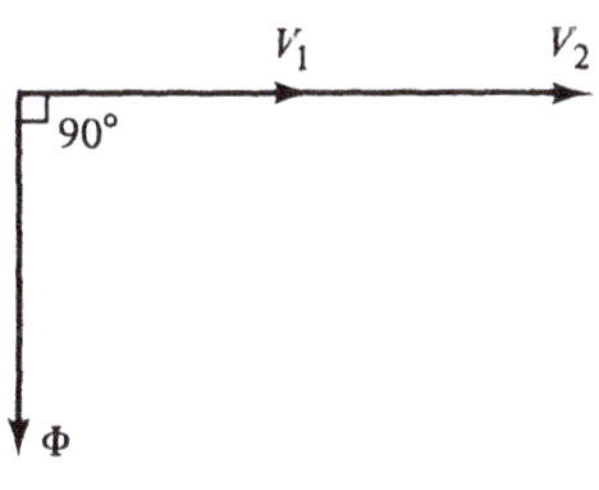

Figure 5.3

[3] The *effective* core area refers to the actual cross-sectional area of the iron (the gross area minus the insulation between the laminations).

$$\frac{|V_1|}{N_1} = \frac{\omega\Phi_{max}}{\sqrt{2}} = \frac{2\pi f}{\sqrt{2}} AB_{max} = 4.44fAB_{max} \qquad [\text{V/t}]. \tag{5.20}$$

The ratio $|V_1|/N_1$ is the *voltage per turn* (VPT), an important transformer design parameter. Note that in view of equation (5.15) we have

$$\text{VPT} = \frac{|V_1|}{N_1} = \frac{|V_2|}{N_2} \qquad [\text{V/t}]. \tag{5.21}$$

Because f is a system constant (60 Hz) and B_{max} a material constant, from equation (5.20)

$$\text{VPT} \propto A \qquad [\text{V/t}]. \tag{5.22}$$

Example 5.1

Find the VPT of a transformer with an effective cross-sectional core area $A = 0.4$ [m^2]. The core is operated at a peak flux density of $B_{max} = 1.5$ [T]. The frequency $f = 60$ Hz.

Solution: Equation (5.20) gives

$$\text{VPT} = 4.44 \cdot 60 \cdot 0.4 \cdot 1.5 = 159.8 \qquad [\text{V/t}]. \tag{5.1.1}$$

5.3.1.4 Transformer Size and Frequency

As mentioned, B_{max} is a material constant. We can write equation (5.20) as

$$A \propto \frac{\text{VPT}}{f} \tag{5.23}$$

This tells us that transformer size is inversely proportional to frequency. This means that European transformers are bulkier than North American transformers of equal power rating, since the frequency is 50 Hz in Europe and 60 Hz in North America. (From equation (5.23) we deduce that a *dc transformer* ($f = 0$) will be of infinite size.)

Example 5.2

An ideal transformer is characterized by the following data: $N_1 = 100$ turns, $N_2 = 300$ turns.

The LV winding is connected to a single-phase generator delivering a sinusoidal voltage of 3 kV rms at 60 Hz.

Find the VPT, core flux, and secondary voltage at no-load.

Solution: The turns ratio is

$$a = \frac{100}{300} = 0.3333. \tag{5.2.1}$$

As the 3-kV generator voltage is applied to the 100-turn primary we get

$$\mathrm{VPT} = \frac{|V_1|}{N_1} = \frac{3000}{100} = 30 \qquad [\mathrm{V/t}] \tag{5.2.2}$$

The secondary voltage follows from (5.15) as

$$|V_2| = \frac{|V_1|}{a} = \frac{3000}{0.3333} = 9000 \qquad [\mathrm{V}]. \tag{5.2.3}$$

The core flux (peak value) is computed from equation (5.19) as

$$\Phi_{\max} = \frac{\sqrt{2} \cdot 3000}{377 \cdot 100} = 0.113 \qquad [\mathrm{Wb}]. \tag{5.2.4}$$

5.3.2 The Ideal Transformer Under Loaded Conditions

5.3.2.1 Voltage, Current, and Flux:

With the circuit breaker B in Figure 5.2 open, both i_1 and i_2 are zero, since the core requires zero mmf to create the magnetic flux in the core. When the circuit breaker B is closed, the secondary transformer voltage v_2 is applied to the "load." The load may be considered to be a single impedance made up of a variety of impedances such as presented by heaters and electric motors. In any case, the result will be a secondary current, i_2, as shown in Figure 5.2. This current will cause a change in the mmf from zero to $N_2 i_2$. Ohm's law for the magnetic circuit must, however, be satisfied. The only way in which this can be achieved is for a finite primary current, i_1, to arise. The magnitude of i_1 will be such as to restore the mmf balance in the core.

In mathematical terms, this mmf balance requires that

$$i_1 N_1 - i_2 N_2 = \mathfrak{R} \cdot \Phi \qquad [\mathrm{A \cdot t}] \tag{5.24}$$

where $\mathfrak{R}$ is the magnetic reluctance of the core. [Compare this to equation (3.90).] However, we had assumed that the core material was characterized by infinite magnetic permeability and therefore $\mathfrak{R} = 0$. Because Φ is finite, equation (5.24) reduces to

$$i_1 N_1 - i_2 N_2 = 0 \qquad [\mathrm{A \cdot t}] \tag{5.25}$$

or

$$\frac{i_1}{i_2} = \frac{N_2}{N_1} = \frac{1}{a}. \tag{5.26}$$

Equation (5.26) applies to the instantaneous value of the currents. If all variables are sinusoidal, the equation must apply also to current phasors, that is,

$$\frac{I_1}{I_2} = \frac{N_2}{N_1} = \frac{1}{a}. \tag{5.27}$$

We had assumed zero resistance in each winding. Consequently, the above currents will cause zero resistive voltage drops. Therefore equations (5.14) and (5.15) are still valid. The relationship between the flux and voltages, as it existed at no-load, remains unchanged.

5.3.2.2 Power

The IT was assumed to be lossless. This means that the input power

$$p_1 = v_1 i_1 \qquad \text{[W]} \tag{5.28}$$

must be equal to the output power

$$p_2 = v_2 i_2 \qquad \text{[W]}, \tag{5.29}$$

hence

$$v_1 i_1 = v_2 i_2 \qquad \text{[W]}. \tag{5.30}$$

[This can also be derived by the multiplication of equations (5.14) and (5.26).]

In words: All the instantaneous power that enters the primary of an ideal transformer must exit from the secondary. The power passing through the device is referred to as the *transformed power.*

As this rule applies to instantaneous power, it will also apply to both real and reactive powers. In the case of sinusoids, we have

$$V_1 I_1^* = V_2 I_2^* \qquad \text{[VA]}. \tag{5.31}$$

Example 5.3

An impedance,

$$Z_2 = 100 + j30 \qquad [\Omega] \tag{5.3.1}$$

is connected across the secondary terminals of the transformer in Example 5.2. Find the resulting currents and also compute the power transformed.

Solution: On the assumption that the current drain will not reduce the generator voltage,[4] the primary and secondary transformer voltages will remain at 3000 V and 9000 V, respectively. Let us choose (similar to Figure 5.3) these voltages as reference phasors. We then have for the secondary current:

[4] In a practical situation a *voltage regulator* is used to maintain the generator voltage constant. If the generator emf were uncontrolled, the current would cause a voltage drop across the synchronous reactance, resulting in a lower terminal voltage.

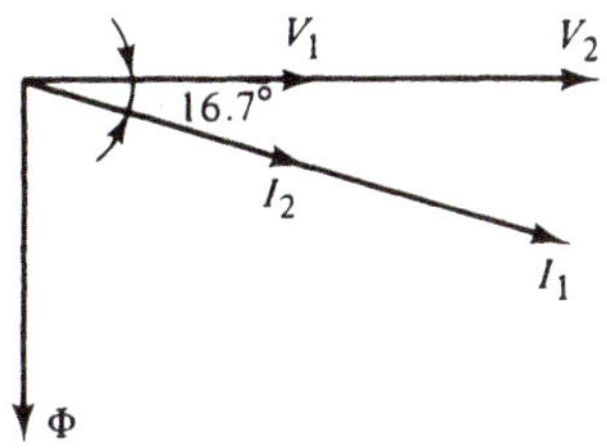

Figure 5.4

$$I_2 = \frac{V_2}{Z_2} = \frac{9000}{100 + j30} = 86.20e^{-j16.70^\circ} \qquad [A]. \qquad (5.3.2)$$

For the primary current equation (5.27) gives

$$I_1 = \frac{1}{0.3333} I_2 = 258.6e^{-j16.70^\circ} \qquad [A]. \qquad (5.3.3)$$

The currents and voltages are shown in the phasor diagram in Figure 5.4.

The complex power supplied to the load impedance is

$$S_2 = V_2 I_2^* = 9000 \cdot 86.2e^{j16.70^\circ} = (743.1 + j222.9) \cdot 10^3 \qquad [VA]. \quad (5.3.4)$$

According to equation (5.31),

$$P_1 = P_2 = 743.1 \qquad [kW] \qquad (5.3.5)$$

and

$$Q_1 = Q_2 = 222.9 \qquad [kVAr]. \qquad (5.3.6)$$

5.3.2.3 The Ideal Transformer as an Impedance Transformer

In the previous example the application of Ohm's law at the secondary gave

$$\frac{V_2}{I_2} = Z_2 \qquad [\Omega]. \qquad (5.32)$$

Using equations (5.15) and (5.27), (5.32) can be rewritten as

$$\frac{V_2}{I_2} = \frac{V_1/a}{aI_1} = Z_2 \qquad [\Omega]. \qquad (5.33)$$

In terms of the primary variables we have

$$\frac{V_1}{I_1} = a^2 Z_2 \equiv Z_2' \qquad [\Omega]. \qquad (5.34)$$

In words: The impedance Z_2 connected across the secondary terminals of the transformer has the same effect (that is, it draws the same amount of power from the generator connected to the primary) as an equivalent impedance $Z_2' = a^2 Z_2$ connected directly across the generator terminals. Z_2' is called *the secondary impedance referred to the primary.*

Example 5.4

Calculate the value of the impedance which will draw the same amount of power, as in Example 5.3., when connected directly across the terminals of the 3-kV generator.

Solution: Equation (5.34) gives

$$Z_2' = 0.3333^2 \cdot (100 + j30) = 11.11 + j3.333 \qquad [\Omega]. \qquad (5.4.1)$$

5.3.2.4 *Equivalent Circuit of the Ideal Transformer*

The basic IT voltage and current relationships [given by equations (5.14) and (5.26)] can be summarized in the *equivalent circuit* shown in Figure 5.5. It is shown later that this diagram is useful in circuit analysis involving transformers.

In the ideal transformer there exists a direct proportionality between primary and secondary voltages. A voltage v_2 applied to the secondary winding will give rise to the same core flux as the voltage av_2 applied to the primary. We refer to av_2 as *the secondary voltage referred to the primary* and give it the special symbol v_2', that is,

$$v_2' \equiv av_2 \qquad [\mathrm{V}]. \qquad (5.35)$$

Similarly, the secondary current i_2 gives rise to the same core mmf as the current i_2/a flowing in the primary. We define

$$i_2' \equiv \frac{1}{a} i_2 \qquad [\mathrm{A}] \qquad (5.36)$$

as *the secondary current referred to the primary.*

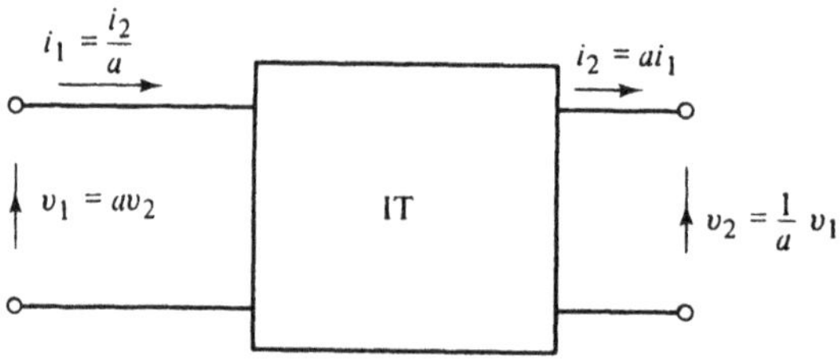

Figure 5.5

Later examples will demonstrate the practical significance of these "referred" variables.

5.3.2.5 A Mechanical Analog of the Ideal Transformer

The IT transforms the voltages in the ratio N_1/N_2 and the currents in the inverse ratio. The transformed power remains invariant. An analogous situation exists in the mechanical gear train shown in Figure 5.6. Let the *gear ratio* be defined as the ratio of the radii:

$$a \equiv \frac{R_1}{R_2}. \tag{5.37}$$

As the "secondary" shaft turns a degrees for every degree of the "primary" shaft, it is clear that the primary and secondary angular speeds ω_1 and ω_2 are related by

$$\frac{\omega_1}{\omega_2} = \frac{R_2}{R_1} = \frac{1}{a}. \tag{5.38}$$

This equation clearly corresponds to (5.26).

As the primary and secondary torques T_1 and T_2 are proportional to the respective radii, we have

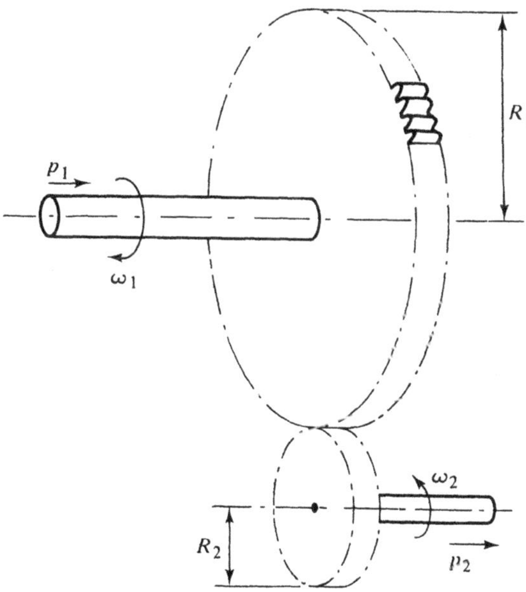

Figure 5.6

$$\frac{T_1}{T_2} = \frac{R_1}{R_2} = a. \qquad (5.39)$$

This equation is the analog of (5.14).

The primary mechanical power

$$p_1 = \omega_1 T_1 \qquad [\text{W}]. \qquad (5.40)$$

For the secondary power p_2 we have

$$p_2 = \omega_2 T_2 = \left(a\omega_1\left(\frac{T_1}{a}\right)\right) = \omega_1 T_1 = p_1 \qquad [\text{W}]. \qquad (5.41)$$

This equation shows that the transformed mechanical power remains invariant (provided the gear train is lossless).

5.4 The Physical Transformer: The Ideal Transformer

The voltage, current, and power relationships for an ideal transformer are given in equations (5.14), (5.26), and (5.30), respectively. Within very good accuracy limits these equations also describe the behavior of a real power transformer. In fact, the errors made by modeling an actual power transformer by this simple set of equations normally amount to a few percent—often less than 1%.

However, the simple IT model, in certain instances, is totally inadequate for describing an actual physical transformer. For example, an ideal transformer is characterized by zero power loss. This means that there is no limit to the power an IT can transform. This, of course, does not apply to a real transformer. The actual losses that occur in the physical device, although relatively small, set a definite limit to its power transformation capability.

We proceed to investigate how to adjust the simple IT model in Figure 5.5 so as to make it more fully representative of the actual device. We do this by removing the idealized (unrealistic) assumptions made earlier, one at a time.

5.4.1 Finite Permeability

5.4.1.1 The Magnetization Current

We first remove the IT assumption of infinite permeability. This assumption led to equation (5.26). If an IT is operated with the secondary open, that is, with $i_2 = 0$, then equation (5.26) tells us that the primary current is also zero. This would imply that the impedance, seen across the primary terminals when the secondary is on open circuit, is infinite. In a practical transformer, the primary current will be small (compared to the rated load current) but not zero. In fact, according to equation (5.23), the primary, open-circuit, or *magnetization current*, will be

$$i_{1m} = \frac{\mathcal{R}\Phi}{N_1} \qquad [\mathrm{A}]. \tag{5.42}$$

When the primary voltage is sinusoidal, the magnetization current, substituting for Φ from equation (5.18), will be

$$i_{1m} = \sqrt{2}\,\frac{|V_1|\mathcal{R}}{\omega N_1^2}\sin\left(\omega t - \frac{\pi}{2}\right) \qquad [\mathrm{A}]. \tag{5.43}$$

5.4.1.2 The Magnetization Reactance

We pointed out in Chapter 3 that the reluctance $\mathcal{R}$, strictly speaking, is *not* a constant because of the variable core permeability, μ. However, let us for the moment disregard this fact and consider it to be a constant. Under that assumption, equation (5.43) tells us that i_{1m} is sinusoidal and lags the voltage by 90°. We therefore represent it by the phasor I_{1m}, having the rms value

$$|I_{1m}| = \frac{|V_1|\mathcal{R}}{\omega N_1^2} \qquad [\mathrm{A}]. \tag{5.44}$$

Equation (5.43) can then (see Appendix B) be expressed in phasor form as

$$I_{1m} = \frac{V_1\mathcal{R}}{j\omega N_1^2} \equiv \frac{V_1}{jX_m} \qquad [\mathrm{A}], \tag{5.45}$$

where

$$X_m \equiv \omega\frac{N_1^2}{\mathcal{R}} \qquad [\Omega] \tag{5.46}$$

is the *magnetization reactance* as viewed from the primary winding.

Example 5.5

Consider the toroidal iron core discussed in Example 3.22, Case 1. The core is wound with primary and secondary coils of 100 turns and 40 turns, respectively. A primary voltage of 16 V rms, at 60-Hz sinusoidal, is applied. What will be the magnetization current?

Solution: In Example 3.22, we computed the reluctance of the flux path as

$$\mathcal{R} = 1.601 \cdot 10^5 \qquad [\mathrm{A \cdot t/Wb}]. \tag{5.5.1}$$

From equation (5.46) we compute the open circuit reactance:

$$X_m = 377 \cdot \frac{100^2}{1.601 \cdot 10^5} = 23.5 \qquad [\Omega]. \tag{5.5.2}$$

The magnetization current will be

$$|I_{1m}| = \frac{16}{23.5} = 0.681 \qquad [\text{A rms}]. \tag{5.5.3}$$

5.4.1.3 Adjustment of the Ideal Transformer Equivalent Circuit to Include the Magnetization Reactance

Equation (5.25) can be replaced by

$$i_1 = \frac{N_2}{N_1} i_2 + \frac{\mathcal{R}\Phi}{N_1} = \frac{i_2}{a} + i_{1m} \qquad [\text{A}]. \tag{5.47}$$

For sinusoids, the equation can be written in phasor form as

$$I_1 = \frac{I_2}{a} + I_{1m} = I_2' + I_{1m} = I_2' + \frac{V_1}{jX_m} \qquad [\text{A}]. \tag{5.48}$$

Clearly, the primary current, I_1 consists of two components:

1. The magnetization current I_{1m}, which is proportional to the voltage V_1.
2. The load current $I_2' = I_2/a$, which is proportional to the secondary current.

As I_{1m} is proportional to V_1 we can "pull out" of the not-so-ideal transformer "box" a reactor connected across the input terminals of the IT circuit, as shown in Figure 5.5, to provide a path for this current. We obtain the adjusted equivalent circuit shown in Figure 5.7. To put things in the proper perspective, the current I_{1m} is normally of the order of a few percent of the rated primary current. It is therefore often neglected, which means that equation (5.48) reduces to the simpler IT equation (5.27).

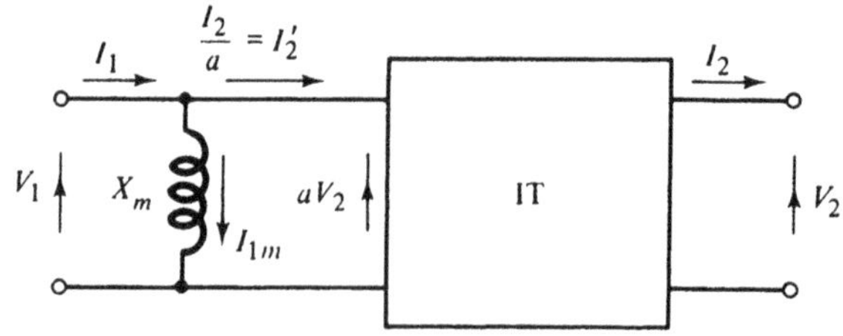

Figure 5.7

5.4.2 Core Losses

Under no-load conditions (that is, with $I_2 = 0$) the model shown in Figure 5.7 gives

$$I_1 = I_{1\text{m}} = \frac{V_1}{jX_\text{m}} \qquad [\text{A}] \tag{5.49}$$

Furthermore (cf. item 2 in Table 4.1), we note that the power drain from the generator is

$$\begin{aligned} P_1 &= 0 \qquad [\text{W}]; \\ Q_1 &= \frac{|V_1|^2}{X_\text{m}} \qquad [\text{VAr}]. \end{aligned} \tag{5.50}$$

In reality, a physical transformer operated under no-load conditions will, in addition to reactive power, draw real power from the source—the *no-load power.* Since the output power is zero, all the no-load input power must be "lost" in the transformer in the form of heat. This loss consist of hysteresis and eddy–current losses. This real power can be accounted for by the addition of a shunt resistance, R_m, as shown in Figure 5.8. R_m is a resistance that dissipates power equal to the hysteresis and eddy–current losses when V_1 is applied across it.

5.4.3 Effects of Core Nonlinearity

The corrections to the model were made on the assumption that the relative permeability of the core iron was constant. In reality, it varies with flux density (see Figure 3.33). This causes nonlinear effects that are of great practical significance in power system operations. We demonstrate this with the following example.

Example 5.6

Obtain the waveform of the no-load current i_1 in the primary winding of a small power transformer with the following specifications: Core cross-sectional area = 10 cm^2. Length of flux path = 60 cm. The B–H characteristics of the core have been found experimentally to be as shown in Figure 5.9. Primary winding:

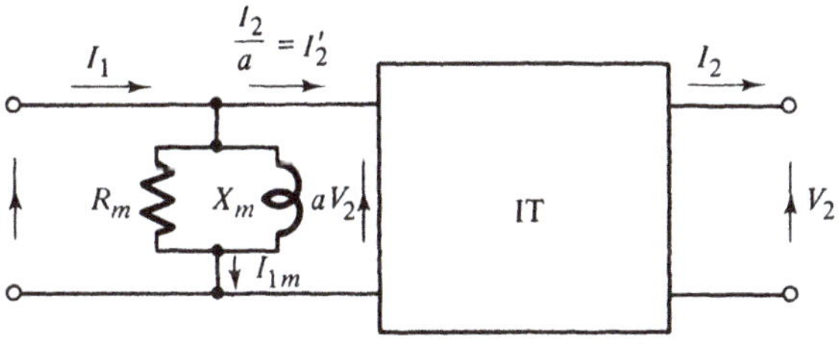

Figure 5.8

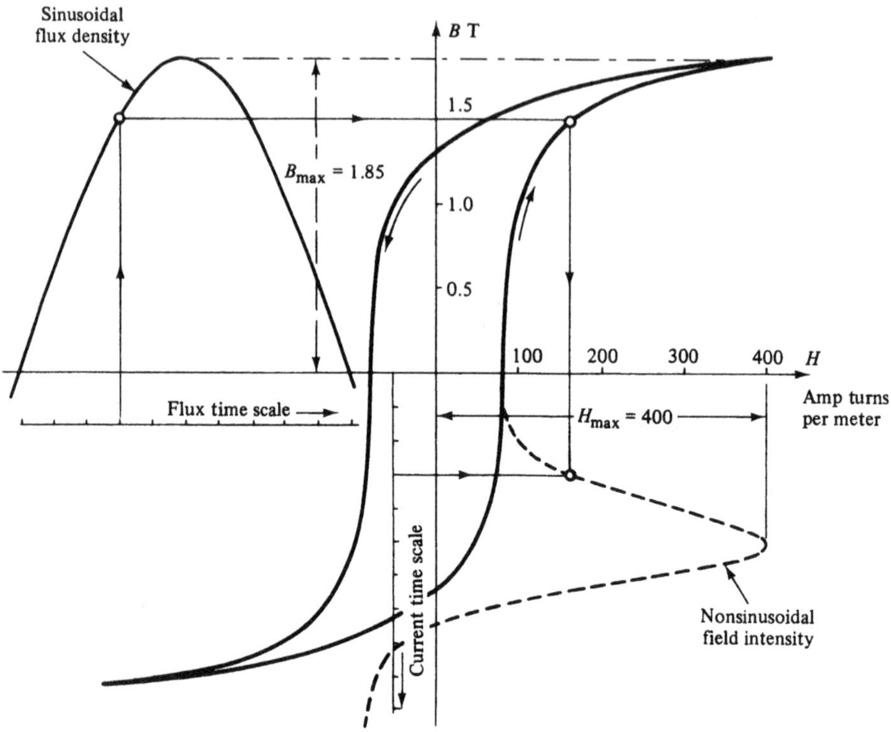

Figure 5.9

$N_1 = 1000$ turns. A sinusoidal 60-Hz voltage of magnitude $|V_1| = 493$ V rms is applied to the primary.

Solution: We first compute the maximum core flux density by using equation (5.20):

$$\frac{|V_1|}{N_1} = \frac{493}{1000} = 4.44 \cdot 60 \cdot 10 \cdot 10^{-4} \cdot B_{max} \qquad [\text{V/t}]. \tag{5.6.1}$$

Thus,

$$B_{max} = 1.85 \text{ [T]}. \tag{5.6.2}$$

From the core data shown in Figure 5.9 we can determine the current waveshape. This must be done graphically as follows:

> From a knowledge of its peak value we can plot the flux density waveform, which we know is essentially sinusoidal. The half-cycle is shown in Figure 5.9. We can construct the corresponding waveshape of the field intensity H (shown in the dashed line) point by point. The construction of one point is shown, and it should be self-explanatory.

Note that the resulting H wave is highly nonsinusoidal. Because H is proportional to i_1 the waveforms of i_1 and H are identical. The constant of proportionality is established from equation (3.95) which gives

$$H = \frac{1000}{0.60} i_1 = 1667 i_1 \; [\text{A} \cdot \text{t/m}] \tag{5.6.3}$$

For example, we note from Figure 5.9 that

$$H_{max} = 400 \qquad [\text{A} \cdot \text{t/m}]. \tag{5.6.4}$$

Thus,

$$i_{1max} = \frac{400}{1667} = 0.24 \qquad [\text{A}]. \tag{5.6.5}$$

This example teaches us the following important facts:

1. The no-load component of the transformer current is highly nonsinusoidal. A harmonic analysis (Appendix B) of the current reveals that the 180- and 300-Hz components can be quite large in magnitude and through induction can cause "noise" in communication networks. (The effect of these higher-frequency harmonic currents in three-phase transformers is discussed in Section 5.8.3.)
2. By focusing attention on the 60-Hz component of i_1 and by accepting some degree of approximation, we can still work with the linear equivalent network. (Expressed differently: In working with the linear equivalent circuits shown in Figures 5.7, 5.8, and 5.10, one is, in effect, neglecting the higher harmonics.)

5.4.4 Modeling of the Winding Losses

The model shown in Figure 5.8, with sufficient accuracy, incorporates the nonideal features of the transformer core. Let us now incorporate the nonideal features of the transformer coils into the model; there are two types.

First, we have the winding resistance, which will cause ohmic losses in the coils and also resistive voltage drops. As the resistance is present in both windings, it is natural to account for its presence by adding series resistances to both the primary and secondary windings to the transformer shown in Figure 5.8.

We learned earlier how impedances can be "referred" from one side to the other by the use of equation (5.34). We find it practical to lump the effects of both winding resistances into one equivalent series resistor R_s, as shown in Figure 5.10. It does not matter on which side of the IT the resistance is placed. Of course, R_s will have different values (differing by the factor a^2), depending on where it is placed.

One can also argue whether R_s should be placed to the left or right of the R_m–X_m shunt. This is of no practical importance because, as we noted above, the series path carries the current I_2/a, which is much larger than I_{1m}.

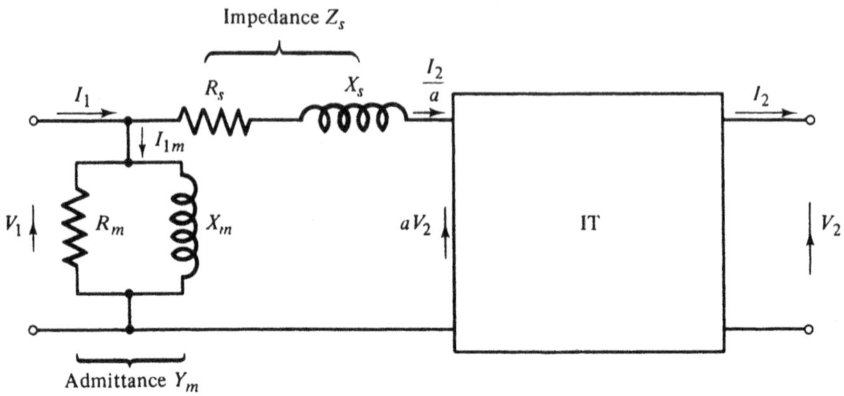

$$Z_s = R_s + jX_s$$
$$Y_m = \frac{1}{R_m} + \frac{1}{jX_m}$$

Figure 5.10

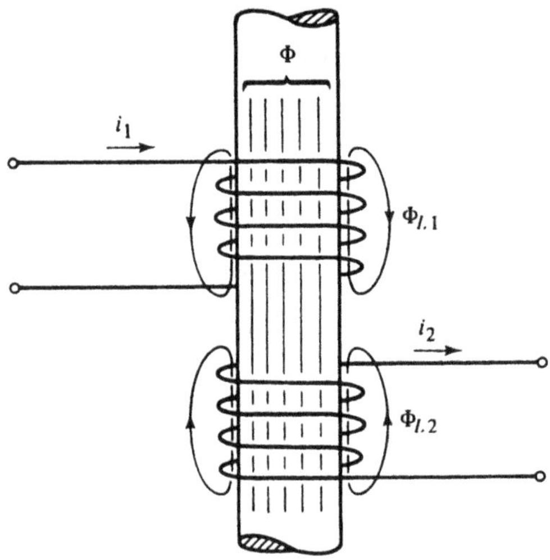

Figure 5.11

We have also added, in series with R_s, the *leakage reactance*, X_s. This equivalent reactance represents the effect of the transformer leakage fluxes, which are shown schematically in Figure 5.11. The leakage fluxes are those portions of the flux that are not mutually linked to the windings. Their magnitudes are very small in relation to the core flux Φ because the paths for leakage flux are predominantly in air. The flux is therefore proportional to the current causing it. The equivalent reactance X_s is therefore with good accuracy, a constant, independent of the magnitude of the current. (The shunt reactance X_m, in contrast, is not constant because of the nonlinearity of the iron.)

In the examples that follow it will be shown that the shunt impedance elements in Figure 5.10 are vastly larger than the series elements, that is,

$$R_s \ll R_m \qquad [\Omega], \tag{5.51}$$

and

$$X_s \ll X_m \qquad [\Omega]. \tag{5.52}$$

In most power transformers it is also true that

$$X_s \gg R_s \qquad [\Omega]. \tag{5.53}$$

5.4.5 Measurement of Transformer Losses

In order to determine the values of the components of the equivalent circuit of the transformer, it is necessary to carry out two tests.

5.4.5.1 The Open-Circuit Test

The full-rated voltage is applied to the primary with the secondary on open circuit. The real and reactive power supplied to the transformer are measured. From Figure 5.10 it is clear that when the secondary is on open-circuit I_2 is zero and therefore no current flows in the series impedance Z_S. It then follows that the real and reactive power taken by the transformer must be dissipated in R_m and X_m, respectively. The values of R_m and X_m can be calculated from the measurements. The open-circuit test can also be carried out with full-rated voltage applied to the secondary with the primary on open circuit. The corresponding values of R_m and X_m will then be referred to the secondary.

5.4.5.2 The Short-Circuit Test

A short circuit (ammeter) is connected across the secondary and a *reduced* voltage is applied to the primary and adjusted to give rated current in the secondary (and in the short circuit). The real and reactive power supplied to the transformer are measured. Since the applied voltage required to drive full-load current in the secondary coil is usually less than 5% of the rated value, it follows that the dissipation in R_m and X_m will be negligible. The real and reactive power taken by the transformer must be dissipated in R_s and X_s, respectively. The values of R_s and X_s can be calculated from the measurements. Similar to the open-circuit test, short-circuit test can also be carried out with a short circuit across the primary and the reduced voltage applied to the secondary. The corresponding values of R_s and X_s will then be referred to the secondary.

We can now demonstrate how to construct a model of a transformer from test data and how to use the model to predict the operating characteristics of the transformer.

Example 5.7

Construct a more accurate model of the transformer discussed in Examples 5.2 and 5.3, using the more complete set of data:

$$\text{rated voltage} = 3\text{ kV primary, } 9\text{ kV secondary;}$$

$$\text{rated frequency} = 60\text{ Hz;}$$

$$\text{rated current} = 333.3\text{ A primary,}^5\ 111.1\text{ A secondary.}$$

Test results are as follows:

$$P_{oc} = 4.31\text{ kW}, \quad Q_{oc} = 9.11 \quad [\text{kVAr}]. \tag{5.7.1}$$

$$P_{sc} = 8.31\text{ kW}, \quad Q_{sc} = 50.3 \quad [\text{kVAr}]. \tag{5.7.2}$$

Solution: As before, the turns ratio:

$$a = \frac{3}{9} = 0.3333. \tag{5.7.3}$$

Assuming that both the series and shunt elements in Figure 5.10 are placed on the 3-kV (primary) side, we can compute the elements as follows:

1. Using Table 4.1 (item 5) we compute the shunt elements from the OC test data as

$$R_m = \frac{3000^2}{4310} = 2088 \quad [\Omega], \qquad X_m = \frac{3000^2}{9110} = 988 \quad [\Omega]. \tag{5.7.4}$$

2. Item 4 in Table 4.1 tells us how to obtain the series elements from the short-circuit test data, namely:

$$R_s = \frac{8310}{333.3^2} = 0.0748 \quad [\Omega], \qquad X_s = \frac{50300}{333.3^2} = 0.453 \quad [\Omega] \tag{5.7.5}$$

Note that the values of R_s, X_s, R_m, and X_m confirm the inequalities (5.51), (5.52), and (5.53). (What values would the four impedances have if placed on the opposite side of the IT?)

Example 5.8

Use the more accurate model obtained in Example 5.7 to repeat Example 5.3. Find the transformer efficiency.

[5] This current rating corresponds to a transformer power rating of 1000 kVA (see Section 5.5.3 for an explanation).

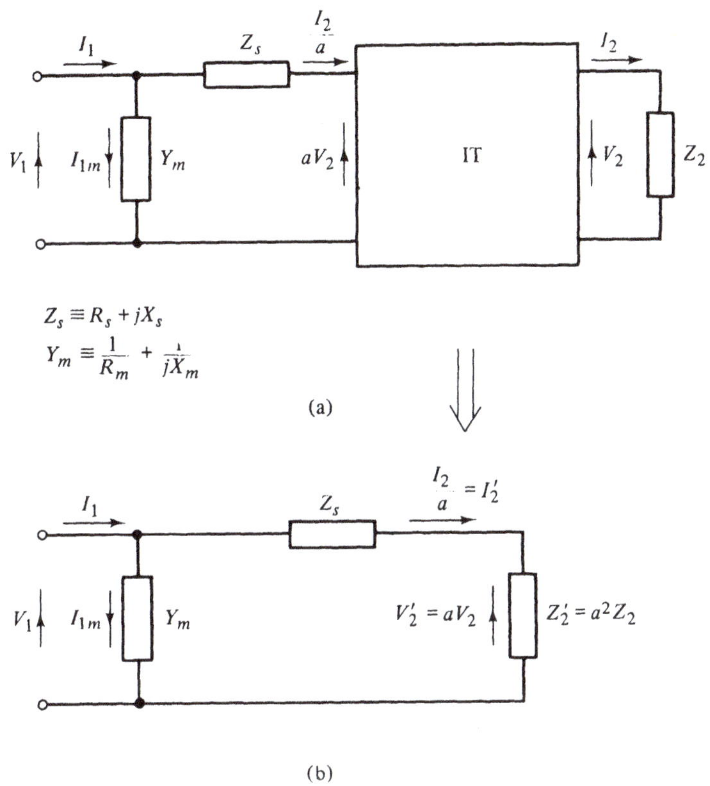

Figure 5.12

Solution: We connect the impedance $Z_2 = 100 + j30\ \Omega$ across the secondary terminals and obtain the equivalent circuit shown in Figure 5.12a. This is further reduced to the simpler circuit shown in Figure 5.12b. The simplification is self-explanatory.

From Figure 5.12b we obtain by inspection,

$$I_2' = \frac{I_2}{a} = \frac{V_1}{Z_s + a^2 Z_2} \qquad \text{[A]}, \tag{5.8.1}$$

$$V_2' = aV_2 = a^2 Z_2 \frac{I_2}{a} = V_1 \frac{a^2 Z_2}{Z_s + a^2 Z_2} \qquad \text{[V]}, \tag{5.8.2}$$

and

$$I_1 = \frac{I_2}{a} + I_{1m} = \frac{I_2}{a} + V_1 Y_m \qquad \text{[A]}. \tag{5.8.3}$$

We introduce the following numerical values from Example 5.7:

$$\begin{aligned} V_1 &= 3000 \angle 0^\circ \quad [\mathrm{V}]; \\ a &= 0.3333; \\ Z_s &= 0.0748 + j0.453 \quad [\Omega]. \end{aligned} \tag{5.8.4}$$

From which we obtain

$$Y_m = \frac{1}{2088} + \frac{1}{j988} = (0.479 - j1.012) \cdot 10^{-3} \quad [\mathrm{S}] \tag{5.8.5}$$

$$\begin{aligned} I_2 &= 84.69 \angle -18.70^\circ \quad [\mathrm{A}], \\ V_2 &= 8841 \angle -2.00^\circ \quad [\mathrm{V}], \\ I_1 &= 256.4 \angle -19.24^\circ \quad [\mathrm{A}]. \end{aligned} \tag{5.8.6}$$

Compare these values with those obtained from Example 5.3. (Note that the errors in the magnitudes of both voltage and current in no case exceed 2%.)

The transformer efficiency is defined as

$$\eta = \frac{P_2}{P_1} 100 = \frac{P_2}{P_2 + P_{\mathrm{loss}}} 100\%. \tag{5.8.7}$$

We have

$$P_2 = R_2 |I_2|^2 = 717.2 \cdot 10^3 \quad [\mathrm{W}]; \tag{5.8.8}$$

$$P_{\mathrm{loss}} = P_{\mathrm{Fe}} + P_{\mathrm{Cu}} = 4310 + 0.0748 \cdot 256.4^2 = 9227 \quad [\mathrm{W}]. \tag{5.8.9}$$

Thus,

$$\eta = \frac{717.2}{726.4} 100 = 98.73\%. \tag{5.8.10}$$

Example 5.9

If the secondary terminals of the transformer in Example 5.8 are accidently short-circuited, how large will the short-circuit current be?

Solution: A "metallic" short circuit across the secondary represents a secondary impedance, $Z_2 = 0$.

From equation (5.8.1) we get

$$I_1 \approx \frac{I_2}{a} = \frac{V_1}{Z_s} = \frac{3000}{0.0748 + j0.453} \quad [\mathrm{A}]. \tag{5.9.1}$$

$$|I_1| = 6540 \quad [\mathrm{A}]. \tag{5.9.2}$$

This is 19.6 times the rated current! A current of such magnitude can be highly destructive, and measures must be taken to quickly interrupt it. (If we had used the IT model in the above analysis we would have come to the absurd conclusion that the short-circuit current would be infinite.)

5.5 Some Practical Design Considerations

In this section we shall comment briefly on some of the more important practical aspects of transformer design.

5.5.1 Core and Coil Design

The transformer core is always laminated to minimize core losses. Figure 5.13 shows a typical arrangement showing three separate windings of a concentric coil design; multiwinding transformers are discussed in Section 5.6. Due to the greater insulation distances needed for increased voltage levels, the high-voltage windings are placed farthest from the core. The whole transformer is immersed in a

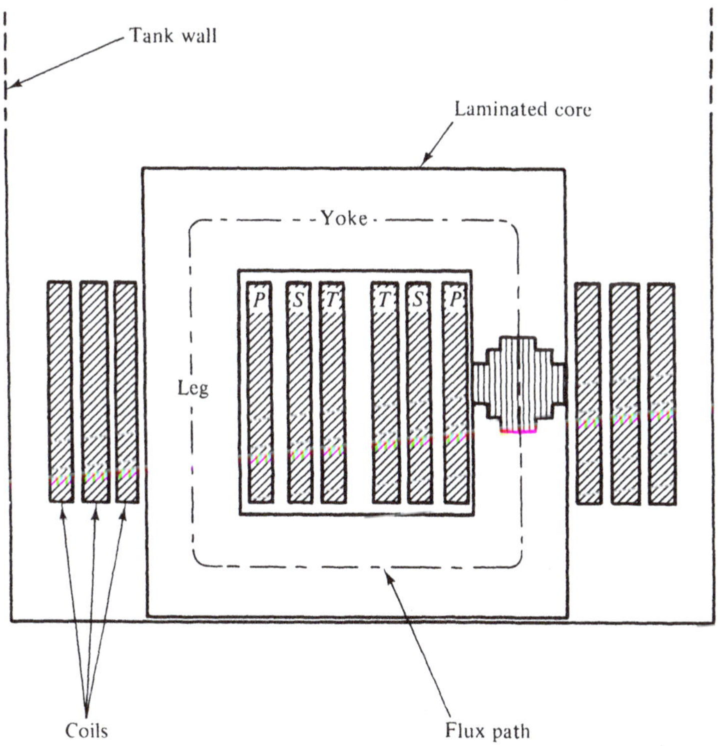

Figure 5.13

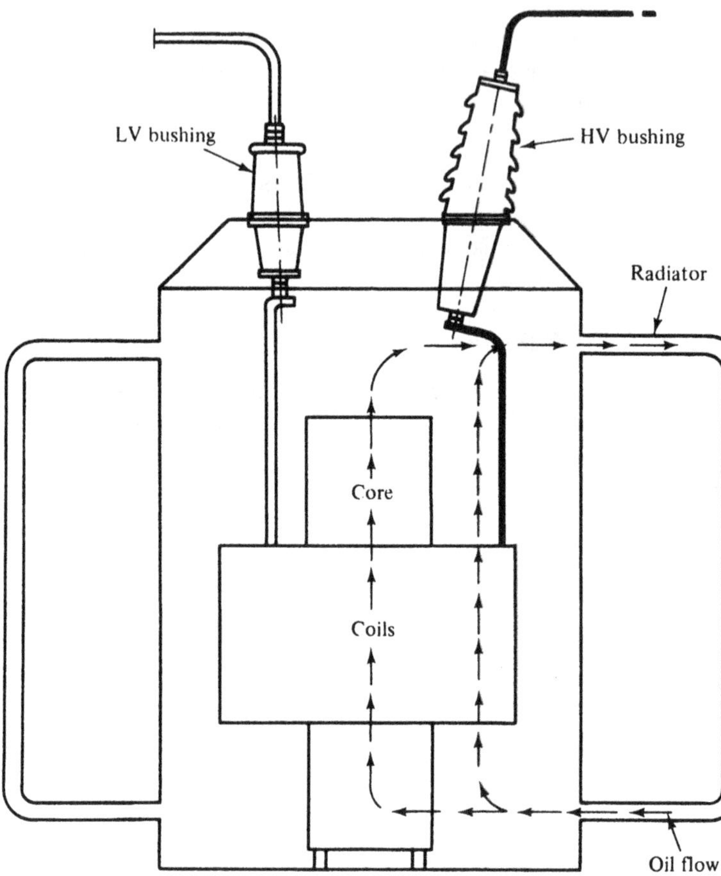

Figure 5.14

tank of oil, which serves a double purpose: it improves the insulation strength and also serves as a transport medium to the outside for the heat generated by the iron and copper losses in the core and coils, respectively. The power is brought in and out of the tank through *bushings* (see Figure 5.14).

5.5.2 Cooling Methods

Required insulation distances and minimum core and coil dimensions determine the overall size of a power transformer. The need to remove the heat has important implications on the size of the device. The core losses may be minimized by laminating the core, but they cannot be entirely eliminated.

The ohmic losses in the windings depend on the resistance. To minimize these losses, power transformer windings are usually made of copper. The choice of the cross-sectional area of the copper is determined by the size of the core and the economics of electric power. Core and ohmic losses are proportional to the volume of the core and the windings, respectively.

If L is a linear dimension of core plus windings, the above reasoning thus tells us that the total power loss is proportional to the volume of the active parts, that is,

$$P_{\text{loss}} \propto L^3 \qquad [\text{W}]. \tag{5.53}$$

The amount of heat that can be conducted through a medium is proportional to the surface area of conduction. Therefore the amount of heat that can be removed via conduction from the core plus windings is

$$P_{\text{removed}} \propto L^2 \qquad [\text{W}]. \tag{5.54}$$

Thus,

$$\frac{P_{\text{loss}}}{P_{\text{removed}}} \propto L. \tag{5.55}$$

This shows the difficulty of heat removal as the size of the transformer increases. In small transformers (see Figure 5.14), the self-induced flow (due to natural convection) of the heated oil, up beside the core and coils, and down the cooling tubes of the tank, is sufficient to transport the heat to the outside and hence maintain an acceptable ambient temperature.

For larger units *forced cooling* is necessary. The oil must be pumped through heat exchangers for efficient heat removal.

5.5.3 Transformer Ratings

The transformer core losses depend on the magnitude of the core flux, which, according to equation (5.18), is proportional to the voltage. The transformer ohmic, or copper, losses depend on the currents in the windings. The total losses determine the maximum transformer temperatures, and therefore its ratings must include information on both voltage and current. The product of current and voltage would evidently be a meaningful figure for rating the transformer. This explains why it is customary to rate a transformer in terms of its kilovolt-ampere (or megavolt-ampere) load. As the operating voltage is kept nearly constant, the maximum permissible kilovolt-amperes set the limits on the load current.

For example, the ratings of a particular transformer may read: voltage = 50/10 kV, 60 Hz; power = 6000 kVA. From these ratings we get the maximum load current as follows:

$$\begin{aligned} I_{\text{HV}} &= \frac{\text{kVA}}{\text{kV}_{\text{HV}}} = \frac{6000}{50} = 120 \qquad [\text{A}]; \\ I_{\text{LV}} &= \frac{\text{kVA}}{\text{kV}_{\text{LV}}} = \frac{6000}{10} = 600 \qquad [\text{A}]. \end{aligned} \tag{5.56}$$

Note that kilowatt ratings would be meaningless. A transformer may deliver zero power (if the load is purely reactive) and still have to withstand maximum permissible core *and* copper losses.

5.6 Multiwinding Transformers

It is quite common practice in electric power systems for power to be transformed to more than one voltage level in a single transformer. Consider, for example, the three-winding transformer shown in Figure 5.15. If the *primary* with a 600-turn winding is connected to a 6-kV source, the 200-turn *secondary* will deliver 2 kV and the 100-turn *tertiary*, 1 kV, to separate loads.

As mmf balance must be maintained, we get

$$I_1 N_1 = I_2 N_2 + I_3 N_3 \qquad [\mathrm{A \cdot t}]. \tag{5.57}$$

(We have neglected the magnetization current as it is very small compared to the load current.)

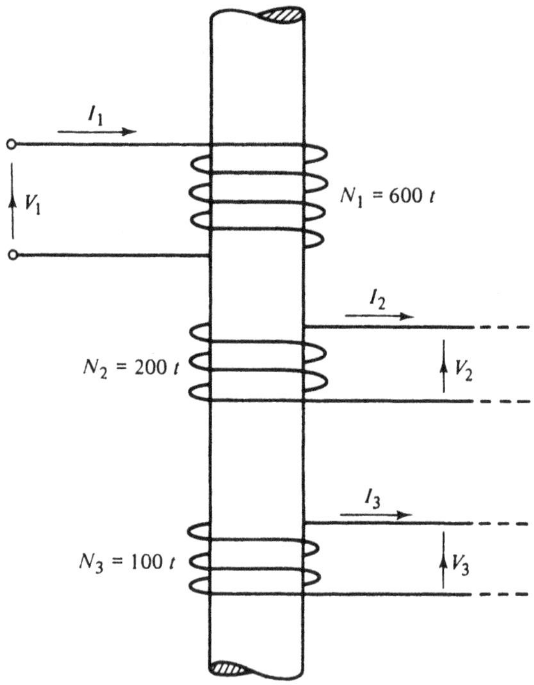

Figure 5.15

Example 5.10

The three windings in the transformer in Figure 5.15 have the following kVA ratings:

$$\begin{aligned} \text{maximum primary kVA} &= 300, \\ \text{maximum secondary kVA} &= 150, \\ \text{maximum tertiary kVA} &= 200. \end{aligned} \tag{5.10.1}$$

Rated voltages:

$$\begin{aligned} \text{primary kV} &= 6 \\ \text{secondary kV} &= 2 \\ \text{tertiary kV} &= 1 \end{aligned} \tag{5.10.2}$$

and

$$N_1 = 600; \qquad N_2 = 200; \qquad N_3 = 100. \tag{5.10.3}$$

The secondary is delivering full-rated kVA to a purely resistive load. The tertiary load consists of a variable reactor. As the reactor load is increased, at what point will the kVA ceilings be reached? Use the IT model in the analysis.

Solution: The three voltages, V_1, V_2, and V_3 will be in phase (neglecting the voltage drops) as indicated in Figure 5.16. The secondary is fully loaded, therefore,

$$|I_2| = \frac{\text{kVA}_2}{\text{kV}_2} = \frac{150}{2} = 75\ [\text{A}]. \tag{5.10.4}$$

Since the secondary load is purely resistive, I_2 will be in phase with V_2. As the tertiary load is purely reactive, I_3 will lag V_3 by 90°. In Figure 5.16 we have shown $N_2 I_2$ and $N_3 I_3$, as equation (5.57) tells us that the mmf's, rather than the currents, determine the loading.

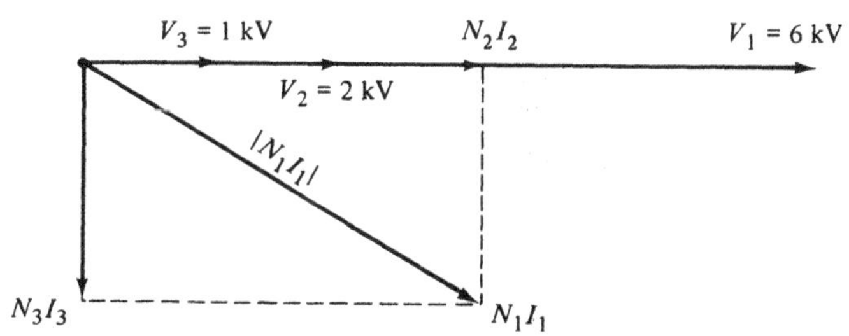

Figure 5.16

Let us assume that we adjust the reactor load until the tertiary is fully loaded. This means that its current, I_3, will have an rms value of

$$|I_3| = \frac{\text{kVA}_3}{\text{kV}_3} = \frac{200}{1} = 200 \qquad [\text{A}]. \tag{5.10.5}$$

According to equation (5.57) we get

$$600I_1 = 200I_2 + 100I_3 \qquad [\text{A} \cdot \text{t}]. \tag{5.10.6}$$

The value $|I_1|$ is given by

$$600|I_1| = |200I_2 + 100I_3| = |200 \cdot 75 + 100 \cdot (-j200)| \qquad [\text{A} \cdot \text{t}] \tag{5.10.7}$$

$$= \sqrt{15{,}000^2 + 20{,}000^2} = 25{,}000 \tag{5.10.8}$$

$$\therefore |I_1| = \frac{25{,}000}{600} = 41.7 \qquad [\text{A}]. \tag{5.10.9}$$

The rated primary current is

$$|I_{\text{rated}}| = \frac{\text{kVA}_1}{\text{kV}_1} = \frac{300}{6} = 50 \qquad [\text{A}] \tag{5.10.10}$$

This means that the primary is not loaded to its capacity. However, we cannot increase the tertiary load any further because it is already at its full-load capacity.

5.7 Autotransformers

Autotransformers can be thought of as a reactive voltage divider (step-down transformer) or multiplier (step-up transformer). Figure 5.17a shows the configuration of the autotransformer and the relationship between the voltages, the number of turns, and the currents.

If a two-winding transformer is reconnected as an autotransformer, its power rating can be increased considerably. To demonstrate this, consider the 30-kVA transformer shown in Figure 5.17b. It has a turns ratio of 250/50, and the voltage rating is 500/100 V. The current rating (based on 30 kVA) is therefore 60/300 A.

If the windings are reconnected as shown in Figure 5.17c, we obtain a new primary with 300 turns. It is possible to energize this new primary from a 600-V source without changing the VPT (and hence the flux). We obtain a new transformer with a voltage rating of 600/500 V. (The alternate voltage rating 600/100 V is clearly also available.)

As shown in Figure 5.17c the new 500-V secondary can supply current to a load with a maximum of 360 A with neither of the two windings being *current-overloaded.* At this point the primary current will be 300 A, and the transformer has a power flow of

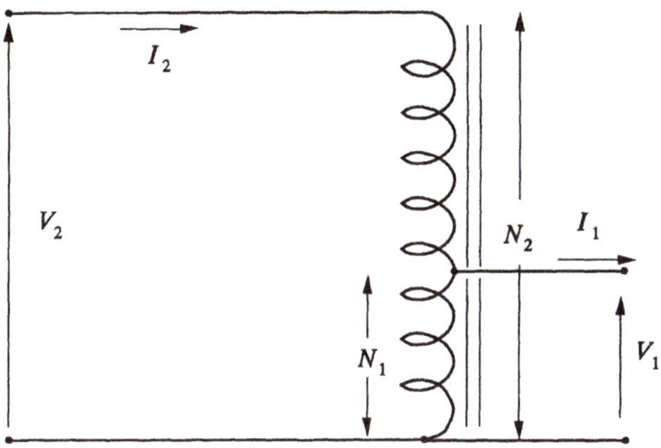

$$\frac{V_1}{V_2} = \frac{N_1}{N_2} = \frac{I_2}{I_1} = a$$

(a)

Figure 5.17

$$300 \cdot 600 = 360 \cdot 500 = 180 \qquad [\text{kVA}]. \tag{5.58}$$

As a result of the change, the transformer rating has changed from 30 kVA to 180 kVA. The explanation of this phenomenon is as follows:

> In the original 30-kVA transformer the two windings had no metallic connections, and so the 30-kVA had to pass through the magnetic flux of the transformer. The new autotransformer has its windings interconnected and hence
>
> $$180 - 30 = 150 \qquad [\text{kVA}] \tag{5.59}$$
>
> will pass through the transformer without being transformed magnetically.

If we connected the autotransformer in the alternate voltage ratio of 600/100 V, the new power rating would have been only 36 kVA, a very modest increase. One can easily show that the best power-rating increase is obtained when the autotransformer voltage ratio is close to unity. In general, autotransformers are used in power systems as links between voltage levels of nearly equal magnitudes (see Figure 6.1).

5.8 Three-Phase Power Transformers

Transformation of three-phase power can be achieved in several ways. Let us briefly study the most important ones.

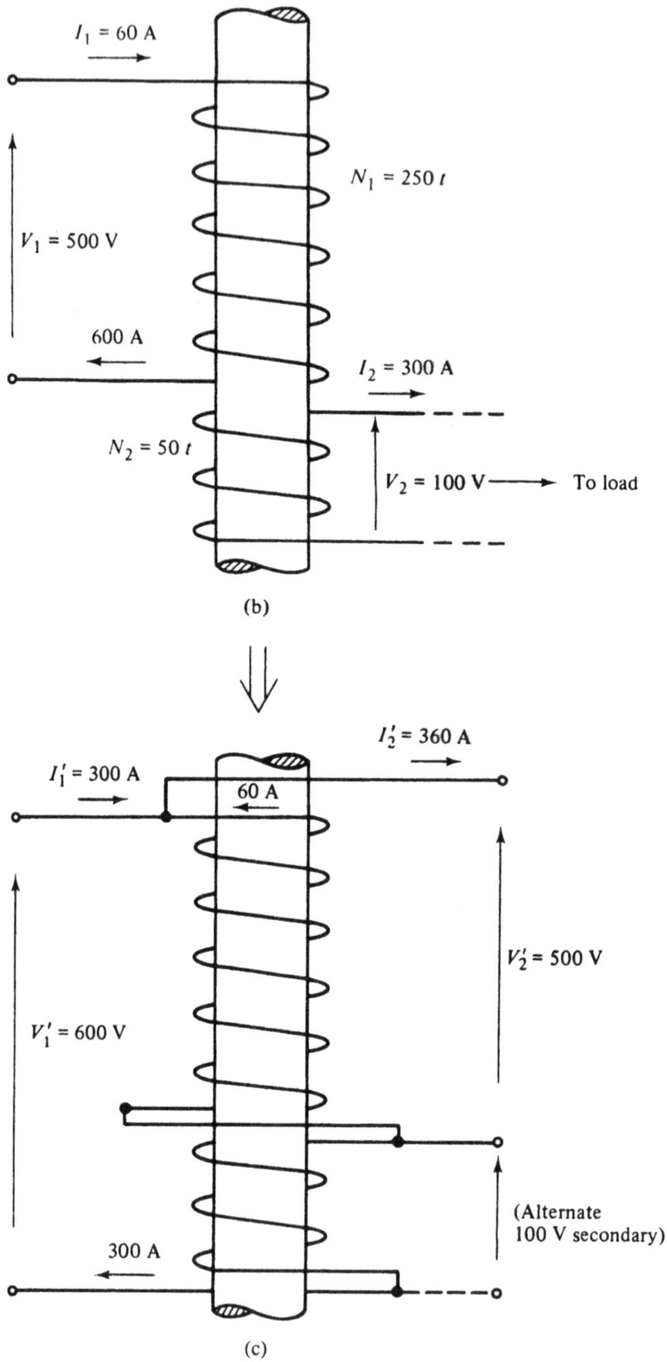

Figure 5.17 ***(cont.)***

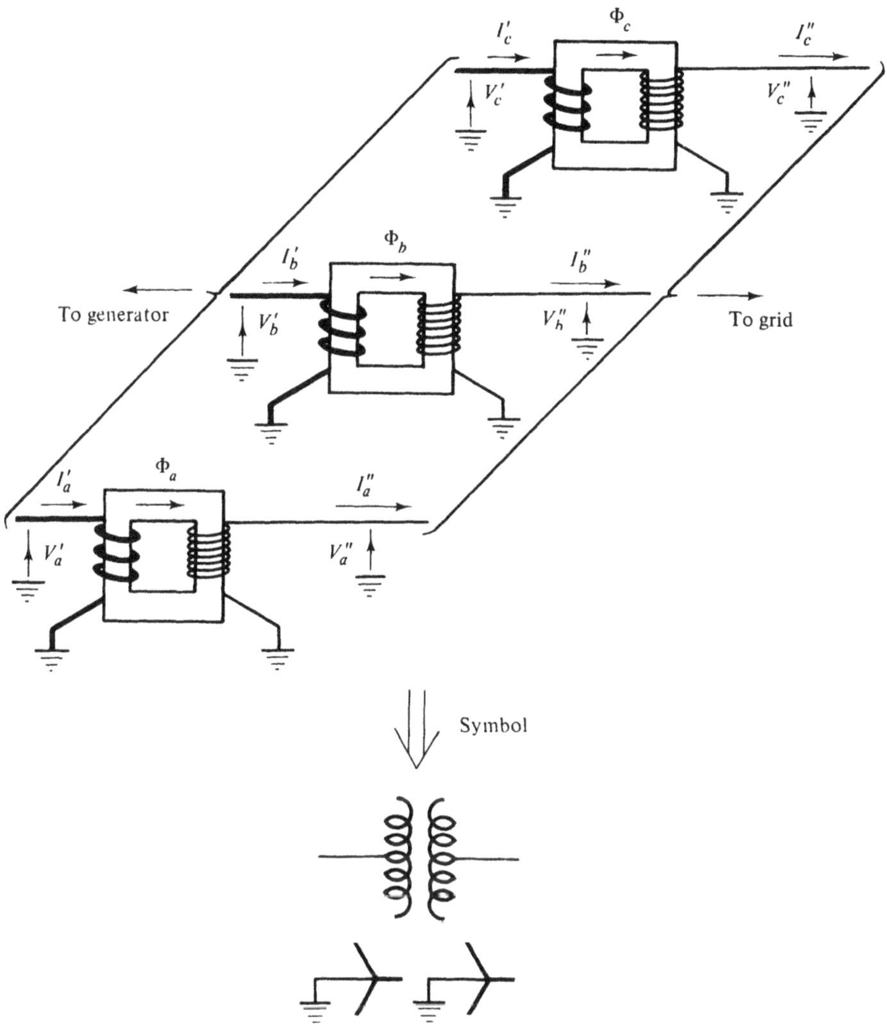

Figure 5.18

5.8.1 Single-Phase Units Connected Y–Y (Bank Arrangement)

The simplest way of transforming three-phase power is shown in Figure 5.18. Three identical single-phase transformer units are connected in parallel—one in each phase. If the HV side is symmetrically loaded, the HV side currents I''_a, I''_b, and I''_c will constitute a symmetrical current set. Because equation (5.27) applies to each transformer unit, the LV side currents I'_a, I'_b, and I'_c will likewise possess

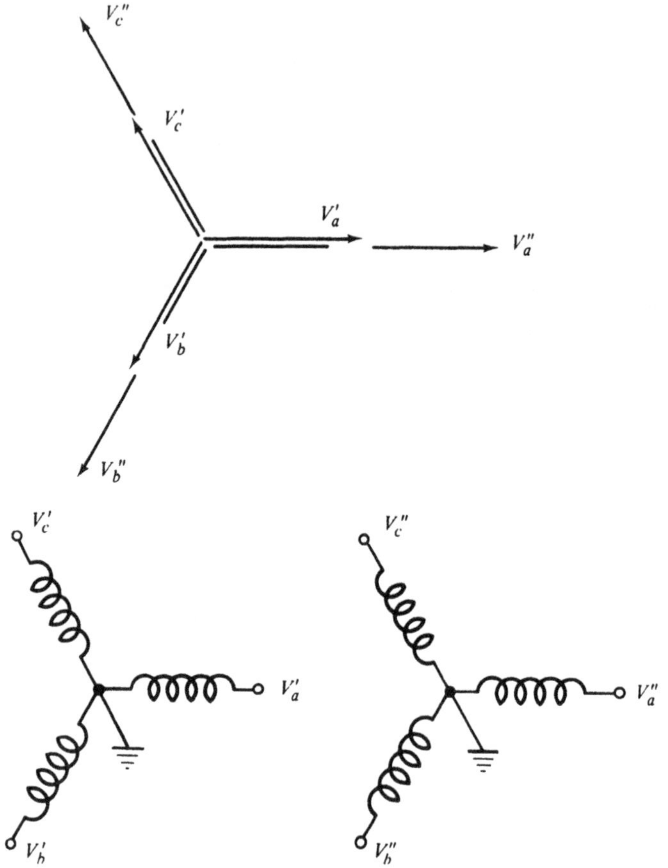

Figure 5.19

three-phase symmetry. Consequently, there will be no ground currents through the neutral [due to equation (4.71)].

Figure 5.19 shows the phasor diagram of the primary and secondary phase voltages. The designation Y–Y is self-evident. The figure also shows a symbolic way of displaying the transformer windings. Each winding is given an angular orientation that corresponds to its voltage phasor.

Example 5.11

The Y–Y connected transformer bank in Figure 5.18 serves as a step-up transformer between a 1000-MVA generator with a 22-kV rating and a 340-kV grid. Determine the voltage and current rating of each individual unit.

Solution: Each transformer must accept, on its LV side, a phase voltage of

$$|V'| = 22/\sqrt{3} = 12.70 \qquad [\text{kV}] \tag{5.11.1}$$

and deliver on the HV side the phase voltage,

$$|V''| = \frac{340}{\sqrt{3}} = 196.3 \qquad [\text{kV}]. \tag{5.11.2}$$

The generator supplies 333.3 MVA per phase, which corresponds to a rated LV side current of

$$|I'| = \frac{333.3}{12.70} = 26.24 \qquad [\text{kA per phase}]. \tag{5.11.3}$$

The turns ratio

$$a = \frac{22/\sqrt{3}}{340/\sqrt{3}} = 0.0647, \tag{5.11.4}$$

the rated HV side current will be

$$|I''| = 0.0647 \cdot 26.24 = 1.698 \qquad [\text{kA per phase}]. \tag{5.11.5}$$

In summary, each unit must have these ratings:

$$\begin{aligned} \text{power rating} &= 333.3 \qquad [\text{MVA}], \\ \text{voltage rating} &= 12.70/196.3 \qquad [\text{kV}], \\ \text{current rating} &= 26.24/1.698 \qquad [\text{kA}]. \end{aligned} \tag{5.11.6}$$

5.8.2 Three-Phase Core Arrangement

The three core fluxes Φ_a, Φ_b, and Φ_c in the bank arrangement shown in Figure 5.18 constitute a symmetrical three-phase set. Consequently, their algebraic sum is identically zero. For this reason, the three separate cores can be replaced by a single core called a "three-phase core" (Figure 5.20), which lacks a flux return path. The advantage of this arrangement is that the core contains less iron, and therefore it is cheaper to manufacture. The disadvantage of the single core arrangement is that a failure of a winding will disable the whole three-phase unit. A similar mishap in the three-bank arrangement disables only one-third of the total three-phase transformer.

Electrically, the arrangements in Figures 5.18 and 5.20 are identical.

5.8.3 The Δ–Y Connection

The Y–Y connection shown in Figures 5.19 and 5.20 would work flawlessly if all currents and voltages in the transformers were at the nominal frequency of 60 Hz.

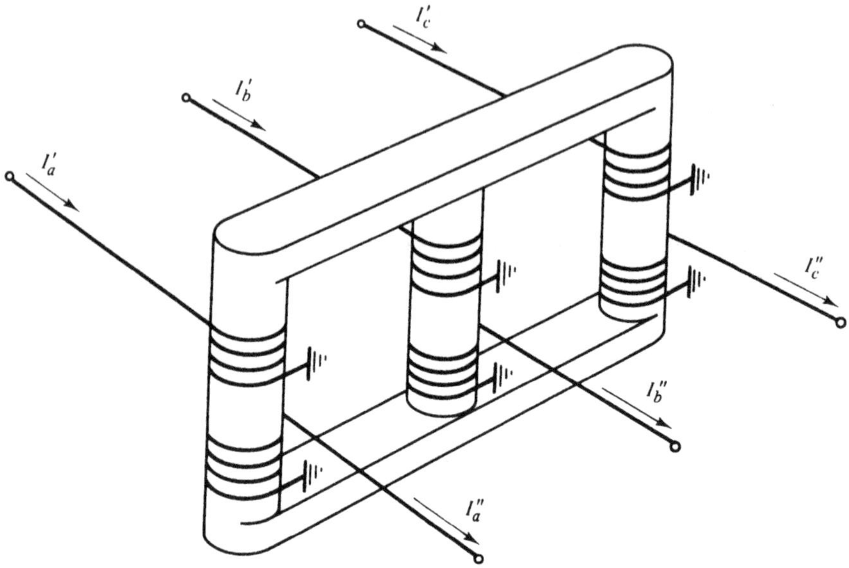

Figure 5.20

However, we have already shown (Figure 5.9) that the magnetization current in a power transformer contains harmonics—particularly the third harmonic of frequency 180 Hz. This harmonic current can cause serious problems in a Y–Y-connected three-phase transformer. Let us examine this.

Figure 5.21 shows the 60-Hz and 180-Hz current waveforms in all three phases as functions of time. Note that the 180-Hz currents in each of the three phases are all in phase. This latter circumstance would mean that the three 180-Hz phase components would add up to a neutral current of a magnitude three times that of each phase component. We can take several remedial measures:

1. Provide a fourth return conductor for the neutral current.
2. Let it exit into ground and "vagabond" back to the generator neutral.
3. Isolate the neutral and prevent the formation of the 180-Hz component.

The first two possibilities are not very attractive, both from the point of view of cost and the generation of electrical noise in communication circuits. The third possibility seems to be the simplest. But when one prevents the formation of the third harmonic current, one will instead cause the flux to become distorted and hence the voltage waveform. (This is easily seen in Figure 5.9. If one removes the 180-Hz component from the current, the flux and the voltage waveforms can no longer remain sinusoidal.)

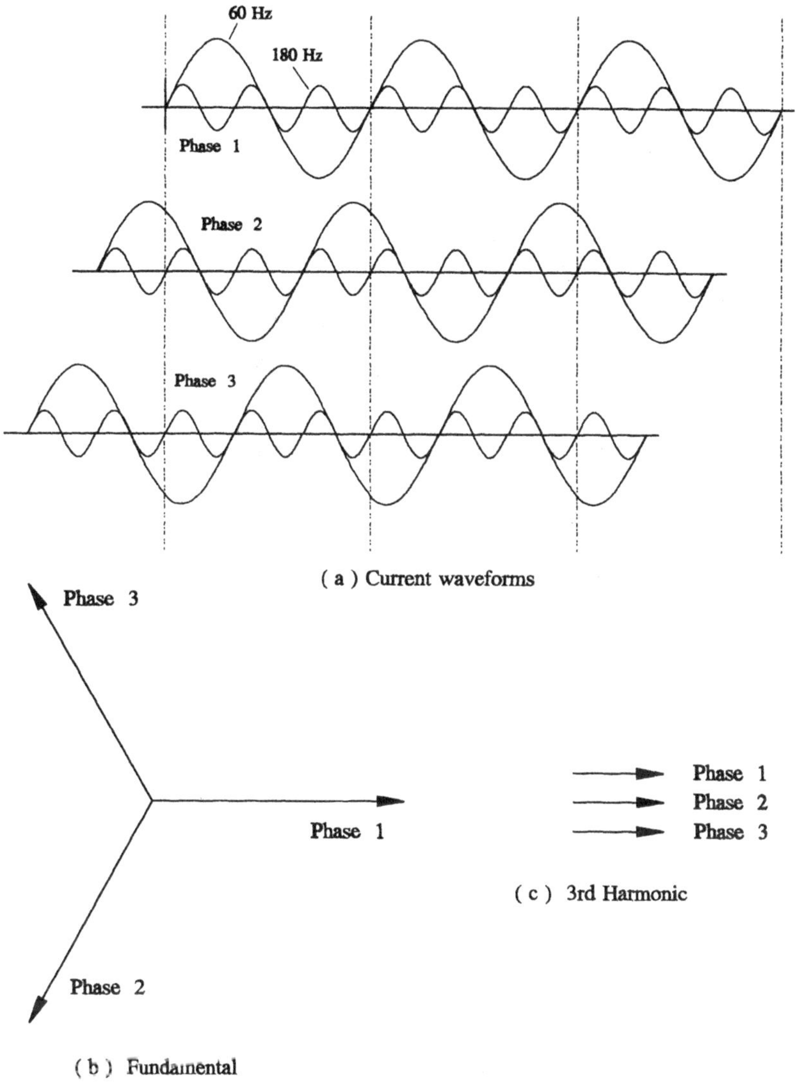

Figure 5.21

The problem is solved by providing a so-called Δ-connected winding, either in the form of a separate tertiary or, as shown in Figure 5.22, by reconnecting the windings in a Δ–Y configuration. The Δ-loop permits the formation of the required 180-Hz current component. It will circulate freely in the Δ, without entering the network. The fluxes can therefore take on the sinusoidal waveshape in all three cores, and no distortion of voltage waveforms take place.

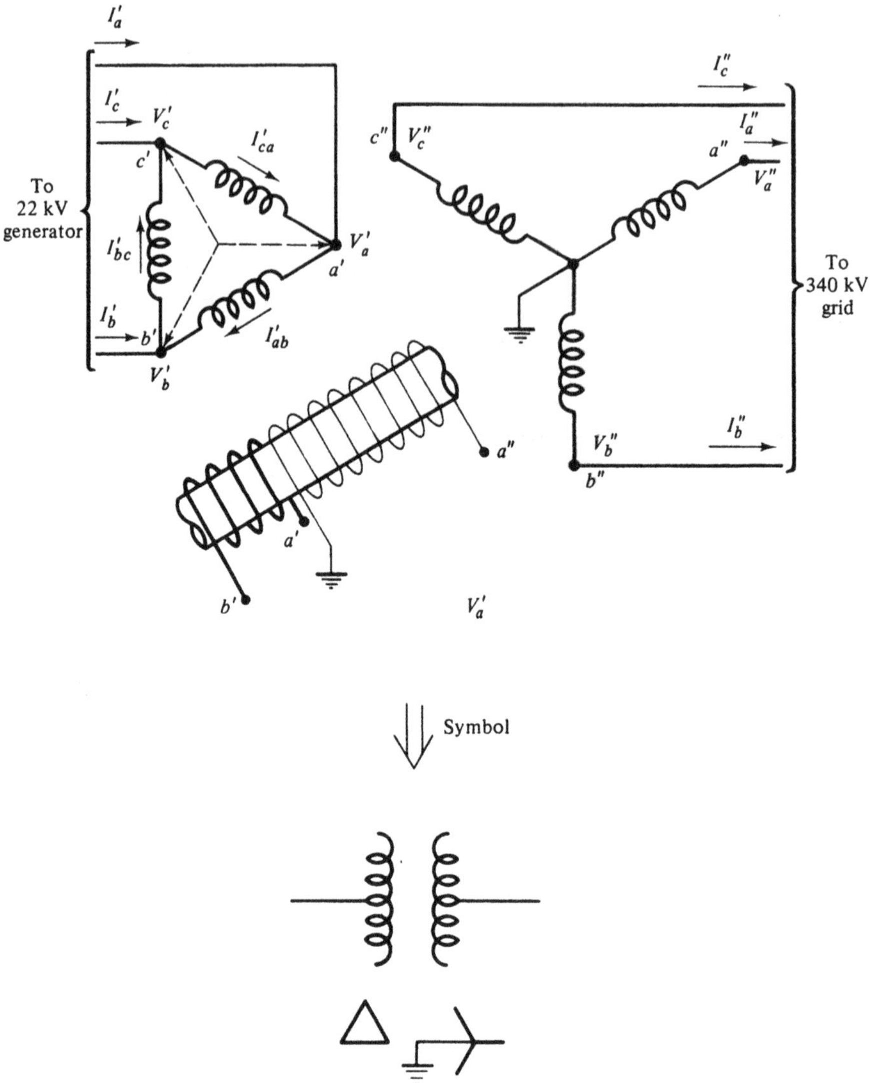

Figure 5.22

It is not immediately clear how the 60-Hz voltages and currents are related. Let us analyze the situation by considering the following specific example.

Example 5.12

The Δ–Y-connected three-phase transformer in Figure 5.22 serves as a step-up transformer between a 22-kV generator and a 340-kV grid. Determine the voltage ratings of each transformer unit and also compute all the currents and voltages that are identified in Figure 5.22. It is assumed that the transformer delivers 700 MW and 90 MVAr to the grid.

Solution: The LV side of each transformer unit is connected to the line voltage of the generator, that is, 22 kV. The HV side of the unit must provide an output equal to the phase voltage of the 340-kV grid, that is, 196.3 kV.

The voltage rating of each transformer unit will be 22.0/196.3 kV, that is, each single-phase unit will have a turns ratio,

$$a = \frac{N_1}{N_2} = \frac{(V_a' - V_b')}{V_a''} = \frac{22.0}{196.3} = 0.1121. \qquad (5.12.1)$$

Voltage Analysis

For an analysis of currents and voltages we shall, for simplicity, assume that each transformer unit is modeled as an IT. The generator neutral is *grounded* and by choosing V_a' as our *reference voltage* we can write for the primary phase voltages:

$$\begin{aligned} V_a' &= 12.70e^{j0°} \quad [\text{kV}], \\ V_b' &= 12.70e^{-j120°} \quad [\text{kV}], \\ V_c' &= 12.70e^{-j240°} \quad [\text{kV}]. \end{aligned} \qquad (5.12.2)$$

These voltage phasors are shown in dashed lines in Figure 5.22.

Now consider the transformer unit (the core of which is shown in Figure 5.22), which delivers the phase voltage V_a'' at its secondary terminals. Because its input voltage is equal to $(V_a' - V_b')$ we obtain from equation (5.15) the secondary phase voltage:

$$\begin{aligned} V_a'' &= \frac{1}{a}(V_a' - V_b') = \frac{1}{0.1121}(12.70 - 12.70e^{-j120°}) \\ &= \frac{1}{0.1121}(12.70 + 6.35 + j10.99) \\ V_a'' &= 196.3e^{j30°} \quad [\text{kV}]. \end{aligned} \qquad (5.12.3)$$

This result could be obtained just as simply by determining graphically the length and direction of the phasor:

$$(V_a' - V_b') = (V_a + [-V_b']) \quad [\text{V}]. \qquad (5.12.4)$$

Similarly, we can find the secondary phase voltages in the remaining two phases. We summarize the results:

$$\begin{aligned} V_a'' &= 196.3e^{j30^\circ} \quad [\text{kV}], \\ V_b'' &= 196.3e^{-j90^\circ} \quad [\text{kV}], \\ V_c'' &= 196.3e^{-j210^\circ} \quad [\text{kV}]. \end{aligned} \tag{5.12.5}$$

Note that the directions of all these voltage phasors agree with the symbolic winding orientation shown in Figure 5.22.

The phase angles of the secondary phase voltages indicate that all the secondary phase voltages are in phase with the Δ-phase voltages of the primary but they lead the corresponding generator phase voltages by 30°. (Remember that in the Y–Y case the primary and secondary phase voltages were all in phase.)

Current Analysis

From a knowledge of secondary power we can compute the secondary phase currents. Equation (4.21) yields, when applied to phase a,

$$V_a''(I_a'')^* = \frac{700}{3} + j\frac{90}{3} = 233.3 + j30.0 \quad [\text{MVA}]. \tag{5.12.6}$$

We solve for the secondary conjugate phase current:

$$(I_a'')^* = \frac{233.3 + j30.0}{196.3 \cdot e^{j30^\circ}} = 1.198e^{-j22.67^\circ} \quad [\text{kA}]. \tag{5.12.7}$$

For the secondary phase current we have

$$I_a'' = 1.198e^{j22.67} \quad [\text{kA}]. \tag{5.12.8}$$

As the secondary phase currents must possess three-phase symmetry (you can easily confirm this by analysis), we have

$$\begin{aligned} I_a'' &= 1.198e^{j22.67^\circ} \quad [\text{kA}], \\ I_b'' &= 1.198e^{-j97.33^\circ} \quad [\text{kA}], \\ I_c'' &= 1.198e^{-j217.33^\circ} \quad [\text{kA}]. \end{aligned} \tag{5.12.9}$$

If we now apply equation (5.27) to the winding shown in Figure 5.22, we get, for the primary Δ-phase current I'_{ab}:

$$I'_{ab} = \frac{1}{a} I_a'' = 10.69e^{j22.67^\circ} \quad [\text{kA}]. \tag{5.12.10}$$

Similarly for the Δ-phase current I'_{ca}:

$$I'_{ca} = \frac{1}{a} I_c'' = 10.69e^{-j217.33^\circ} \quad [\text{kA}]. \tag{5.12.11}$$

From Figure 5.22 we note that the primary line current I'_a is

$$I'_a = I'_{ab} - I'_{ca} \qquad [\text{kA}]. \tag{5.12.12}$$

Thus,

$$I'_a = 10.69e^{j22.67°} - 10.69e^{-j217.33°} = 18.52e^{-j7.33°} \qquad [\text{kA}]. \tag{5.12.13}$$

The primary line currents also possess three-phase symmetry, therefore,

$$\begin{aligned} I'_a &= 18.52e^{-j7.33°} \qquad [\text{kA}], \\ I'_b &= 18.52e^{-j127.33°} \qquad [\text{kA}], \\ I'_c &= 18.52e^{-j247.33°} \qquad [\text{kA}]. \end{aligned} \tag{5.12.14}$$

We have summarized the results of the above analysis in the phasor diagram in Figure 5.23. The primary and secondary currents and voltages are shown. Because complete three-phase symmetry exists, we have drawn only the phasors belonging to phase a. The following are important observations:

1. The ratio of the primary Δ-phase voltage (which is equal to the generator-line voltage) to the secondary phase voltage is

$$\frac{|V'_a - V'_b|}{|V''_a|} = \frac{N_1}{N_2} = \frac{22}{340/\sqrt{3}} = 0.1121. \tag{5.60}$$

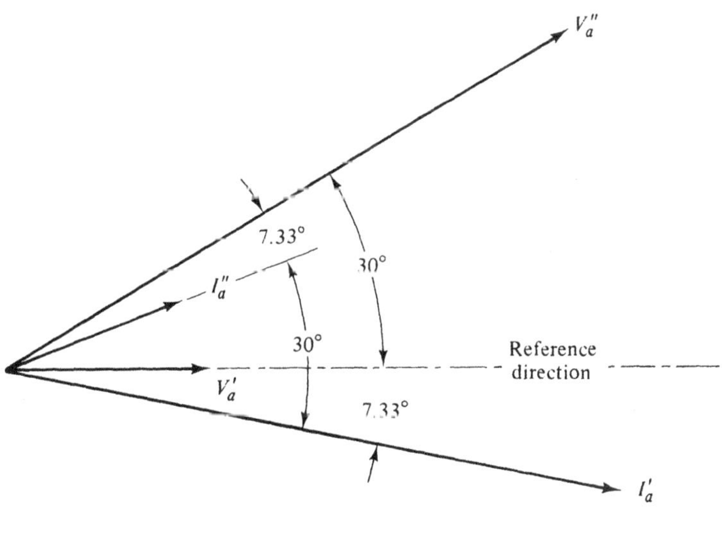

Note: Phasors are *not* shown in proper magnitude scale.

Figure 5.23

2. The Δ-phase currents in the primary and the secondary phase currents are transformed inversely in the same ratio:

$$\frac{|I'_{ab}|}{|I''_a|} = \frac{1}{0.1121} = 8.923. \tag{5.61}$$

3. The secondary phase voltages (V''_a, V''_b, and V''_c) and currents (I''_a, I''_b, and I''_c) are advanced 30° relative to the corresponding generator-phase voltages (V'_a, V'_b, and V'_c) and currents (I'_a, I'_b, and I'_c).
4. The magnitude of the primary line current is $\sqrt{3}$ times the primary Δ-phase current. (Note that the Δ-phase voltage (equal to the generator-line voltage) is $\sqrt{3}$ times the generator-phase voltage.)
5. It should be noted that the primary transformer windings can be interconnected to form a Δ in two ways. They are shown in Figure 5.24. (The first was analyzed in Example 5.12). If we perform a similar analysis of the second, we find that the only

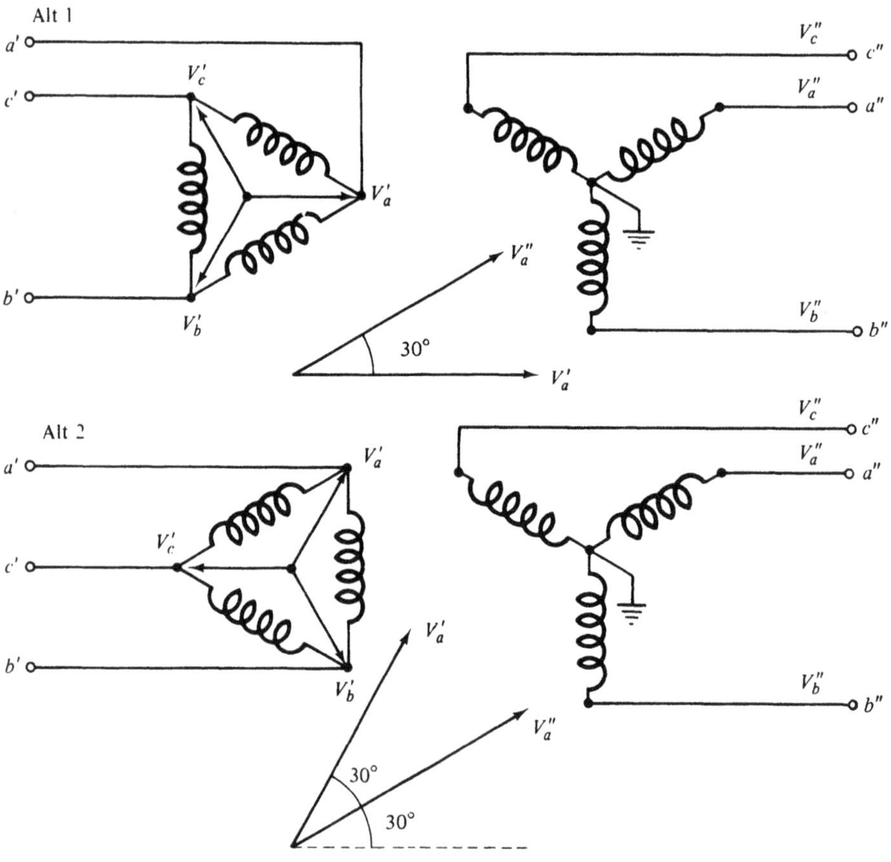

Figure 5.24

difference is a change of phase shift from $+30°$ to $-30°$ between primary and secondary variables.

6. The fact that the Δ winding introduces a phase shift of $\pm 30°$ has one very important consequence: One must be careful in connecting three-phase transformers in parallel. For example, one cannot operate a Y–Y- and Δ–Y-connected transformers in parallel, even if their line-voltage ratios are identical. Why?

5.8.4 Equivalent Circuits for Three-Phase Transformers

When we compare the results of the analysis in Examples 5.11 and 5.12 we conclude that both Y–Y and Δ–Y transformers are identical in one important respect: In both transformer configurations the magnitudes of the primary and the secondary phase voltages and currents are transformed in the same ratio—this ratio is computed as the ratio of the phase voltages or the number of turns in the phases. The transformers are not identical because of the $\pm 30°$ phase shift introduced by the Δ–Y transformer configuration.

In power system studies it is always of great importance to know how three-phase transformers affect the phase values of current and voltage. The power systems engineer will find it helpful to draw an equivalent circuit that shows the relationship between primary and secondary phase variables.

In view of the above findings we can conclude that the equivalent circuits for Y–Y and Δ–Y transformers should be identical in all respects except that the Δ–Y equivalent circuit, must reflect the $\pm 30°$ phase shift. The two circuits shown in Figure 5.25 satisfy these requirements. The equivalent circuit in Figure 5.25a represents a Y–Y connected transformer, the one in Figure 5.25b a Δ–Y connected transformer.

The box labeled "phase shifter" can be thought of as a circuit capable of rotating both the input current and voltage phasors through the angle $\pm 30°$ (sign depending on the polarity according to Figure 5.24). We have also added to the diagram (dashed line elements) series and shunt impedances that will account for the nonideal features.

Example 5.13

Consider the 1000-MVA three-phase transformer discussed in Examples 5.11 and 5.12. What voltage must be maintained at the generator terminals if the secondary voltage level is to be held at 340 kV? We assume the same power delivered as in Example 5.12. From a short-circuit test the transformer series impedance was found to be

$$Z_s = 0.00130 + j0.0711 \qquad [\Omega \text{ per phase}], \tag{5.13.1}$$

referred to the LV side.

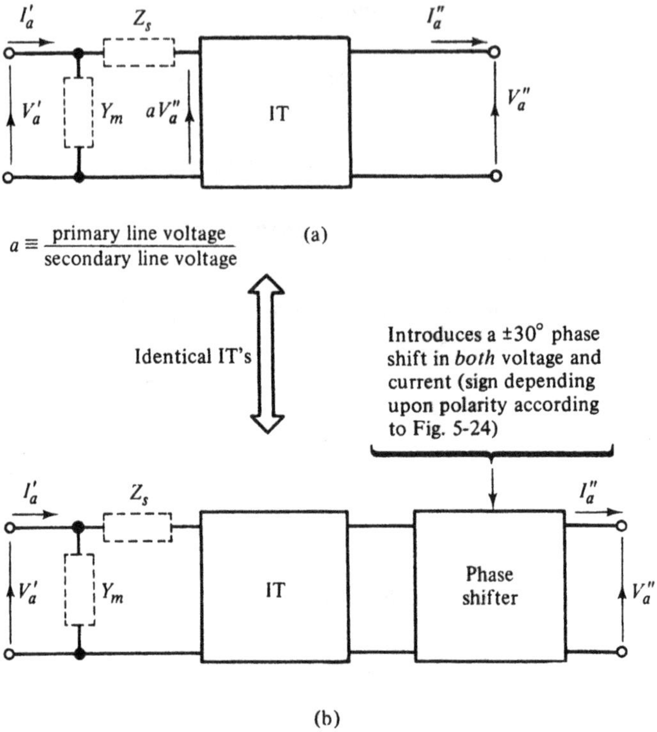

Figure 5.25

Before we proceed with the solution, we make two observations:

1. If the transformer were simply modeled as an IT we would give the answer as 22 kV. Clearly, we are (as in real life) interested in the voltage drop caused by the series impedance Z_s.
2. Because we are interested in the magnitude of the generator voltage we are not concerned whether this is a Δ–Y or Y–Y transformer. We therefore choose to work with the Y–Y configuration which is the simpler of the two circuit (Figure 5.25a).

Solution: We choose the secondary phase voltage in phase *a* as our reference phasor:

$$V''_a = \frac{340}{\sqrt{3}} e^{j0°} = 196.3 \qquad [\text{kV}]. \tag{5.13.2}$$

From a knowledge of the secondary power we compute the secondary phase current I'_a as follows:

$$V''_a (I''_a)^* = \frac{700}{3} + j\frac{90}{3} \qquad [\text{MVA}]. \tag{5.13.3}$$

Solution for the current yields

$$I''_a = 1.198e^{-j7.33^\circ} \qquad \text{[kA]}. \tag{5.13.4}$$

The line-to-line or phase-to-phase voltage turns ratio is

$$a = \frac{22.0}{340.0} = 0.0647. \tag{5.13.5}$$

A knowledge of this ratio permits us to compute the primary phase current:

$$I'_a = \frac{1}{0.0647} \cdot 1.198e^{-j7.33^\circ} = 18.52e^{-j7.33^\circ} \qquad \text{[kA]}. \tag{5.13.6}$$

The equivalent circuit diagram (Figure 5.25a) gives the primary phase voltage,

$$V'_a = aV''_a + I'_a Z_s \qquad \text{[kV]}. \tag{5.13.7}$$

Numerically,

$$\begin{aligned} V'_a &= 0.0647 \cdot 196.3 + 18.52e^{-j7.33^\circ}(0.00130 + j0.0711) \\ V'_a &= 12.96e^{j5.77^\circ} \qquad \text{[kV per phase]}. \end{aligned} \tag{5.13.8}$$

We conclude that we must increase the generator terminal voltage from 22 kV to

$$\sqrt{3} \cdot 12.96 = 22.45 \qquad \text{[kV]} \tag{5.13.9}$$

if we wish the transformer to deliver 340 kV to the grid.

5.9 Summary

Modern power systems span continents and thousands of megawatts of electric power must often be transmitted over distances measured in hundreds and even thousands of kilometers. Synchronous generators cannot generate power at voltage levels in excess of about 25–30 kV. We have seen that it is impossible to transmit bulk energy at such low-voltage levels. We need to transform the power to and from voltage levels in the hundreds of kilovolts. This is the job of the power transformer.

In this chapter we first learned the basic characteristics of the single-phase transformer. We also developed mathematical models that can be used to predict the transformer behavior in system studies. The first model was developed for the "ideal transformer." Corrections were then made to account for realistic nonideal core and winding behavior. A brief discussion of multiwinding and autotransformers was included.

All power systems are operated in the three-phase mode and it is necessary therefore to use transformers in this mode. We have discussed the basic Y–Y and Δ–Y connections. Numerous examples have been used to illustrate the theory.

EXERCISES

5.1 A single-phase transformer is designed to operate at 60 Hz: Voltage ratings: primary, 500 V; secondary, 200 V. The maximum permissible load is 30 kVA.

- **a)** What will be the magnitudes of the primary and secondary currents when the device is fully loaded?
- **b)** Loading is accomplished by an impedance connected across the 200-V terminals. How many ohms will correspond to full load of the transformer? (Use the IT model.)

5.2 Assume that you were to use the transformer in Exercise 5.1 in a 50-Hz power network. If you thoughtlessly connected it to a 500-V source, what values would you measure for (i) core flux? (ii) secondary voltage? Answer in percent of design values. Would this unintended use damage the transformer, in your opinion? Explain your answer.

5.3 The 30-kVA transformer in Exercise 5.1 is made the subject of a short-circuit test. One winding is short-circuited, and the other is fed from a 60-Hz voltage source. The voltage is raised until rated current is circulated in the windings, which occurs when the applied voltage is equal to 5.11% of rated winding voltage. The transformer consumes 290 W during the test.

- **a)** Compute the series impedance $Z_s = R_s + jX_s$ of the transformer referred to the primary and secondary sides.
- **b)** Compute the core flux during the short-circuit test. (Express its magnitude in percent of normal operating flux.)
- **c)** Why is it permissible to assume that all of the 290 W constitute ohmic losses in R_s and no part of it is core loss?

5.4 Assume that the transformer in Exercise 5.3 is operated from an ideal voltage source. If a short circuit occurs on the secondary, what would be the winding currents? Express your answer in amperes and also as a percentage of the rated currents.

5.5 The transformer in Exercise 5.3 is fed from a 500-V source. A load of impedance $Z_L = 1.03 + j0.72\ \Omega$ is connected across the secondary.

- **a)** Find the currents in both windings and the secondary voltage by using the IT model.
- **b)** Same as in part a, but now include the transformer impedance Z_s in your analysis. Comment on the change in your answers.
- **c)** Is the transformer "current-overloaded"?

5.6 The 30-kVA transformer in Exercise 5.1 is made the subject of an open-circuit test. It is fed from a 500-V source with the secondary terminals open circuit. The transformer consumes 230 W.

Based on the short-circuit and open-circuit test data compute the efficiency of the transformer when loaded with the impedance, Z_L as specified in Exercise 5.5.

5.7 The 500/200-V, 30-kVA transformer in Exercise 5.1 is reconnected as a 700/500-V autotransformer. Compute the new kVA rating of the device.

5.8 The terminals of the 500/200-V transformer windings in the previous exercise can be interconnected in four different ways, two of which will result in a 700/500-V autotransformer. Assume that you have interconnected the windings in the wrong way, but you believe that you did it the right way. In other words, you think that you have a 700/500-V autotransformer when in fact you have something else.

As you now connect the "700-V terminals" of your device to a 700-V source, you expect to obtain 500 V between what you presume to be "500-V terminals." To your surprise you get an entirely different voltage.

a) What voltage would you get?

b) What will happen to your transformer under these conditions?

5.9 Small power transformers used as variable voltage supplies (in laboratories, for example) are often connected as autotransformers as shown in Figure 5.26. Compute the ratio between the two winding currents, I' and I'':

a) When the sliding contact is adjusted to give 50% of the input voltage.

b) When it is adjusted to give 10% of the input voltage. (Note that the maximum voltage obtainable is 100%.)

c) Can you envision any danger of burning the winding in the extreme contact position?

5.10 The "hydraulic transformer" shown in Figure 5.27 transforms mechanical power between a high-force primary piston to a low-force secondary piston. The piston velocities are in inverse ratio to the forces. Thus the primary and secondary piston powers (force times velocity) are equal. This mechanical transformer can often be used as a good analog of an electric one.

Consider the odd-looking electric transformer in Figure 5.28a. The winding 1 is connected to a generator. The magnetic flux divides equally between the two outer legs. One now attempts to load one of either windings 2, 3, and 4. Can this be done? Use the hydraulic analog in Figure 5.28b to predict the results. Specifically, what happens with the voltages on windings 2 and 4 if you attempt to load 2?

5.11 Repeat Example 5.10 with one change: The variable tertiary load, like the fixed secondary load, is purely resistive. At what value of the tertiary load will the kVA ceilings be reached?

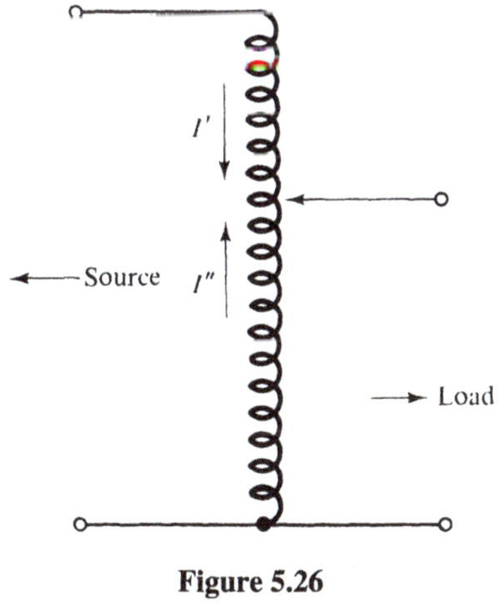

Figure 5.26

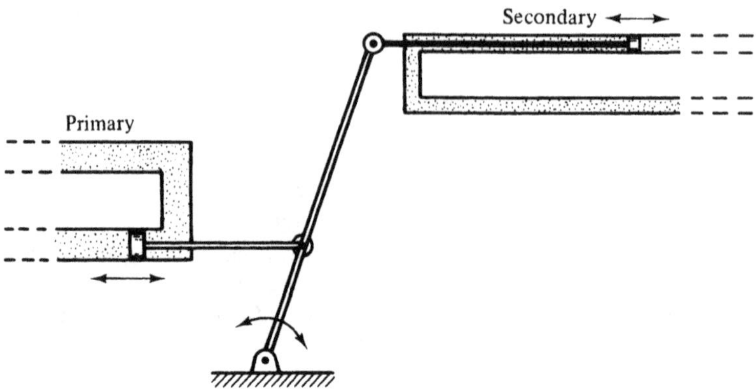

Figure 5.27

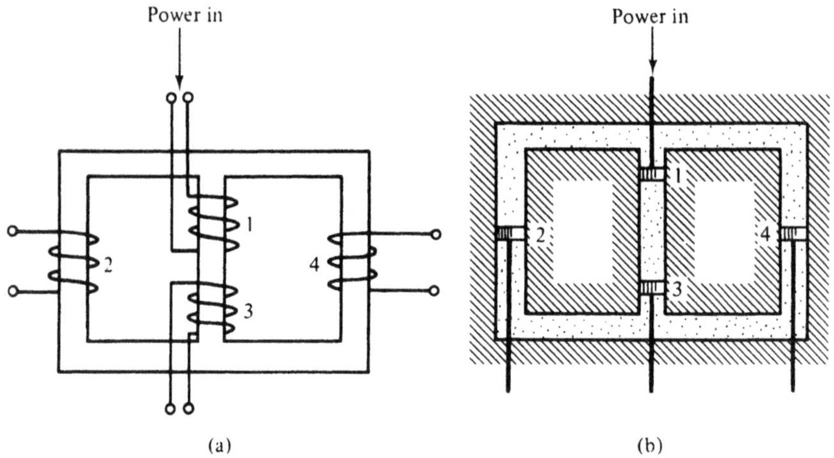

Figure 5.28

5.12 Three identical single-phase transformers, each having the ratings 5.500/23.20 kV, 60-Hz, 3000 kVA, are used to form a Y–Y connected three-phase bank of total power rating of 9 MVA. The bank is fed from a three-phase generator with a terminal voltage of 9.526 kV (line value). The load consists of three equal resistors R, connected in Δ. Use the IT models exclusively in your analysis to find:

a) the value of R which will result in 9-MVA load on the transformer bank;

b) the current in each load resistor;

c) the current in each transformer, primary and secondary.

5.13 The three identical single-phase transformers in Exercise 5.12 are connected Δ–Y and fed from a 5.5-kV three-phase generator. The load consists of three identical reactors connected in Y. Each reactor has an inductance L [H]. Again, use the IT model to find:

a) The value of L to give a 9-MVA load.
b) The voltage across each reactor.
c) All line currents.
d) The Δ-phase current in each transformer primary.

5.14 Consider Exercise 5.12 and keep the load resistances R unchanged. In this exercise we take into account the transformer series reactance, which is

$$Z_s = 0.711 + j6.98 \qquad [\Omega] \tag{5.62}$$

for each single-phase unit (based upon ratings). What is the percentage increase in the generator voltage, from the previous setting of 9.526 kV, required to keep the secondary voltage at 100% exactly?

5.15 The three-phase autotransformer shown in Figure 5.29 serves as a link between the 500 kV and 340 kV (line-to-line) portions of a power system. Determine the rms values of the currents indicated by the arrows when the transformer transforms a total power (three-phase) of 900 + j100 (MW and MVAr). Model the transformer as an IT. (The "idle" Δ-winding will not carry any 60-Hz currents. It accommodates the circulating harmonic components thus preserving the sinusoidal flux.)

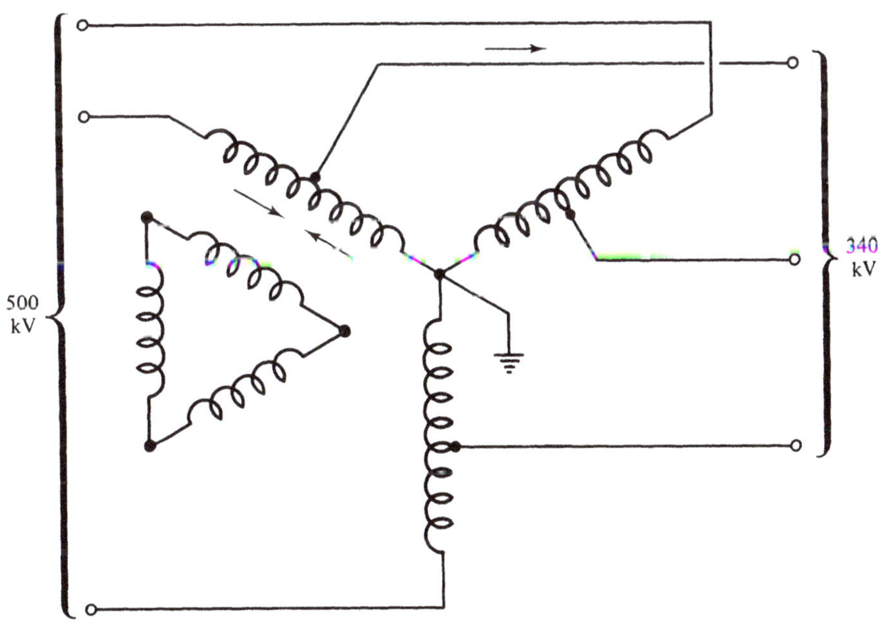

Figure 5.29

References

Del Toro, V. *Electric Machines and Power Systems*. Englewood Cliffs, NJ: Prentice-Hall, 1985.

Elgerd, O. *Electric Energy Systems Theory: An Introduction*. New York: McGraw-Hill, 1971.

McPherson, G., Laramore, R.D. *An Introduction to Electrical Machines and Transformers*, 2nd ed. New York: John Wiley & Sons, 1990.

Schultz, R.D., Smith, R.A. *Introduction to Electric Power Engineering*. New York: HarperCollins, 1985.

Slemon, G.R. *Electric Machines and Drives*. Reading, MA: Addison-Wesley, 1992.

6

The Electric Power Network

Local generation and consumption of electricity is unacceptable for economic, environmental, and reliability reasons. Consequently, electricity is generated in bulk quantities in *power stations* or *centers* and, as the customers are located over a vast geographic area, the electric energy must be transmitted over an *electric power network* connecting the power stations to the customers.

The locations of the power stations are dictated by a number of factors. Fuel availability is one important factor. For example, coal-fired generating stations are often located in the vicinity of coal mines. It is obviously more practical and economical to transport the energy to the user in electric form over electric transmission lines than in the form of coal shipments.

Environmental factors are becoming increasingly important in the choice of sites for power stations. So is the availability of water for cooling the condensers in thermal and nuclear power stations. The sites of nuclear power stations and nuclear power itself continue to be controversial.

Electric generators are becoming increasingly larger and power ratings in excess of 1000 MW are not uncommon. The electric power network must be designed so as to meet the needs of every customer and at the same time permit the transmission of power in the 1000-MW range. Immediately, one notices the analogy between an electric power network and a well-designed transportation system. The interstate superhighway system handles huge trucks, buses, and long-distance automobile traffic. State highways take care of medium-distance traffic. The small and lightly traveled urban and rural roads and lanes serve local traffic needs.

In a power network, huge quantities of electric power move on the *grid* or *transmission links*. From the grid, the power is subdivided into smaller units and fed into the *subtransmission* network systems. Finally, the individual customers are serviced from the *distribution* networks.

6.1 The Structure of the Power Network

From an operational point of view, the power network can be divided into several substructures based on *operating voltage levels*. Highest on the voltage scale is the *transmission* or the *power grid*. The continent-spanning grid is the "interstate power highway system" over which the electric power is "transported" in large quantities. Figure 6.1 shows, in a "single-line diagram," a part of the grid. The

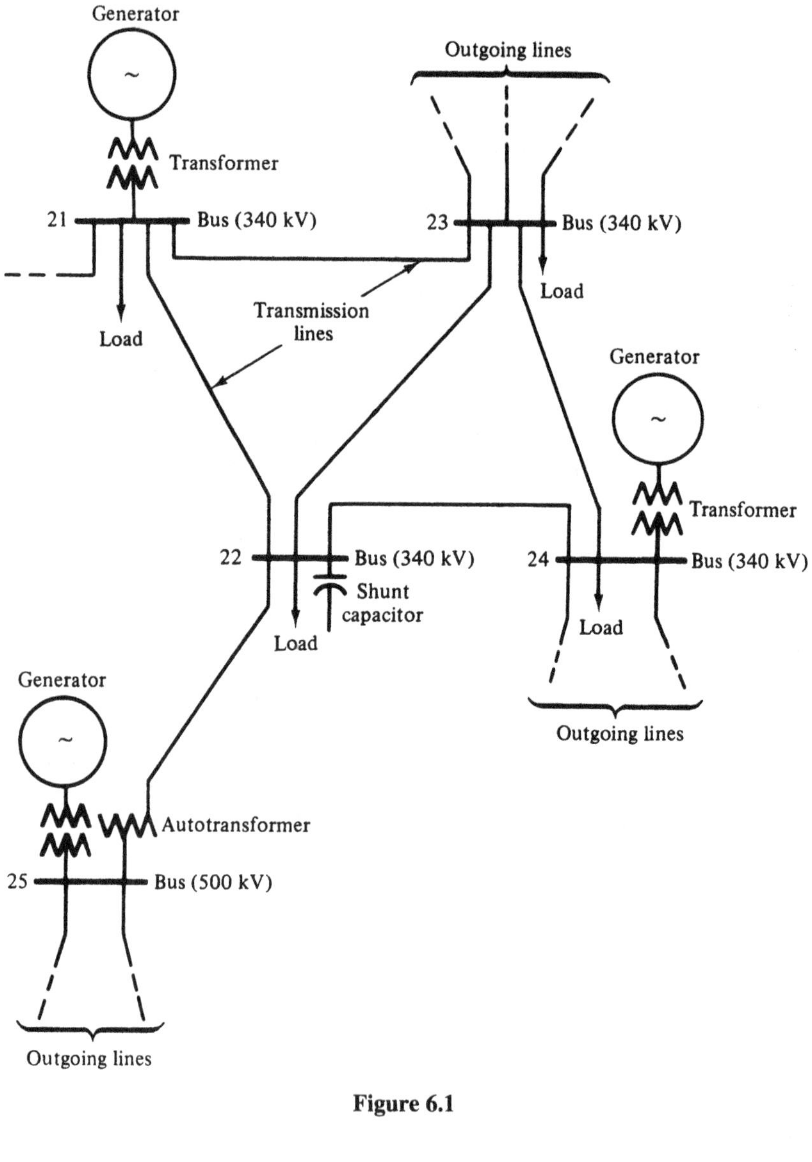

Figure 6.1

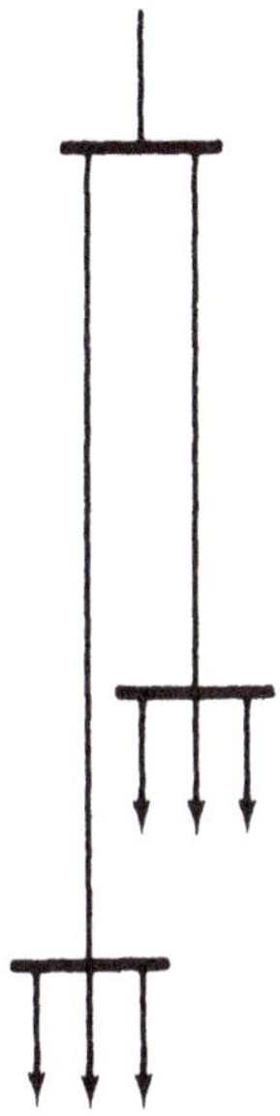

Figure 6.2

generator, transformer, and line symbols should be self-explanatory. (Note that an autotransformer is used to intertie the 340-kV and 500-kV parts of the grid.)

A typical feature of the grid is its *loop structure.* Such a structure, as distinguished from a *radial*-type network (see Figure 6.2), offers many alternate paths. The transmission lines form the coarse meshes in the grid. The lines are interconnected at *switching stations* or at the *generating stations.* In power engineers' lingo, these network nodes are referred to as *buses*. The term *bus* emanates from the bus bars used for the actual intertie between the various components.

Due to the physical size of the equipment, these "nodes" may cover acres of ground. Figure 6.3 shows the physical appearance of a three-phase interconnection between a transformer and two outgoing lines. This is highly simplified—for example, no switches or circuit breakers are shown.[1]

The arrows labeled "load" in Figure 6.1 refer to power taps. They indicate service to a single huge industrial user or to a whole city. Figure 6.1 also includes the symbol for a shunt capacitor. Such a device is often found on a bus to which no generator is connected. It is used to "support" the voltage of the bus in question.

[1] A power *switch* is a network isolation device that can be operated only under no-load condition. It is used to block power flows during repair or may also be used to segregate parts of the system. A *circuit breaker* is a device that can interrupt a circuit under load. Its operation is particularly crucial under fault conditions when large currents may have to be interrupted. The actual circuit interruption is performed by high-pressure gas or oil.

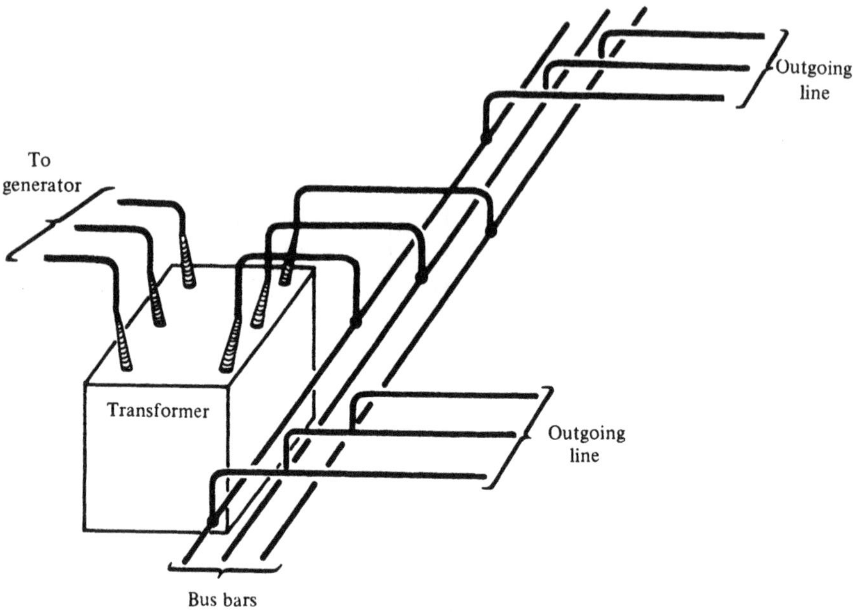

Figure 6.3

The grid is not operated by a single agency. For political, historical, and economic reasons, the grid is divided into individual *power systems*. Most of these are privately owned; some are municipal operations; part of the grid is controlled by government agencies. The sizes of these individual systems vary greatly. Many of them operate as "power pools" for mutual economic and technical benefit.

The high-voltage (HV) power grid does not operate at a single voltage level. In the United States, the highest voltage levels are about 750 kV. The grid does not have a uniform strength; in some regions, its links are quite weak. By "strength," we mean the ability to transmit power (see Section 6.5.4). Some large industrial customers may draw their power directly from the grid. More commonly, the power is transformed to a lower voltage in *power substations* and fed into the *subtransmission system*, the voltage levels of which may vary typically from 50 kV to 150 kV. The subtransmission system forms an intermediate and more fine-meshed link between the grid and the distribution circuits. It may be partly loop-structured, partly *feeder-structured*.

Finally, the power is fed into the fine meshes of the *distribution system* via *distribution substations*. Figure 6.4 shows, in schematic form, the voltage level division of the power system. It should be pointed out that no sharp demarcation lines exist between the various levels. It should also be understood that a power system, like a road system, is continuously undergoing changes. What today may constitute a transmission link may be part of tomorrow's subtransmission system.

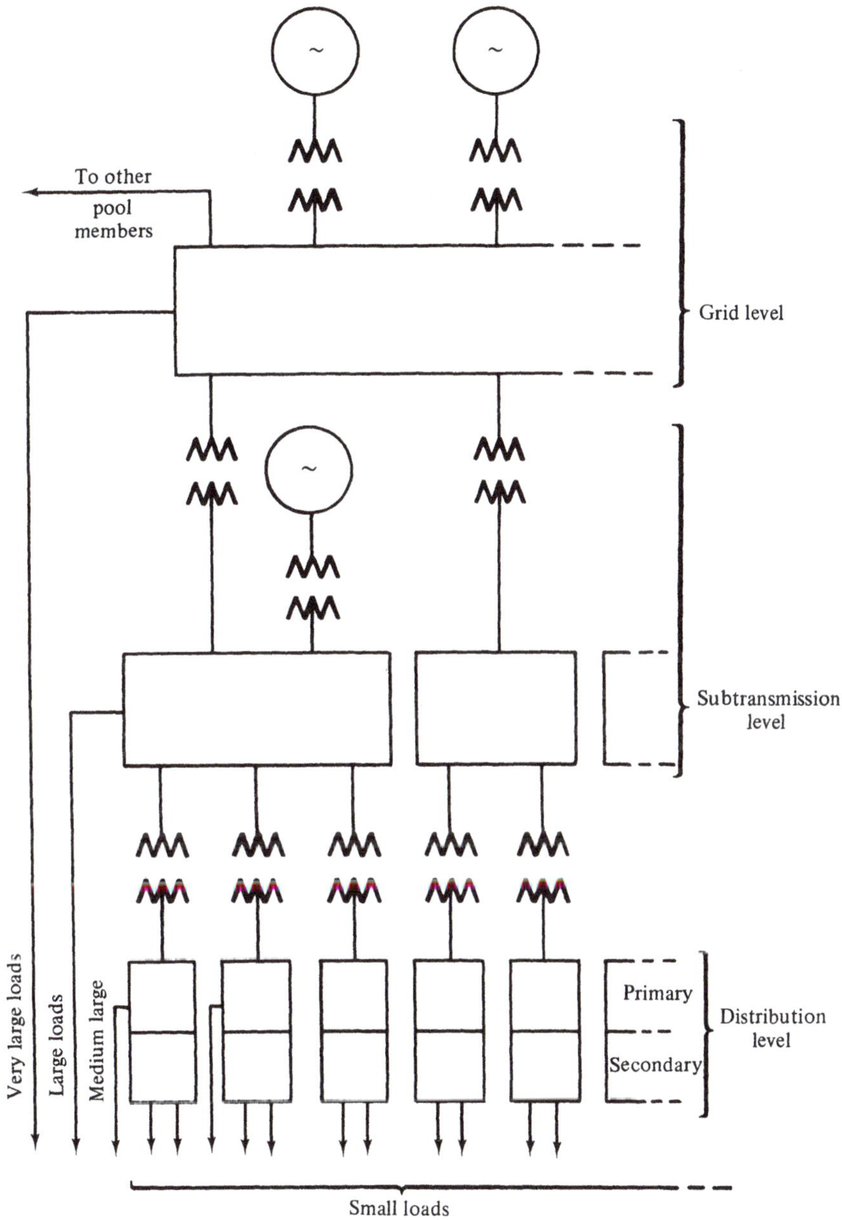
To other
pool
members
Grid level
Subtransmission
level
Primary
Secondary
Distribution
level
Very large loads
Large loads
Medium large
Small loads

Figure 6.4

6.2 Objectives of Power System Operation

We have pointed out that a grid may span an entire continent. Theoretically, turning on a light switch in Florida will affect the current flow in the California portion of the grid. In practice, however, the effects of network changes (such as a step change in loads or generations) will be most strongly felt locally. The remote effects diminish rapidly with increasing distance.[2] This makes it possible to operate the individual power systems, although interconnected, on an individual basis. However, certain functions—for example, frequency control and pooling operations—must be done by mutual agreements between all power stations.

We now identify the most important objectives that must be met in the normal operation of a grid and hence of the individual power systems which supply power to the grid:

1. Maintenance of real power balance
2. Control of frequency
3. Maintenance of reactive power balance
4. Control of the voltage profile
5. Maintenance of an "optimum" generation schedule
6. Maintenance of an "optimum" power routing

We stress that these are objectives to be met in a normal system operation. Under abnormal or *fault* conditions, the effects of system disturbances must be minimized—that is, we wish to operate with maximum security. It is not possible and indeed undesirable to cover all aspects of system operation in this book. Our aim is to give the reader an understanding of the most basic operational problems encountered under normal system operation conditions.

6.3 Real Power Balance: The Load-Frequency Control Problem

The six main objectives that were stated above are not necessarily mutually exclusive. For example, the problem of keeping the frequency constant at 60 Hz is closely intertwined with the problem of real power balance. The term *load-frequency control* (LFC) describes this joint task. No doubt the LFC problem is the most basic one that confronts the power systems engineer. We shall presently turn our attention to its solution. Before that, however, we need to explain the term *load.*

[2] Physically this can be attributed to the "diffusion effect" that is always present in a huge power system. To demonstrate this effect consider what will happen in the case of a sudden load increase on bus 22 in Figure 6.1. There are four incoming lines at this bus and the increased load current will thus be divided between these individual lines. The current in the line that terminates in bus 23 will again "subdivide" into five new lines, and so on.

6.3.1 Load Characteristics

On several occasions in this and the preceding chapters, we have referred to system "load." Let us now define more carefully what we mean by this term and, more important, study the factors that influence its characteristics.

As indicated in Figure 6.4, power can be drained from a power network at all voltage levels. Large industrial loads tend to consist of big individual units—for example, a 300-hp mine elevator motor or a 10,000-hp drive motor for a steel mill. Such huge load objects are invariably of three-phase design—that is, they represent a *balanced* three-phase loading on the system.

Domestic-type loads are quite different. They are small but numerous and generally tend to be single-phase. Figure 6.5 shows the typical connection of a small

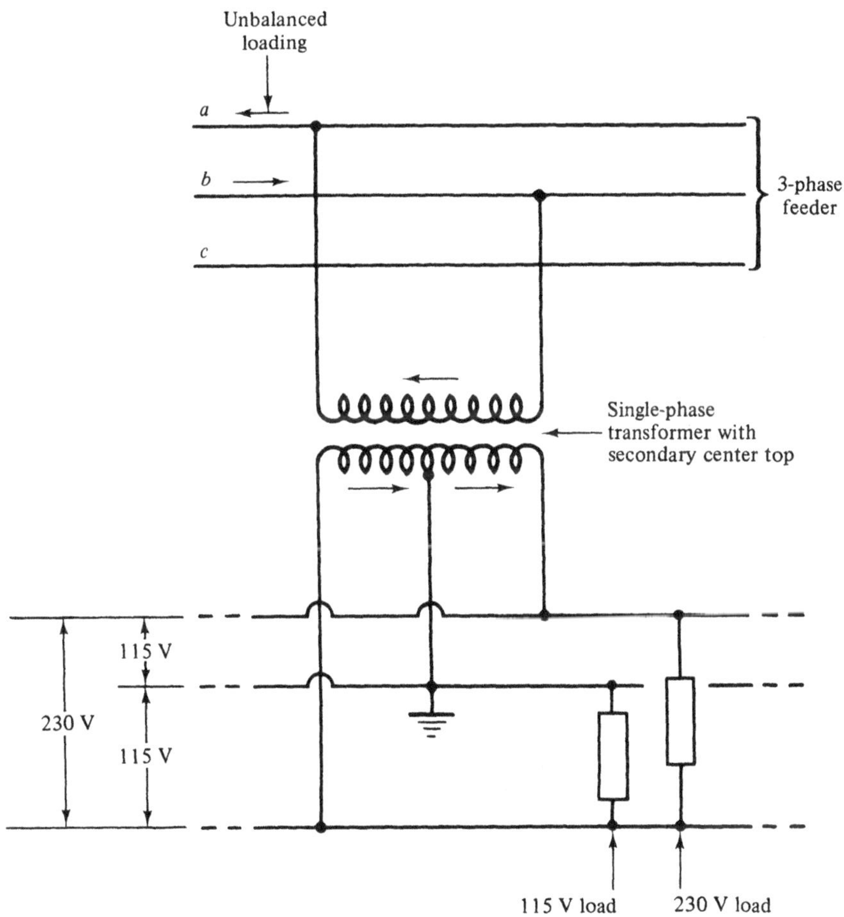

Figure 6.5

single-phase distribution transformer. It has a 230-V single-phase secondary with a midpoint tap that is normally grounded. This type of secondary connection permits the loading of, for example, a 115-V television set and a 230-V heater—both single-phase.

The primary and secondary transformer currents are indicated in the figure, and it is quite clear that this represents an *unbalanced* load on the three-phase network. However, by distributing the individual loads between the three phases, the power authority can achieve a balanced effect. For example, of the 60 houses in a subdivision, 20 may be connected between phases a and b, 20 between phases b and c, and the remaining 20 between phases c and a. (See Exercises 6.1 and 6.2.) This balancing effect is more pronounced the further up the voltage ladder we go. Thus, the composite load as represented by a whole city is, for all practical purposes, totally balanced between the three phases.

All individual loads, whether they are a single bread toaster or a 10,000-hp motor run by a night-shift mill operator, have a random time character. However, because of the large number of individual loads, we can always rely on the averaging effect of the laws of statistics. As a result, although the individual loads are entirely unpredictable, the total load is a highly predictable function of time.

Figure 1.9 shows load variation as a function of time for a typical city. The power demand shifts from hour to hour in a smooth and predictable manner. The power system operator can usually predict, with good accuracy, from one day to the next what the load will be at a given hour, and hence schedule appropriate generation. What we have said about *real power* loads can also be applied to *reactive* loads. Practically all electric loads consume[3] reactive power, usually in proportion to the real power.

6.3.2 Load-Frequency Dynamics

Load demand curves, of the type shown in Figure 1.9, are characterized, by a fairly smooth and predictable pattern if we view them on an hour-by-hour basis. If we look at them on a second-by-second or minute-by-minute basis, they are quite different. Figure 6.6 shows a "microscopic" look at the load demand over a 15-minute interval. The system frequency is also recorded over the same time interval.

We note that the load demand has an average of about 1061 MW, and it is characterized by a random fluctuation of about ±2 MW. This "noise" component is entirely unpredictable. Similarly, the frequency has an average of about 60.0 Hz, but superimposed on this, is a "noise" component. There is, however, a definite correlation between the load and frequency curves. For example, a 3-MW load peak occurs around 9:40. We note the resulting dip in the frequency graph.

[3] Motors are some of the most important parts of any industrial load. Motors always have iron cores which require magnetization current from the power source. (See the equivalent circuit of the induction motor in Chapter 8.)

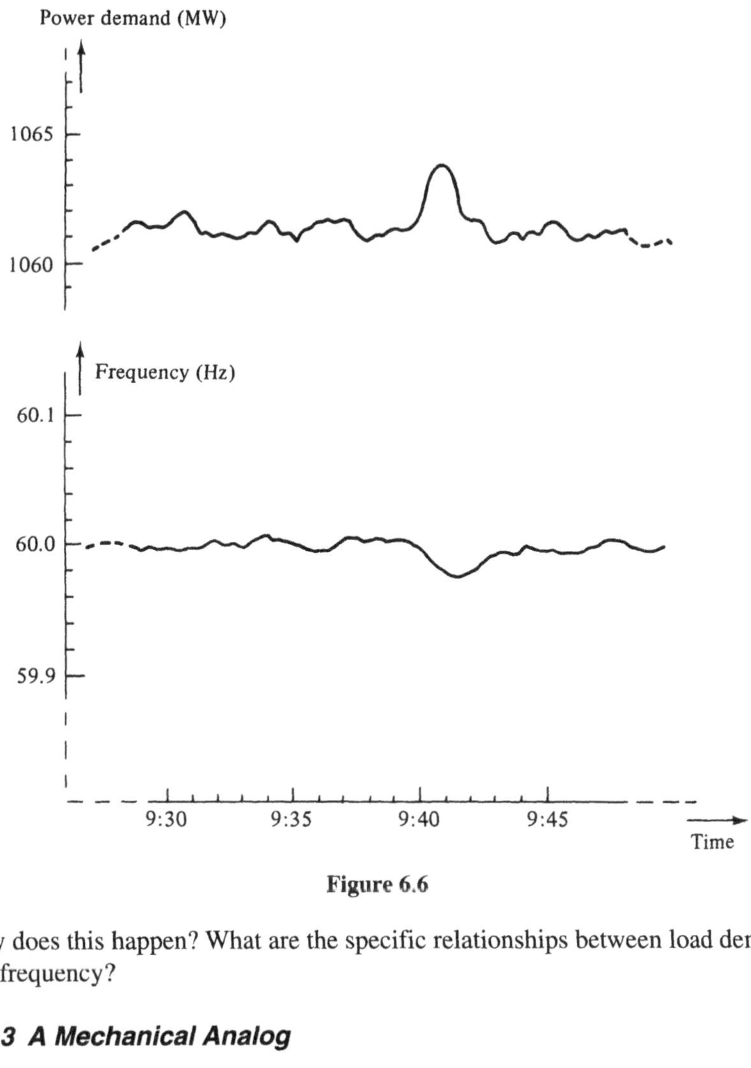

Figure 6.6

Why does this happen? What are the specific relationships between load demand and frequency?

6.3.3 A Mechanical Analog

The following mechanical analog illustrates the basic features of load-frequency dynamics in an interconnected electric power system.

A freight train consisting of many locomotives and many freight cars is supposed to travel at a constant speed of 100.0 kph. The locomotives represent the individual generators. The freight cars are the analogs of the electric loads. The elastic couplings between the engines and the freight cars represent the electric transmission lines—the "couplings" between the electric generators and loads.

Consider what will happen if the train encounters a sudden change in load in the form of an uphill grade. If the throttle settings of the engines remain unchanged, the speed will, no doubt, drop. Similarly, if the load in a power system suddenly increases, but the generator power output remain fixed, the frequency will drop.

In both the train analog and the power system, the increase in load without a corresponding upward adjustment in generation would result in a net power deficiency. The power must be taken from somewhere. The only available source of energy is the kinetic energy of the moving masses. As the kinetic energy is being consumed, the speed must drop.

Example 6.1

The total kinetic energy of the spinning rotors and turbines in a power system is equal to 600 MJ (measured when the frequency is at 60 Hz). The system is running at constant frequency in perfect power balance, when a sudden power-load increase of 2 MW sets in. What deceleration, measured in hertz per second, will the system experience if the turbine power output remains unchanged?

Solution: As the power deficiency must be equal to the rate of change of kinetic energy we have the equation:

$$-2 \cdot 10^6 = \frac{d}{dt}(w_{\text{kin}}) \qquad [\text{W}]. \tag{6.1.1}$$

Because the kinetic energy is proportional to the square of velocity (or frequency) we can write:

$$w_{\text{kin}} = 600 \cdot 10^6 \left(\frac{f}{60}\right)^2 \qquad [\text{J}]. \tag{6.1.2}$$

Substituting and performing the differentiation, we obtain for the frequency change:

$$\frac{df}{dt} = -0.1 \qquad [\text{Hz/s}]. \tag{6.1.3}$$

6.3.4 Automatic Load-Frequency Control

In Example 6.1, if the power imbalance were sustained, the frequency would clearly drop to 59 Hz in 10 seconds. A frequency change of such a magnitude would be entirely unacceptable in a modern power system, the frequency of which is normally kept within a tolerance of 60.0 ± 0.1 Hz. There are many reasons why the frequency must be controlled to within these narrow limits of accuracy. Generally, it can be said that the tighter control we have over the frequency, the better control we have over the entire system.[4]

[4] There are millions of electric clocks running off the power system. The accuracy of these clocks depends on the frequency of the system. A sudden shift in frequency is one of the surest signs of a system in trouble. If the system control mechanism can keep the frequency deviation very small, we can

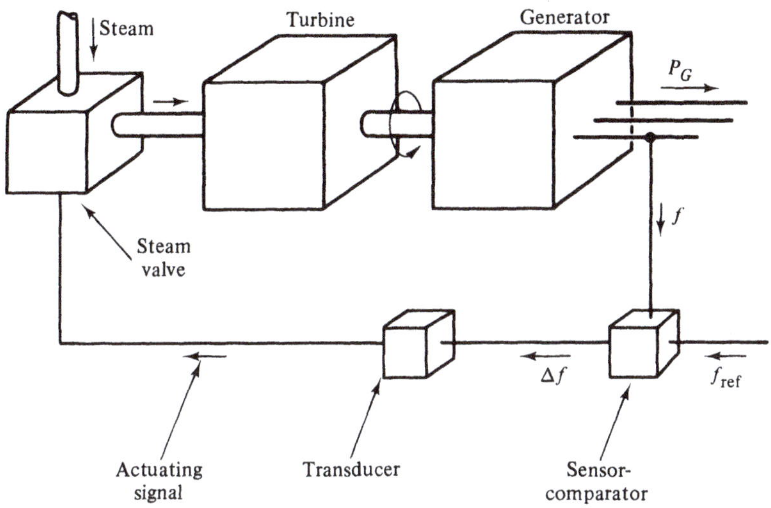

Figure 6.7

In most modern power systems automatic regulation is used to control the frequency. Figure 6.7 shows schematically the operation of an automatic load-frequency control (ALFC) system. A frequency "sensor-comparator" senses the system frequency f and compares it to a reference frequency f_{ref} (60.00 Hz). A *frequency error* signal,

$$\Delta f = f - f_{\text{ref}} \qquad [\text{Hz}], \tag{6.1}$$

is generated, which is a measure of the frequency deviation. An amplifier converts the error signal into an actuating command, which is sent on to the steam turbine control valve to change its setting.

A positive error signal would indicate too high a frequency and the actuating signal would in that case issue a "lower" command to the steam valve and hence a lower generator output P_G. A negative Δf would result in a "raise" command—that is, an opening of the steam valve.

Of necessity, this description is very superficial.[5] Many interesting questions arise in connection with the actual operation of an ALFC system, such as:

1. How "responsive" should the control loop be? Clearly, it is not wise to let the generators "chase" every load excursion, however short it may be. This will cause unnecessary wear and tear on the equipment.

detect trouble more easily in the system when one starts—one can detect the slightest perturbation (caused by a falling stone) on a perfectly smooth body of water.

[5] For detailed analysis of ALFC, see Elgerd, 1971.

2. Which generators in a given station should participate in the ALFC? In our previous mechanical analog, when the train encounters the uphill grade, it may not be necessary to raise the power in all the engines equally or even proportionately.

Likewise, in a power system, we delegate the ALFC to those generators most suitable for the job. We had earlier noted that it is much easier to control the power output in a hydroturbine than in a steam-driven generator. Consequently, if we have a mix of prime movers in the system, hydroturbines are the natural candidates for the ALFC role.

6.4 Optimum Generation

The ALFC system maintains real power balance within the system on both a second-by-second and a minute-by-minute basis. This being accomplished, the power system operator must make sure that the generators divide—over a longer time spans—the total system load in a manner that guarantees minimum operating costs.

In a power system, a given load demand can be met in a number of ways. As support for this statement, consider the simple two-bus system shown in Figure 6.8. A real-life power system is never that simple; however, it will serve the purpose of demonstrating the principles of *optimum power dispatch.*

Assume that the system is operating in the power configuration shown in Figure 6.8a. The total output of the system is 500 MW, with 60% (300 MW) of it tapped from bus 2 and 40% (200 MW) from bus 1. Because the fuel is cheaper at the location of bus 1, the generator $G1$ supplies more power to the system than $G2$. Note that the line losses amount to 2 MW. Note also that power balance exists at each bus—that is, the power entering the bus is equal to the power leaving.

Now let us assume that the load on bus 2 increases by 50 MW. Where should this additional power be generated? The first possibility might be to let $G1$ handle the entire increment because of its cheaper fuel. The *load flow* would then be as shown in Figure 6.8(b). The added line power causes the line losses to increase to 5 MW—a 3-MW increase (see equation (6.17) for P_Ω). Instead, if we let generator $G2$ assume 20 MW of the added load, then the line power would be less and so will the losses, say 3 MW—an increase of only 1 MW. The load flow is now as shown in Figure 6.8c.

Which of the two generation strategies is better? We cannot know until we add up the generation costs in the two cases. In spite of the higher fuel costs at bus 2, the second alternative may prove cheaper overall because of the lower line losses. It would seem intuitively obvious that the overall costs will be a minimum at some appropriately chosen load division between $G1$ and $G2$. In fact, a careful analysis confirms that in general, there is one and only one power configuration that is cheaper than all the others. If we adjust our generators accordingly, then our system is said to be on *optimum power dispatch.*

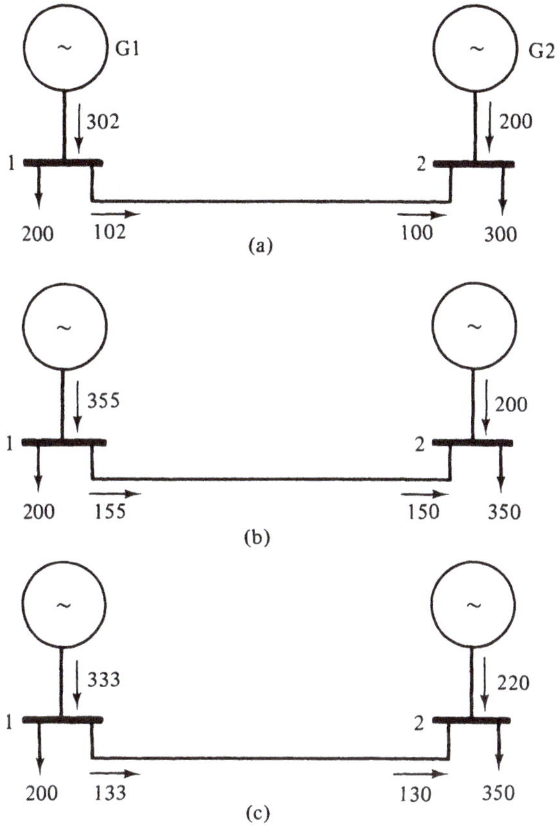

Figure 6.8

It is not particularly difficult to find the optimum situation in the simple two-bus system used in the example. It is considerably more difficult in a system containing hundreds of lines and dozens of generating stations. The job would be impossible without modern computers. Most modern power systems are *computer dispatched*.

It is beyond the scope of this book to enter into a discussion of the mathematics of optimum dispatch. The reader will find a more detailed discussion in Elgerd, 1971.

6.5 Line Power and Its Control

The transmission lines permit us to dispatch surplus power from one grid bus to another, as was demonstrated in the previous example. They constitute the important network links that make it possible to choose alternate power flow configurations for optimum economy and security. In this section, we study the factors that affect the line power flow and, particularly, the method for controlling the flow. In order to do this, we have to develop a model of the transmission line.

6.5.1 Line Parameters

A three-phase transmission line is usually of the overhead variety (Figure 5.1). (In densely populated urban areas, underground cables are often used when overhead lines present an unacceptable hazard.) Typically, the bare conductors consist of a steel core (for mechanical strength) and an outer shell made of aluminum, which is a better electrical conductor than steel. To obtain a more flexible conductor, both the steel and aluminum portions are stranded. The conductors are hung on insulator strings from crossbars of the wooden, steel, or concrete transmission towers. Each of the three conductors in a three-phase line (Figure 6.9a) is characterized by resistance. The current in each conductor generates a magnetic field, resulting in a self-inductance. Finally, as shown in Figure 6.9a, there exists capacitance between each conductor. As shown in Figure 6.9b, these are equivalent to a set of capacitances between each phase and a neutral node. (This equivalence can be verified by the Δ–Y transformation of impedances.)

Electrically, the transmission line can be characterized by an equivalent circuit with both *series* and *shunt* elements. The line resistance belongs to the former group and so does the *self-inductance* caused by the magnetic flux surrounding each conductor. The capacitance that exists between the conductors represents a *shunt* or parallel admittance. (There is also a shunt *conductance* that represents a path for the leakage current along the insulator strings shown in Figure 5.1. For normal weather conditions this leakage current can be neglected.)

All of the above circuit parameters are *distributed* and can be expressed in ohms-, henry-, and farad-per-meter of line. If the line is designed to be *phase symmetrical* (similar to the one shown in Figure 6.9a), all three capacitances of the Y-equivalent in Figure 6.9b are equal. Consequently, the neutral node will have zero (or ground) potential. It is then possible to represent the three-phase line on a per-phase basis, as shown in Figure 6.10a. The distributed parameters of the line can be shown to be equivalent to that of the *lumped* circuit in Figure 6.10b. If the line is "electrically short" (less than about 160 km at a system frequency of 60 Hz), the values of the lumped equivalent circuit elements can be obtained from the ohm-, henries-, farads-per-meter values by multiplication. In an "electrically long" line, the rules for calculating the elements of the lumped equivalent circuit are more complicated.[6] For more details, see Elgerd, 1971.

[6] The discussion given here is actually somewhat simplified. In addition to the self-inductance per phase there is also a mutual inductance between phases. In addition to capacitance between phases, there exists also capacitance between each phase and ground. Finally, (as demonstrated by the line in Figure 5.1), a practical conductor arrangement is not always phase symmetrical. It can be shown, however, that when all those factors are considered, a practical line can still be represented by the per-phase equivalent circuit of Figure 6.10b.

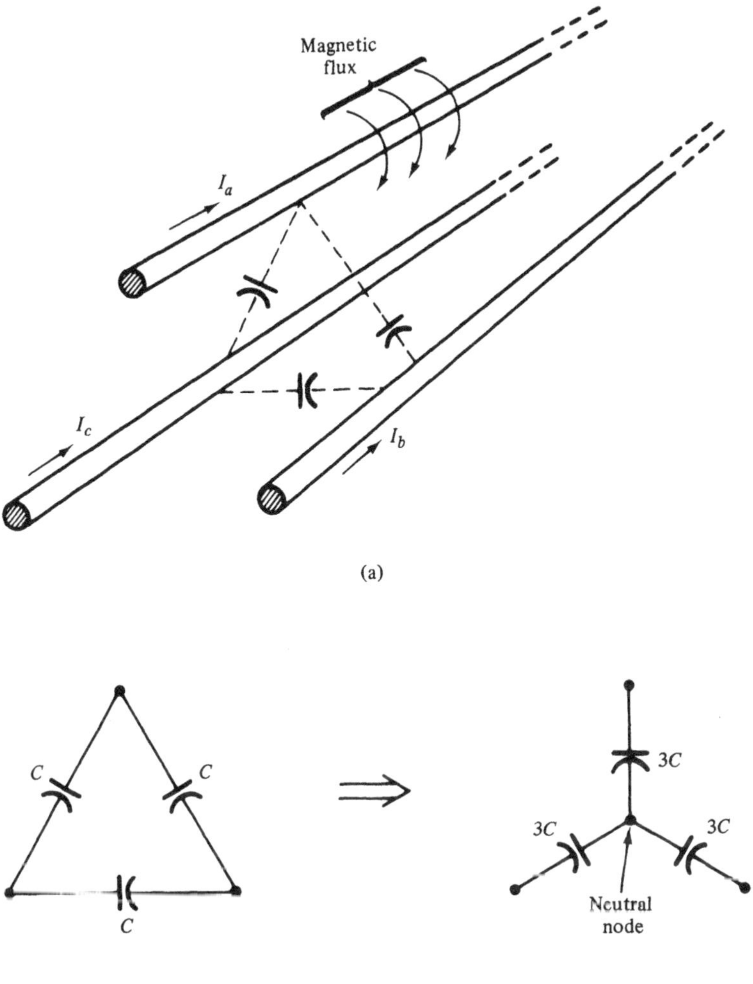

Figure 6.9

Normally, the magnitude of the series reactance is much larger than that of the resistance. In general, all three network parameters (R, L, and C) must be taken into account when describing a transmission line. However, for short lines, one may disregard both the resistance and capacitance, and work with the simplified line model shown in Figure 6.11.

Example 6.2

A three-phase 140-kV, 100-km transmission line consists of three conductors arranged as shown in Figure 5.1. From measurement, the line parameters are

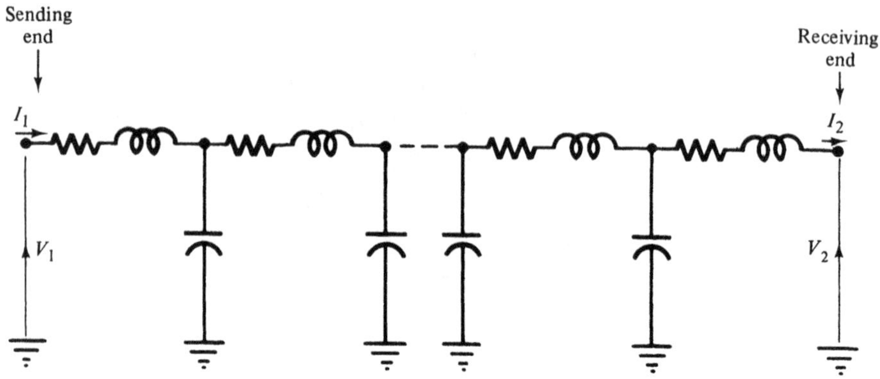

(a)

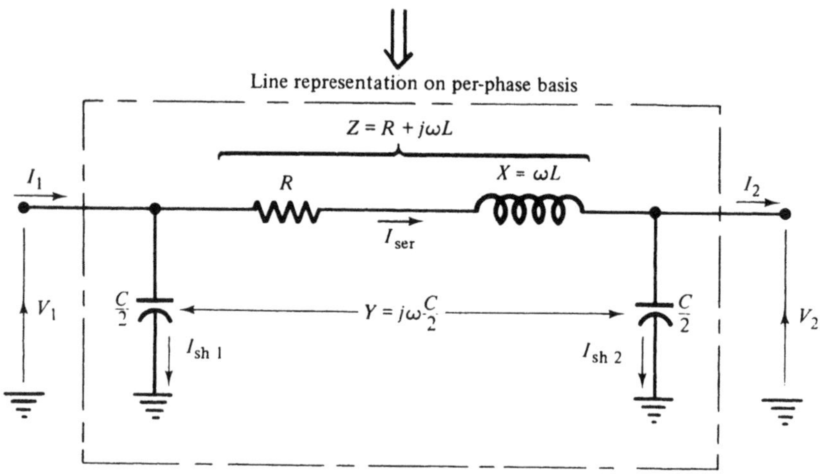

(b)

Figure 6.10

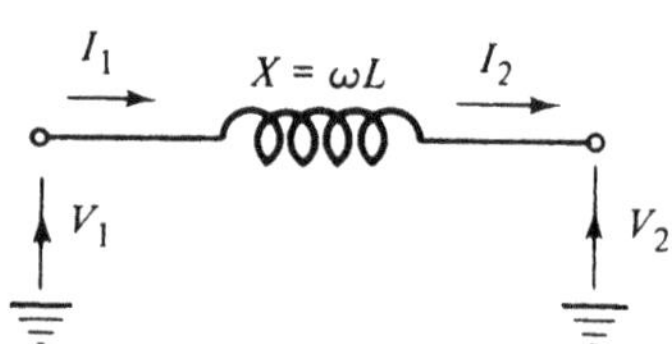

Figure 6.11

$$\text{resistance} = 0.0910\ \Omega/\text{km and phase}$$

$$\text{inductance} = 1.34\ \text{mH/km and phase}$$

$$\text{capacitance} = 8.85 \cdot 10^{-9}\ \text{F/km and phase}$$

Compute the sending-end power if the receiving-end voltage, $|V_2| = 140$ kV and the receiving-end power, $S_2 = 100 + j60$ (three-phase megavalues). Assume the line to be "electrically short."

Solution: We first compute the circuit parameters of the equivalent lumped circuit (assuming 60-Hz frequency):

$$R = 100 \cdot 0910 = 9.10 \qquad [\Omega \text{ per phase}]; \tag{6.2.1}$$

$$X = 100 \cdot 377 \cdot 1.34 \cdot 10^{-3} = 50.5 \qquad [\Omega \text{ per phase}]; \tag{6.2.2}$$

$$Y = j50 \cdot 377 \cdot 8.85 \cdot 10^{-9} = j1.67 \cdot 10^{-4} \qquad [\text{S per phase}]. \tag{6.2.3}$$

Using these impedance and admittance values in the equivalent circuit in Figure 6.10b, we compute the sending-end power as follows:

Step 1. Find the receiving-end current.

From equation (4.21), we get

$$S_2 = V_2 I_2^* = \frac{100}{3} + j\frac{60}{3} = 33.33 + j20.00 \qquad [\text{MVA}]. \tag{6.2.4}$$

If V_2 is chosen as the reference phasor, we have

$$I_2 = \frac{33.33 - j20.00}{140/\sqrt{3}} = 0.4124 - j0.2474 \qquad [\text{kA per phase}]. \tag{6.2.5}$$

Step 2. Find the sending-end voltage.

The current[7] I_{sh2} (see Figure 6.10b) is

$$I_{sh2} = V_2 Y = \frac{140}{\sqrt{3}}(j1.67 \cdot 10^{-4}) = j0.0135 \qquad [\text{kA per phase}] \tag{6.2.6}$$

Next, we compute the current[7] I_{ser} (see Figure 6.10b):

$$I_{ser} = I_2 + I_{sh2} = 0.4124 - j0.2339 \qquad [\text{kA per phase}]. \tag{6.2.7}$$

The voltage drop across the series branch is

$$\Delta V = I_{ser}(R + jX) = 15.56 + j18.70 \qquad [\text{kV per phase}]. \tag{6.2.8}$$

[7] The currents I_{sh1}, I_{sh2}, and I_{ser}, *cannot* be measured physically. Why not?

The sending-end voltage is then obtained as

$$V_1 = V_2 + \Delta V = 96.39 + j18.70 \qquad \text{[kV per phase]}. \tag{6.2.9}$$

Step 3. Find the sending-end current.

The shunt current[7] I_{sh1} (see Figure 6.10b) is

$$I_{sh1} = V_1 Y = -0.0031 + j0.0161 \qquad \text{[kA per phase]}. \tag{6.2.10}$$

The sending-end current is

$$I_1 = I_{sh1} + I_{ser} = 0.4093 - j0.2178 \qquad \text{[kA per phase]}. \tag{6.2.11}$$

Step 4. Finally, find the sending-end power,

$$\begin{aligned} S_1 &= V_1 I_1^* = (96.39 + j18.70)(0.4093 + j0.2178) \\ &= 35.38 + j28.64 \qquad \text{[MW and MVAr per phase]} \end{aligned} \tag{6.2.12}$$

or

$$S_1 = 106.14 + j85.93 \qquad \text{[MW and MVAr]} \qquad \text{(3-phase)}. \tag{6.2.13}$$

By comparison with the receiving-end powers, we conclude that there is a loss of 6.14 MW and 25.93 MVAr in the transmission line. We note also that the receiving-end current is 481 [A]. However, the sending-end current is 464 [A]. Why is the magnitude of the receiving-end current greater than that of the sending-end?

6.5.2 Control of the Line Voltage Profile

In operating a system such as the one shown in Figure 6.8, it is important to keep the magnitudes of the voltages $|V_1|$ and $|V_2|$ at each end constant. The reason is that all electric loads are voltage-rated—that is, they are designed to operate at a specified voltage. For example, the light intensity from an incandescent light bulb is very sensitive to changes in voltage and, in general, fluctuations in the light flux can cause an annoyance to the customers.

This type of voltage control is achieved by *automatic excitation control* (AEC) of the individual generators. We remember from Chapter 4 that the generator emf is proportional to the rotor field current, which suggests that an AEC system could be designed as shown in Figure 6.12.

The generator bus voltage $|V|$ is sensed and compared in a voltage sensor-comparator to a reference voltage $|V|_{ref}$. The error voltage,

$$\Delta|V| \equiv |V| - |V|_{ref} \qquad \text{[V]}, \tag{6.2}$$

is amplified and sent on as an actuating signal to the *field current source*. The latter will adjust the field current of the generator to minimize the magnitude of the error voltage.

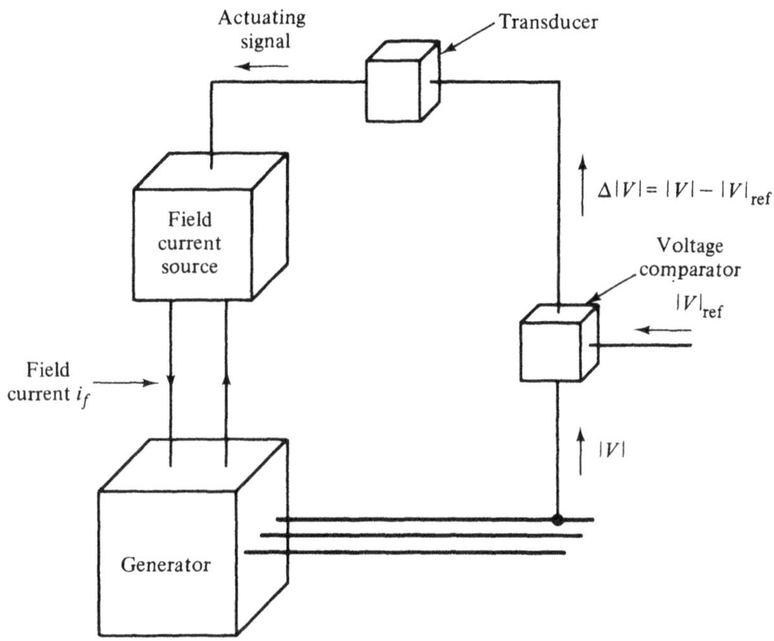

Figure 6.12

The automatic excitation control (AEC) loop shown in Figure 6.12, together with the automatic load-frequency control (ALFC) loop shown in Figure 6.7, constitute the two basic control systems for synchronous generators. They are essentially noninteracting; the action of one does not significantly affect the other. Because the AEC loop involves only electrical variables, it has a much faster response time than the ALFC loop, which includes electrical as well as mechanical variables (steam valves, mechanical inertia, etc.)

6.5.3 Control of Real Line Power

We are now ready to discuss how the real power in a transmission line can be controlled in both magnitude and direction. To avoid unnecessary complications, we assume a *lossless* transmission line—we make the very reasonable and practical assumption that the line resistance $R = 0$. If we choose the simplified line representation in Figure 6.11, the sending-end and receiving-end currents are equal. The complex powers at each end can then be computed from

$$\begin{aligned} S_1 &= P_1 + jQ_1 = V_1 I^* \qquad [\text{VA}] \\ S_2 &= P_2 + jQ_2 = V_2 I^* \qquad [\text{VA}]. \end{aligned} \tag{6.3}$$

Let the transmission line impedance, $Z = jX$.

From Figure 6.11, the line current I is

$$I = \frac{V_1 - V_2}{Z} \qquad [\text{A}]. \tag{6.4}$$

Substituting into equation (6.3) we get

$$\begin{aligned} P_1 + jQ_1 &= V_1 \frac{V_1^* - V_2^*}{Z^*} = \frac{|V_1|^2 - |V_1||V_2|e^{j\delta}}{-jX} \\ &= \frac{|V_1||V_2|}{X}\sin\delta + j\frac{|V_1|^2 - |V_1||V_2|\cos\delta}{X} \qquad [\text{VA}]. \end{aligned} \tag{6.5}$$

where

$$\delta \equiv \angle V_1 - \angle V_2 \tag{6.6}$$

By separating the real and imaginary parts we get

$$\begin{aligned} P_1 &= \frac{|V_1||V_2|}{X}\sin\delta \qquad [\text{W}]; \\ Q_1 &= \frac{|V_1|^2 - |V_1||V_2|\cos\delta}{X} \qquad [\text{VAr}]. \end{aligned} \tag{6.7}$$

Similarly,

$$\begin{aligned} P_2 &= \frac{|V_1||V_2|}{X}\sin\delta \qquad [\text{W}]; \\ Q_2 &= \frac{|V_2||V_1|\cos\delta - |V_2|^2}{X} \qquad [\text{VAr}]. \end{aligned} \tag{6.8}$$

These equations are very important in the determination of the limits of the transmission line to carry power. We remind the reader that the *power angle*, δ, was defined as the phase angle between V_1 and V_2 [equation (6.6)]. Note that the voltages and reactance must be given in per-phase values to yield per-phase values of powers. Note also that because we neglected the line resistance, the real line powers at each end are equal. (For this reason, we discard the subscripts and refer henceforth to the real line power P.) Here, X is a *fixed* line parameter. We assume automatic excitation control of the generators at both ends of the line—that is, the magnitudes of the voltages $|V_1|$ and $|V_2|$ are kept constant. Thus, P is a function of only δ.

Figure 6.13 shows how P varies as a function of δ. As δ increases in a positive sense (V_1 leading V_2), the power increases to a maximum value:

$$P_{\max} = \frac{|V_1||V_2|}{X} \qquad [\text{W per phase}], \tag{6.9}$$

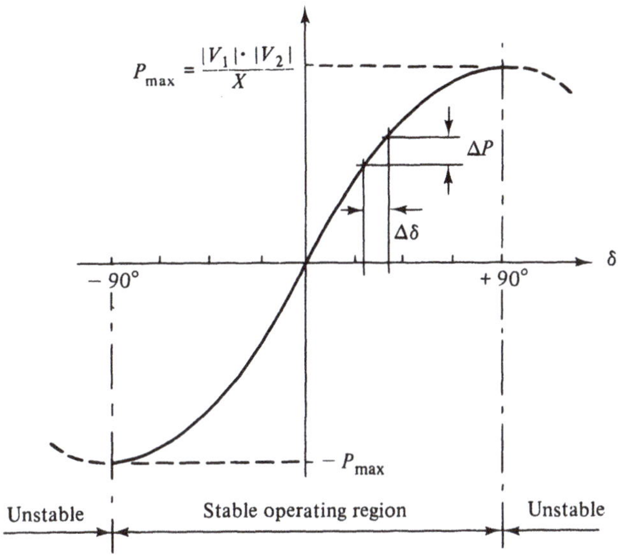

Figure 6.13

which occurs when $\delta = 90°$. If we attempt to increase P by further increasing δ, the power will in fact decrease. At this point the transmission collapses—the two parts of the system steps out of synchronism. We have reached the *transmission limit* or *static stability limit* for the line. "Stepping out of synchronism" means that $G1$ and the bus 1 load will run at one frequency and $G2$ and the bus 2 load at another frequency. This phenomenon is similar to synchronous machine pullout discussed in Section 4.7.5.

If δ increases in a negative sense (V_2 leading V_1), the power becomes negative. We now transmit power from right to left in Figure 6.11. Note the difference between dc and ac transmission. In a dc system, the magnitude of the sending-end voltage must exceed that of the receiving-end voltage in order for power to flow in the proper direction. In an ac system, the magnitudes of the voltages do not determine the direction of power flow; the end which has a *leading phase angle* transmits power to the end with the *lagging phase angle*.

From equation (6.9) it can be seen that the maximum power that can be transmitted is a function of the square of the line voltage. Earlier, we had found, from equation (3.32), that the power loss was a function of the inverse of the square of the voltage. These are very good reasons for using high voltage in transmission systems.

Example 6.3

How much power can be transmitted over the line shown in Figure 6.8? We assume that the line designed for 140 kV, is 80 km long, and has a reactance of 40 Ω per phase. We neglect its resistance.

Solution: Assuming that the line voltage is kept at 140 kV at both ends. From equation (6.9), the maximum three-phase power:

$$P_{max} = 3\,\frac{(140/\sqrt{3})\cdot(140/\sqrt{3})}{40} = \frac{140^2}{40} = 490 \qquad [\text{MW}]. \qquad (6.3.1)$$

(Note that the answer is in three-phase megavalues, if we use line voltages in kilovolts.)

Example 6.4

In the introduction to Chapter 5, we concluded that we would encounter difficulties in transmitting 1000 MW along a 20-km line at a transmission voltage of 20 kV. Let us find, in light of equation (6.9), the maximum power that this line can, in fact, transmit.

Solution: In Chapter 5, we computed the line reactance $X = 8.72\ \Omega$ per phase. If we assume 20 kV at both ends and neglect the resistance, we have

$$P_{max} = \frac{20^2}{8.72} = 45.9 \qquad [\text{MW}] \qquad (\text{3-phase}). \qquad (6.4.1)$$

(No wonder we had difficulties squeezing 1000 MW through this line!)

6.5.4 Synchronization Coefficient

From Figure 6.13, it can be seen that the power that can be transmitted over a line approaches its limit as δ approaches $\pm 90°$. This means that if there were two synchronous machines, one at each end of the transmission line, they could not keep in synchronism beyond $\delta = 90°$. If we think of the transmission line as a "coupling" between the machines, we can define a "coupling coefficient" or the "electrical stiffness" or the "synchronizing coefficient," T_{sync}, of the power transmission line as

$$T_{sync} \equiv \frac{\Delta P}{\Delta \delta} \qquad [\text{W/rad}]. \qquad (6.10)$$

It is a measure of the elemental change in power, ΔP, resulting from an elemental change in the power angle, $\Delta\delta$. By noting that the definition of T_{sync} is identical (in the limit as $\Delta\delta \to 0$) to the derivative $dP/d\delta$, we get

$$T_{sync} = \frac{dP}{d\delta} = \frac{|V_1||V_2|}{X}\cos\delta \qquad [\text{W/rad}]. \qquad (6.11)$$

Note that the stiffness approaches zero as we approach the stability limit ($\delta = 90°$). For this reason a transmission line is never operated close to its power limit.

Example 6.5

Consider again the 140-kV lossless transmission line in Example 6.3. If we operate it with a "flat"-rated voltage profile—140 kV at both ends—and the real power flow is 100 MW, what is the power angle δ and T_{sync}?

Solution: From equation (6.7),

$$P = \frac{140/\sqrt{3} \times 10^3 \ 140/\sqrt{3} \times 10^3}{40} \sin\delta = \frac{100 \times 10^6}{3} \ [\text{W}]. \qquad (6.5.1)$$

Therefore,

$$\delta = 11.78^\circ. \qquad (6.5.2)$$

The "electrical stiffness" is given by equation (6.11) as

$$T_{sync} = \frac{140/\sqrt{3}\times 10^3 \ 140/\sqrt{3}\times 10^3}{40} \cos 11.78^\circ = 159.89 \quad [\text{MW/rad}] \quad (6.5.3)$$

or

$$479.67 \quad [\text{MW/rad}] = 8.37 \quad [\text{MW/deg}] \quad (\text{3-phase}). \qquad (6.5.4)$$

This shows that if we increase the power angle from 11.78° to 12.78°, the three-phase power will increase approximately by 8.37 MW.

6.5.5 Control of Reactive Line Power

In Section 6.5.3, it was noted that real power flow in a transmission line takes place from the end with a leading phase angle to the end with a lagging phase angle. In this section we investigate the parameters that affect the flow of reactive line power.

From equations (6.7) and (6.8) we can write:

$$\begin{aligned} Q_1 &= \frac{|V_1|}{X}(|V_1| - |V_2|\cos\delta) \quad [\text{VAr}]. \\ Q_2 &= \frac{|V_2|}{X}(|V_1|\cos\delta - |V_2|) \quad [\text{VAr}]. \end{aligned} \qquad (6.12)$$

Under normal operating conditions, cos δ is fairly close to unity (refer to Example 6.5). The terms inside the parentheses in both Q_1 and Q_2 are approximately proportional to the difference $(|V_1| - |V_2|)$.

We draw the conclusion that reactive power flows from the end of the transmission line that has the higher voltage to the end with the lower voltage, and it is proportional to the difference in the magnitudes of the voltages.

Example 6.6

Calculate the reactive power flow in the transmission line shown in Figure 6.8, assuming the same real power flow (100 MW) as in Example 6.5, under the following conditions:

1. $|V_1| = |V_2|$ ("flat" voltage profile)
2. $|V_1|$ is 20% higher; $|V_2|$ has the initial value
3. $|V_2|$ is 20% higher; $|V_1|$ has the initial value

Solution:

1. The value of δ was calculated earlier ($\delta = 11.78°$). The phase voltage is

$$|V_1| = |V_2| = \frac{140}{\sqrt{3}} \times 10^3 = 80.83 \times 10^3 \quad [\text{V}]. \tag{6.6.1}$$

From equations (6.7) and (6.8) we get

$$\begin{aligned} Q_1 &= \frac{80.83 \times 10^3}{40}(80.83 \times 10^3 - 80.83 \times 10^3 \times \cos(11.78°)) \\ &= 3.44 \quad [\text{MVAr per phase}] \quad \text{or} \quad 10.32 \quad [\text{MVAr}] \quad \text{3-phase.} \end{aligned} \tag{6.6.2}$$

$$\begin{aligned} Q_2 &= \frac{80.83 \times 10^3}{40}(80.83 \times 10^3 \times \cos(11.78°) - 80.83 \times 10^3) \\ &= -3.44 \quad [\text{MVAr per phase}] \quad \text{or} \quad -10.32 \quad [\text{MVAr}] \quad \text{(3-phase).} \end{aligned} \tag{6.6.3}$$

Note that when the voltage profile is kept flat, reactive power flows into the line from both ends. Clearly, the line (i.e., its reactance) consumes 20.64 MVAr.

2. The value of δ is recalculated by using equation (6.7):

$$P = \frac{1.20 \times (80.83 \times 10^3)^2}{40} \sin\delta = \frac{100 \times 10^6}{3} \quad [\text{W}]. \tag{6.6.4}$$

Therefore,

$$\delta = 9.79° \tag{6.6.5}$$

and

$$|V_1| = 1.20 \times 80.83 \times 10^3 = 96.99 \times 10^3 \quad [\text{V}]. \tag{6.6.6}$$

Hence

$$\begin{aligned} Q_1 &= \frac{96.99 \times 10^3}{40}[96.99 \times 10^3 - 80.83 \times 10^3 \times \cos(9.79°)] \\ &= 42.05 \quad [\text{MVAr per phase}] \quad \text{or} \quad 126.16 \quad [\text{MVAr}] \quad \text{(3-phase).} \end{aligned} \tag{6.6.7}$$

$$\begin{aligned} Q_2 &= \frac{80.83 \times 10^3}{40}[96.99 \times 10^3 \times \cos(9.79°) - 80.83 \times 10^3] \\ &= 29.81 \quad [\text{MVAr per phase}] \quad \text{or} \quad 89.44 \quad [\text{MVAr}] \quad \text{(3-phase).} \end{aligned} \tag{6.6.8}$$

3. The angle δ remains the same as in part 2, and from equations (6.7) and (6.8) we get

$$Q_1 = -89.4 \qquad [\text{MVAr}] \tag{6.6.9}$$

and

$$Q_2 = -126.2 \qquad [\text{MVAr}]. \tag{6.6.10}$$

Figure 6.14 shows the three cases.

By changing the voltage levels of the two buses, the reactive power flow changes substantially, but it has no effect at all on the real power flow. However, it has a slight effect on the power angle δ.

6.5.6 Real Power Losses

Figure 6.14 shows a rather substantial reactive power loss on the line. Under condition 1 of Example 6.6, the line absorbs a reactive power of 20.6 MVAr

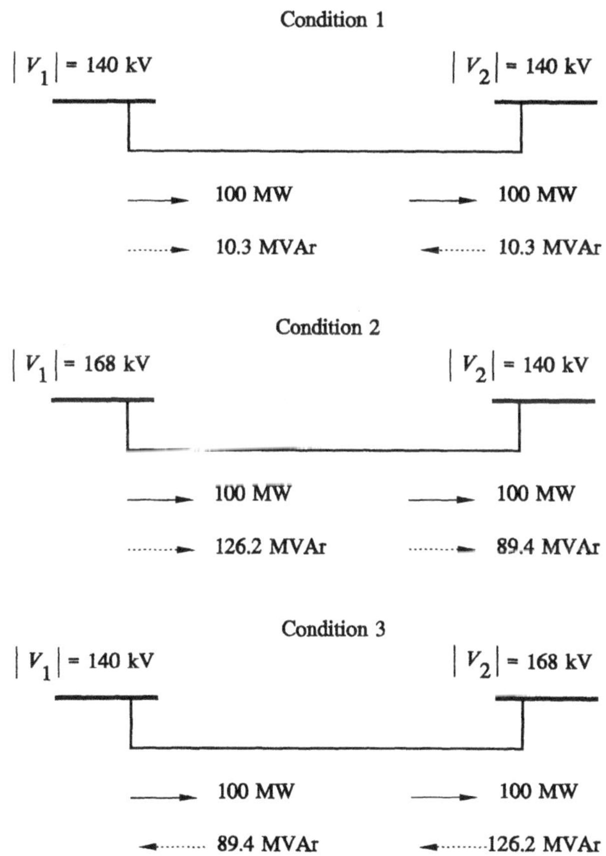

Figure 6.14

(10.3 + 10.3). Under conditions 2 and 3, the loss increased to 36.8 MVAr (126.2 − 89.4). This loss is, of course, consumed by the series line reactance. Equations (6.7) and (6.8) were derived on the assumption that the transmission line had zero resistance, and therefore the real power loss was zero. A real line will, of course, have a series resistance R, which will cause an ohmic power loss P_Ω. This power loss, which can be expressed as

$$P_\Omega = |I|^2 R \qquad \text{[W per phase]}, \tag{6.13}$$

is, in practice, of greater importance than the reactive loss.

We had noted earlier that the voltage, current, and power vary along the line. However, let V, I, P, and Q represent average values measured—for example, at midline. The following relationship must exists between the variables:

$$P + jQ \approx VI^* \qquad \text{[VA]}. \tag{6.14}$$

We then have

$$\begin{aligned} I^* &\approx \frac{P + jQ}{V} \qquad \text{[A]}; \\ I &\approx \frac{P - jQ}{V^*} \qquad \text{[A]}. \end{aligned} \tag{6.15}$$

Multiplication of the currents in equation (6.15) gives

$$I \cdot I^* = |I|^2 \cong \frac{P - jQ}{V} \cdot \frac{P + jQ}{V^*} \cong \frac{P^2 + Q^2}{|V|^2}. \tag{6.16}$$

Substitution of $|I|^2$ into equation (6.13) gives the approximate loss,

$$P_\Omega \approx |I|^2 R \approx \frac{P^2 + Q^2}{|V|^2} R \qquad \text{[W per phase]}. \tag{6.17}$$

This approximation is important because it tells us that the real and reactive line powers contribute equally to the real power loss in the line. Therefore, to minimize power loss during transmission, it is necessary to minimize both real and reactive power flow. In practice, this is accomplished by generating the reactive power at the bus where it is needed. If a generator is not available (remember that an overexcited generator produces reactive power), one can install shunt capacitors to serve the same purpose.

Example 6.7

Use equation (6.17) to find the real power loss in the transmission line in Example 6.2. Compare your answer to the result obtained in Example 6.2.

Solution: From Example 6.2, we compute the following average values for line voltage and line power:

$$|V|_{\text{ave}} = \frac{170 + 140}{2} = 155 \qquad [\text{kV}] \qquad (\text{line to line});$$

$$P_{\text{ave}} = \frac{106.1 + 100}{2} = 103 \qquad [\text{MW}] \qquad (3\text{-phase}); \qquad (6.7.1)$$

$$Q_{\text{ave}} = \frac{85.9 + 60}{2} = 73.0 \qquad [\text{MVAr}] \qquad (3\text{-phase}).$$

From equation (6.17), we get:

$$P_{\Omega} \approx 9.10 \cdot \frac{103^2 + 73^2}{155^2} = 6.04 \qquad [\text{MW}] \qquad (3\text{-phase}). \qquad (6.7.2)$$

The approximation produces less than 2% error.

Example 6.8

In Example 6.2, if the 60 MVAr needed at the receiving end of the line were produced locally, what would be the real transmission loss?

Solution: From equation (6.17), we get:

$$P_{\Omega} \approx 9.10 \frac{103^2}{155^2} = 4.0 \qquad [\text{MW}] \qquad (3\text{-phase}). \qquad (6.8.1)$$

The real losses would decrease from 6 MW to about 4 MW—that is, a 33% reduction. (Note that by using equation (6.15), we assumed the same average voltage as in Example 6.8. This is not quite correct, of course, but we must remember that equation (6.17) is an approximation.)

Example 6.9

Three equal shunt capacitors, C, are to be connected phase to ground at the receiving-end bus of the 140-kV line in Example 6.2 to produce locally, the 60 MVAr needed at this bus. Find the size of the capacitors.

Solution: The voltage across each capacitor is $140/\sqrt{3}$ kV. From Table 4.1, item 3, we get

$$Q_{\text{produced}} = \frac{60}{3} = 20 = \omega C |V|^2 \qquad [\text{MVAr}]. \qquad (6.9.1)$$

$$\therefore\ C = \frac{20}{377 \cdot (140/\sqrt{3})^2} = 8.12 \cdot 10^{-6} = 8.12 \qquad [\mu\text{F per phase}]. \qquad (6.9.2)$$

6.5.7 Summary of Interesting and Important Observations

In this section we summarize the most interesting and important observations from the discussion above and try to place them in the proper perspective.

OBSERVATION 1. An interconnected, synchronously run power system is operated at a constant frequency, which is controlled by an automatic load-frequency control system. ALFC implies a continuous process of real power balance within the system, using the frequency deviations as the indicator that balance is achieved.

OBSERVATION 2. In general, the load is balanced between the three phases and its magnitude is fairly predictable with changes taking place relatively slowly throughout the day.

OBSERVATION 3. In a transmission line, real power flows from the end that has a leading phase angle to the end that has the lagging phase angle. The magnitude of the real power flow is a function of the difference between the sending-end and receiving-end phase angles.

OBSERVATION 4. In a transmission line, reactive power flows from the end with the higher voltage to the end with the lower voltage. The magnitude of the reactive power flow is a function of the difference in the endpoint voltages.

OBSERVATION 5. The real power loss in the line is proportional to the sum of the squares of the real and reactive powers flowing in the line, and inversely proportional to the magnitude of the voltage squared.

OBSERVATION 6. Three equal shunt capacitors connected to a bus will have the same effect on the voltage and reactive power flow as an overexcited generator. Three equal shunt reactors will have the effect of an underexcited synchronous machine.

OBSERVATION 7. A bus lacking both a generator and shunt capacitors and/or reactors has an "uncontrollable" voltage. Its voltage (magnitude and phase) are determined solely by the effects of ALFC and AEC at the other network buses.

6.6 Load Flow Analysis

In the previous sections of this chapter we analyzed the factors that influence the flow of real and reactive power on a single transmission line. It is considerably more difficult to analyze the power flow in a complex system with many interconnections. For example, assume that 200 MW of power is demanded by the load on bus 22 in Figure 6.1. This power will be delivered via the four incoming lines. How will these lines share the load? What portion of the load will be supplied by the various generators in the system?

Load flow analysis (LFA) is the collective term used for a number of computer-aided analysis procedures aimed at determining the actual power flow patterns in a given system and, even more important, control these patterns. The most important objectives of LFA are as follows:

1. Determination of the real and reactive power flow in the transmission lines of a system based on certain *a priori* assumptions regarding loads and generators.
2. Computation of the voltages at all system buses.
3. Determination that no transmission line is overloaded. "Overload" can mean operation too close to its power transmission limit or (in the case of underground cables) overheating.
4. Rerouting of power in case of emergencies.
5. Determination of the specific power flow patterns that results in "optimum dispatch."

LFA of a power system consisting of hundreds of buses and transmission lines is a rather complex procedure, far beyond the objectives of this book.[8] It is possible, however, to demonstrate some of the basic features of LFA by considering very simple networks.

As an example, we choose the simple two-bus system shown in Figure 6.15. We simplify the system further by assuming that generation is available only at bus 1 and the load exists at bus 2. Figure 6.15 also shows the equivalent circuit. (The subscript "*D*" refers to load "demand.")

6.6.1 Load Flow Analysis Is Not a "Standard" Circuits Problem

An electrical engineer would immediately identify the LFA problem as an electric circuits problem. The "standard" procedure in solving such problems is to first represent the active sources as either voltage or current sources. Network equilibrium equations can then be written in which either the network voltages or the currents are the unknowns. The "loads" are invariably represented by impedances, and if these and other circuit parameters are assumed to be known and to be constants, the resultant network equations will be linear.

For example, assume that the load in Figure 6.15 could be specified in terms of a load impedance Z_D. We could then model our two-bus system as shown in Figure 6.16. (For simplicity, we have neglected the shunt impedance elements for the transformer and the transmission line). The current in the circuit is, then,

[8] For a detailed presentation of LFA when applied to large-scale systems, see Elgerd, 1971 or Stevenson, 1975.

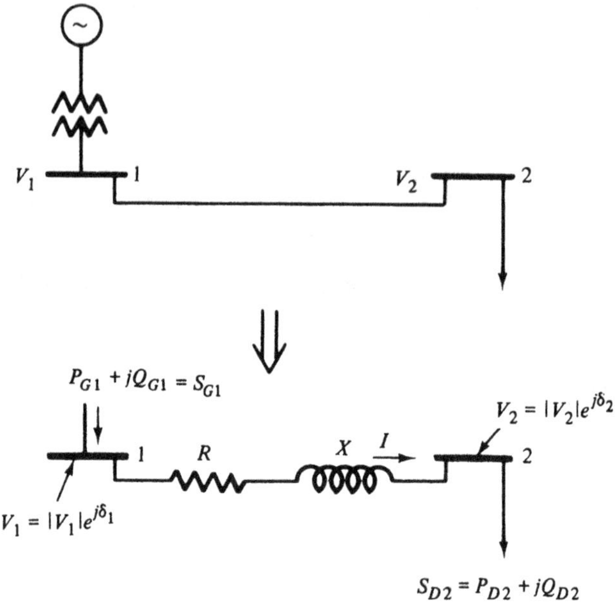

Figure 6.15

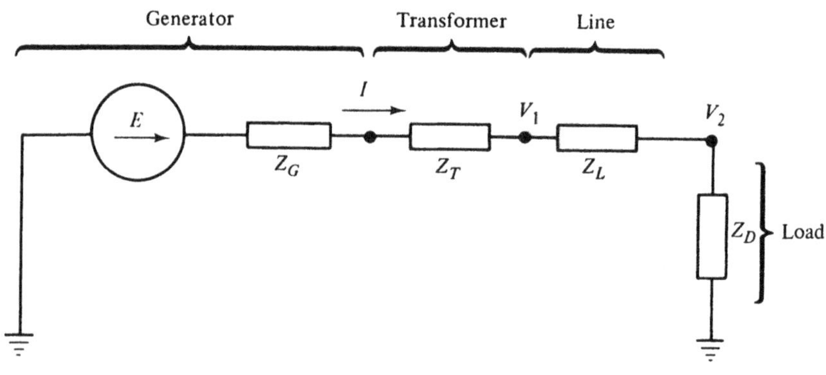

Figure 6.16

$$I = \frac{E}{Z_G + Z_T + Z_L + Z_D} \quad [A]. \tag{6.18}$$

If E were known, equation (6.18) would give the current. With a knowledge of the current, we could then compute the bus voltages and hence all powers of interest.

In summary, the analysis would be straightforward, simple and linear.[9] The linearity feature would still prevail if we were to extend the analysis to a multibus

[9] Linearity implies that I is proportional to E [see equation (6.18)].

system with many generators and loads. Instead of the single linear current equation (6.16), we would now have a system of linear equations.

LFA in a power system can never be performed in such a simple manner for the following reasons:

1. The behavior of a load in a power system is such that it cannot be represented by a constant impedance.
2. In a real power system, the generator emf, E, is never explicitly known.

Instead, we must write network equations in terms of variables that can be measured easily and that have practical significance. In power systems work, these variables are as follows:

1. The real and reactive powers.
2. The magnitudes of the bus voltages.

A typical LFA involves network equations written in terms of voltages and powers, not voltages and currents. This causes the LFA equations to be nonlinear. This fact eliminates the possibility of an analytical solution of the power flow equations in most cases. We can, however, always arrive at a numerical solutions using computer-aided techniques.

6.7 Summary

The main function of an electric power network is to connect the generating stations to the individual customers. It must be designed so it can transmit the appropriate amount of power to its customers, large and small. We have learned that the transmission capacity of a line grows as the square of the line voltage, and inversely as the magnitude of the line reactance.

The primary concern of the power systems engineer is to generate power at a constant frequency. This job is normally assigned to an automatic control system that maintains, at all times, real power balance within the system. A mismatch in real power results in a frequency deviation.

The next most important function is to maintain a proper voltage profile throughout the system. This is accomplished by proper flow of reactive power in the various lines. One may express the problem as follows: if proper reactive power balance can be maintained, the voltage profile remains under control. If the reactive power balance is not maintained, the voltage profile will drift (just as the frequency will drift if real power balance is not maintained).

This is particularly noticeable during night hours. The reactive power generation represented by the shunt capacitors of the lines tend to provide a surplus of reactive power during the night hours. This is not a problem during normal work hours, as the reactive power is consumed by the motor loads in factories etc. Consequently, the bus voltages tend to increase during night hours.

We have also presented, in the simplest of terms, the all-important load-flow analysis problem. By controlling the flow of both real and reactive power on the electric grid, it is possible to control the transmission losses which affect the economic operation of the system.

EXERCISES

6.1 Consider the transformer shown in Figure 6.5. The three-phase feeder voltage measures 11.5 kV between lines. The 115-V load consists of a total of 0.95 kW of single-phase inductive motor load of power factor $\cos\varphi = 0.8$. The 230-V load consists of 3.1 kW heaters at $\cos\varphi = 1.0$. Model the transformer as an IT and compute all currents indicated by the arrows. [HINT: The total mmf on the transformer core must be zero.]

6.2 The single-phase load in the previous exercise is 4.05 kW connected between phases a and b. Assume now that we have three identical single-phase loads of this type. The remaining two are connected between phases b and c, and c and a, respectively. Show that the total set of currents in the three-phase feeder constitutes a symmetrical three-phase set. Find the rms value of the current in each phase.

6.3 A power company does not normally try to exert any control over the amount of power that its customers drain from its network. However, in times of energy crises, the load may exceed the generating capacity of the plant. The load must therefore be reduced. If voluntary means fail, the company, in the end, may have to disconnect its customers on some priority basis. Before this drastic "solution" is adopted, the company can reduce the customer's load gradually by reducing the voltage.

Consider an industrial heating load consisting of three identical 12-Ω resistors connected in Y to a 12-kV, three-phase feeder.

a) What load in MW, do these resistors represent if the 12-kV bus is exactly 12 kV?

b) What load do the same resistors represent if the power company lowers the voltage of the 12-kV bus by 5%?

If you solved the problem correctly, you would have found that the voltage reduction causes a 9.75% reduction in the load. This would seem to be a saving for the customer. Why is she or he not happy?

6.4 The previous exercise demonstrates the dependency of the load upon the voltage. The load may also depend upon the frequency. Consider a load consisting of three identical impedances, again connected in Y to the 12-kV, 3-phase bus in the previous exercise. Each impedance consists of a 20-Ω resistor in series with a 40-mH reactor. Show that the load drawn by this set of impedances will increase if the frequency drops. In particular, by what percent will the load increase if the frequency drops from 60.0 Hz to 59.5 Hz?

6.5 In Example 6.1 the total kinetic energy of a power system was 600 MJ. To get a feel for the magnitudes involved, place all this kinetic energy in an equivalent cylinder made of solid steel, having a diameter equal to its length and rotating at 1000 rpm. How big will this cylinder be? The density of steel is 7800 kg/m^3.

6.6 Consider the 3-phase, 140-kV, 100-km transmission line, the line parameters of which were given in Example 6.2. The sending-end voltage is kept at 145 kV. The three-phase sending-end power is 120 MW at a 0.8 power factor (voltage leading current); $f = 60$ Hz.

a) Compute the current at each end of the line.
b) Compute the voltage at the receiving end.
c) Compute the three-phase power at the receiving end.
d) Find the transmission efficiency.

6.7 Consider the transmission line in Exercise 6.6. The load at the receiving end consists of three equal 200-Ω resistors connected in Y.

a) It is required that the receiving-end voltage $|V_2|$ be kept at 140 kV exactly. What voltage $|V_1|$ must be maintained at the sending end to make this possible? $f = 60$ Hz.
b) If the voltage condition in part a is maintained, what will be the sending-end and receiving-end powers, real and reactive?

6.8 The 100-km line in Exercises 6.6 and 6.7 has its sending-end voltage $|V_1| = 140$ kV. The opposite end of the line is on open-circuit, and $f = 60$ Hz.

a) What will be the sending-end current?
b) Show that the line consumes real power (how much?), but generates reactive power (how much?). Explain this in physical terms.
c) What is the voltage $|V_2|$ at the open end?

6.9 The transmission line in Exercise 6.8 has been deenergized for repair. A sleet storm has put a layer of ice on the line. Before putting the line back into operation, we wish to melt the ice by sending an estimated 100-A current into each phase. For this purpose, we short-circuit all three phases at one end and apply a symmetric three-phase voltage at the other. How much voltage must be applied in order to inject 100 A into each phase? $f = 60$ Hz.

6.10 We represent a 160-km line by the simplified model shown in Figure 6.11. The voltage at each end is 100 kV. The line reactance, $X = 60\ \Omega/\text{phase}$. According to equation (6.9), this line can at most transmit:

$$P_{\max} = 167 \qquad [\text{MW}]. \tag{6.19}$$

By putting three equal series capacitors C at the midpoint in each phase, we seek to reduce the series reactance from 60 Ω to 35 Ω and thus increase the transmittable power to

$$P_{\max} = 286 \qquad [\text{MW}]. \tag{6.20}$$

a) What size capacitor is required? $f = 60$ Hz.
b) Compute the voltage across each capacitor if the line is transmitting 200 MW. The voltages in both ends are kept at 100 kV.

6.11 Although we would never do this in reality, assume that we operate the 140-kV line in Example 6.6 at a power angle of 75°. (The voltages are assumed to be equal to 140 kV at both ends.)

a) Compute the synchronizing coefficient, expressed in MW/degree.
b) If the line power increases by 8.37 MW, what is the corresponding change in the power angle? Compare this to the results in Example 6.6.

6.12 Assume that you operate the line in Exercise 6.11 at a power angle $\delta = 89.9°$ (theoretically, of course). Show that an additional line load of one single horsepower (0.746 kW) will bring about the collapse of the transmission.

6.13 Consider the 2-bus system shown in Figure 6.17. The line connecting the two buses is modeled according to Figure 6.11. The line reactance, X is 50 Ω/phase.
There are synchronous generators at each bus. The generated powers indicated in Figure 6.17 are measured at the HV terminals of the step-up transformers. A particular load is as follows:

$$\begin{aligned} P_{D1} + jQ_{D1} &= 250 + j150 \quad [\text{MVA}]; \\ P_{D2} + jQ_{D2} &= 50 + j50 \quad [\text{MVA}]. \end{aligned} \tag{6.21}$$

The two bus voltages must be maintained at $|V_1| = |V_2| = 175$ kV. The generators divide the real load equally; $P_{G1} = P_{G2} = 150$ MW. This means that 100 MW must be transmitted from bus 2 to bus 1.

Find the reactive power flow at each end of the line and also the reactive powers Q_{G1} and Q_{G2}. What will be power angle δ of the line?

6.14 Consider the system in Figure 6.17. All power specifications are unchanged from Exercise 6.13. The magnitude of the line voltage at bus 1 is increased by 10%:

a) Find the real and reactive line power flow at each end.
b) Find the reactive powers generated.
c) Find the power angle, δ, of the line.

(Note from your results how the higher line voltage $|V_1|$ now requires a higher reactive power generation at bus 1.)

6.15 Consider the system in Figure 6.17. The following powers and voltages are specified:

$$\begin{aligned} P_{D1} + jQ_{D1} &= 250 + j150 \quad [\text{MVA}]; \\ P_{D2} + jQ_{D2} &= 50 + j50 \quad [\text{MVA}]; \\ P_{G1} &= 0 \quad [\text{MW}]; \\ P_{G2} &= 300 \quad [\text{MW}]. \end{aligned} \tag{6.22}$$

$|V_2| = 175$ kV and $|V_1|$ is 10% higher.

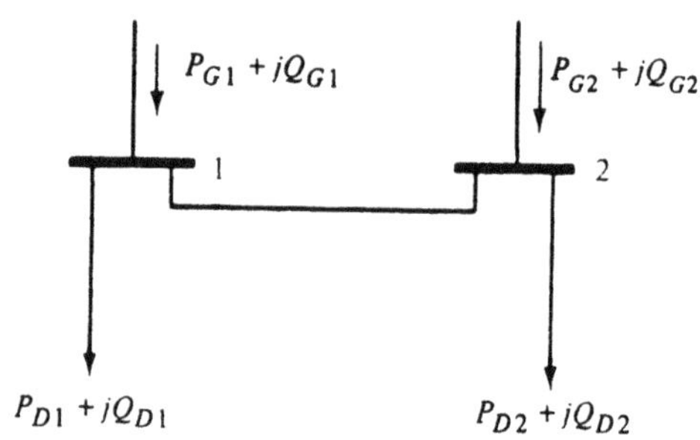

Figure 6.17

a) Find the real and reactive power flow in the line.
b) Find the reactive powers generated.
c) Find the power angle, δ, of the line.

References

Bergen, A.R. *Power System Analysis*. Englewood Cliffs, NJ: Prentice Hall, 1986.

Del Toro, V. *Electric Power Systems*. Englewood Cliffs, NJ: Prentice Hall, 1992.

Elgerd, O. *Control Systems Theory*. New York: McGraw-Hill, 1967.

Elgerd, O.I. *Electric Energy Systems Theory - An Introduction*, New York, McGraw-Hill, 1971.

Elgerd, O.I. *Electric Energy Systems Theory*, 2nd ed. New York: McGraw-Hill, 1982.

Grainger, J.J., Stevenson, W.D. Jr. *Power System Analysis*. New York: McGraw-Hill, 1994.

Gross, C.A. *Power System Analysis*, 2nd ed. New York: John Wiley & Sons, 1986.

Stevenson, William T. *Elements of Power Systems Analysis*, New York, McGraw-Hill, 1975.

Stevenson, S.D. Jr. *Elements of Power Systems Analysis*, 4th ed. New York: McGraw-Hill, 1982.

7

The Direct Current Machine

The system loads that were discussed in the previous chapters consist of a multitude of electrical devices ranging from bread toasters, lights, electric ovens, and vacuum cleaners in the domestic sector to huge motors and lift magnets used in industry.

The vast majority of the electric motors are of ac design (Chapter 8). The dc motor constitutes a distinct minority. Not only is a dc motor more expensive than an ac motor of equivalent size, it has more parts that can go wrong and it requires a dc supply system. However, the dc motor survives because it has some very special characteristics to which no other motor can lay claim.

7.1 Torque–Speed Requirements of Motors

There is a large variety of motors, electrical and nonelectrical, in use today. The internal combustion engine (ICE) is the most important drive system in the transportation sector. The hydroturbine—a descendant of the waterwheel—still finds important applications, particularly as a prime mover in hydroelectric power plants. The steam turbine drives most of the world's electric generators.

The electric motor is by far the most versatile drive system available. Electric motors come in sizes varying from a few watts to thousands of horsepower. The ac induction motor has a simple and rugged design, and it meets the requirements of a vast spectrum of applications. The dc motor is the most versatile and the easiest motor to control and is used when no other motor can do the job. The dc motor is used to drive hoists, cranes, elevators, and road vehicles, and also as an actuator in a number of industrial control systems.

The single most important requirement that a motor—any motor—must meet is matching the torque–speed (TS) characteristics of the load it pulls. Consider for example the TS demands put on an automobile engine.

Figure 7.1a shows a normal duty cycle of an automobile engine, including start, run, stop, and reverse. The torque requirements are most severe during acceleration and deceleration. For example, during acceleration, the motor drive

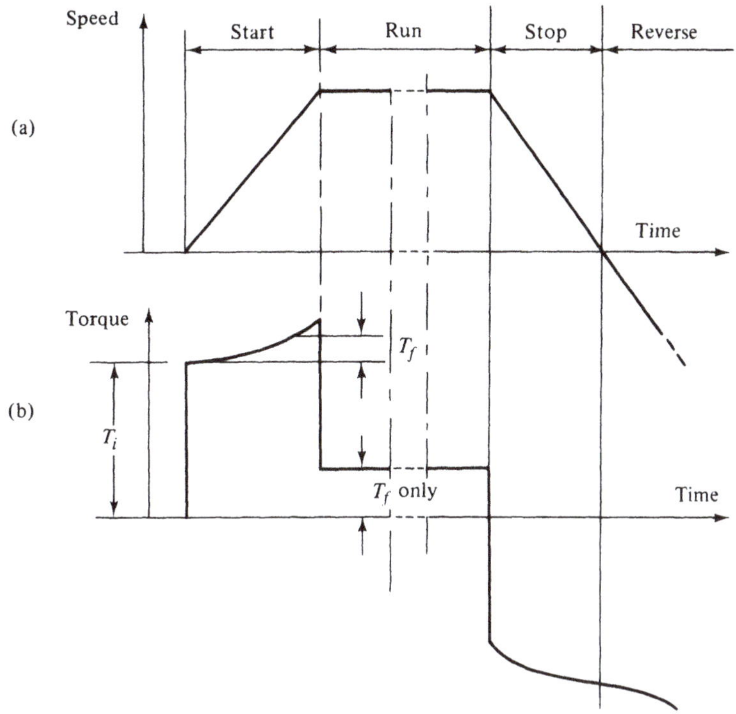

Figure 7.1

torque T_m must be equal to the sum of the inertial torque T_i and the frictional torque T_f:

$$T_m = T_i + T_f = ma + T_f \qquad [N]. \tag{7.1}$$

If the acceleration a is assumed constant, then T_i will also be constant. Usually T_f is a function of the square of the speed. The motor torque requirement, T_m, is shown in Figure 7.1b. The torque–speed characteristics of the ICE are a very poor match to these requirements. It has zero starting torque—in fact it takes a dc motor to get the ICE itself to start. It cannot deliver the smooth torque needed from zero to full speed. Either a gearbox and a clutch or a complicated torque converter (automatic transmission) must be installed between the engine and the road wheels to obtain the desired TS characteristics.

The ICE easily delivers the torque needed during "run," but during deceleration and "stop," most of the negative torque required is obtained from mechanical brakes that convert the kinetic energy of the vehicle into heat. During the "reverse" part of the duty cycle, the torque and speed are both negative. Since the ICE can run in only one direction, we must call upon the torque converter again to provide the proper TS characteristics.

By contrast, a dc motor can easily provide the TS required—without the assistance of either a torque converter or (to some extent) even brakes. Furthermore, a dc motor is relatively quiet, nonpolluting, and cheap, and it has only one moving part—the rotor.

Why, then, is the ICE still dominant in the automobile market? Clearly, it is the unique ability of the ICE to run 300–400 km on a tank of gasoline. When a battery is developed that can store—on the basis of comparable weight—as much energy as a tank full of gasoline, we shall most probably see the "electrics" rapidly capture the automobile market. In the meantime we see them making inroads into the low-speed, short-range (urban), heavy-duty sectors.

7.2 A Direct Current Motor Prototype

The simple arrangement in Figure 7.2 shows a "linear" dc motor. Even in this simple form, it demonstrates most of the typical characteristics of a normal rotating dc motor. A rod of length L can move freely along two supporting horizontal

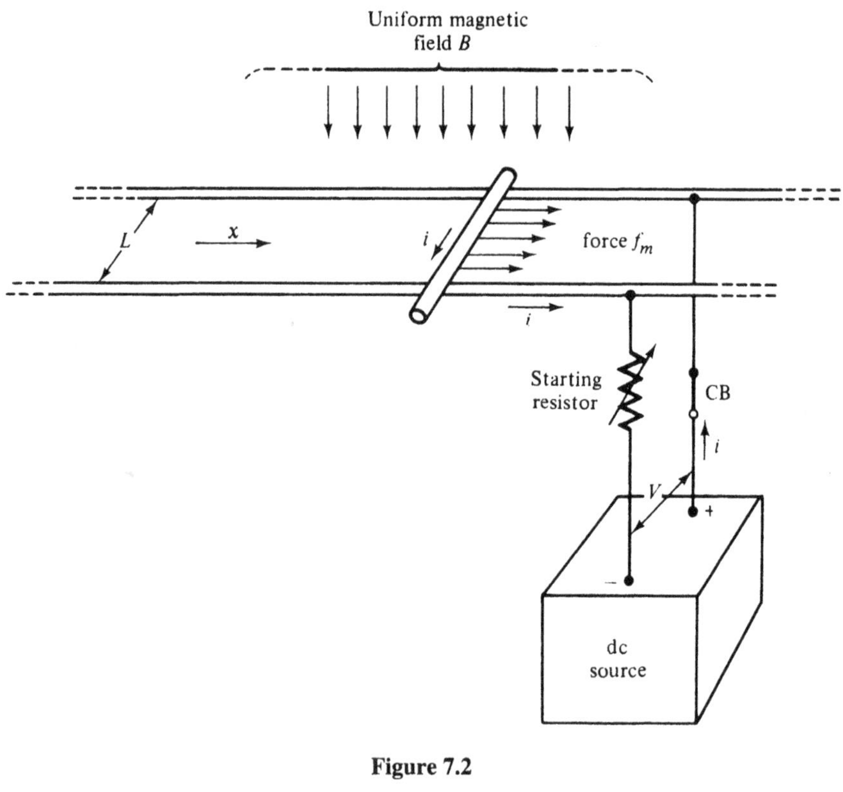

Figure 7.2

rails, perpendicular to a uniform, vertical magnetic field of density B. A dc voltage source, V, supplies current to the rod via a *starting resistor.*

When the circuit breaker (CB) is closed, the source will cause a current i to flow in the direction indicated. According to equation (3.57), the rod will be subjected to an electromechanical or *motor force*, f_m of magnitude,

$$f_m = BiL \qquad [\text{N}], \tag{7.2}$$

acting in the positive x direction. The motor force will cause the rod to accelerate, that is, to overcome the inertia force f_i. It can also pull a cart that we may attach to the rod, which is represented by the force f_L. Mathematically, this can be stated as

$$f_m = f_i + f_L = m\dot{s} + f_L \qquad [\text{N}], \tag{7.3}$$

where s is the velocity of the rod, m/s; m is the mass of the rod and its load, in kg. As the rod picks up speed (the rod is now moving in the magnetic field), an emf, e, will be induced which according to equation (3.47) will be of magnitude,

$$e = BLs \qquad [\text{V}]. \tag{7.4}$$

According to Lenz's law, the polarity of the emf will be such as to oppose the current, that is, the cause of the motion. The total loop resistance R is made up of the resistance of the rod, the contact[1] resistance between the rod and rails and the external current-limiting "starting" resistor. By application of Ohm's law, the current will have the value

$$i = \frac{V - e}{R} = \frac{V}{R} - \frac{BL}{R}s \qquad [\text{A}] \tag{7.5}$$

If we combine equations (7.2), (7.3), and (7.5) we obtain the following first-order, linear differential equation in terms of the speed s:

$$\dot{s} + \frac{B^2L^2}{mR}s - \frac{BLV}{mR} + \frac{f_L}{m} = 0. \tag{7.6}$$

Included in the load force f_L are all friction forces acting on the rod and cart.

7.2.1 Steady-State Speed Under No-Load Conditions

If we assume the absence of friction the load force f_L will be zero. This is described as the *no-load* condition. The rod will now accelerate until the back emf is equal to the source voltage, V, at which time the current will be zero. The force f_m, and hence the acceleration $\dot{s}$, will then be zero, and the rod velocity will remain constant.

[1] The contact resistance is not constant, and we neglect it in our analysis. It is small in comparison to the resistance of the starting resistor.

The steady-state *no-load* velocity, s_0, can be obtained from equation (7.6) by setting both $\dot{s}$ and f_L equal to zero. We get

$$s \equiv s_0 = \frac{V}{BL} \qquad [\text{m/s}]. \tag{7.7}$$

Note that the velocity, s_0 increases as the magnetic field flux density, B, decreases. This is a somewhat surprising situation and one that always confuses the novice.

7.2.2 Energy Transformation in Direct Current Motors

The linear motor converts electric power drawn from the source, into useful motor power plus ohmic power loss in the loop resistance. The motor power goes into additional kinetic energy of the rod and any useful work done by the rod. We can obtain a mathematical statement of this mechanism by considering equation (7.5) which can be written as

$$V = iR + e \qquad [\text{V}]. \tag{7.8}$$

If we multiply it by i, we have

$$Vi = i^2R + ei \qquad [\text{V}]. \tag{7.9}$$

The left-hand side of equation (7.9) represents the power delivered by the source. The first term on the right-hand side is the ohmic power loss. The last term must be the *motor power*, p_m, where

$$p_m = ei \qquad [\text{W}]. \tag{7.10}$$

7.2.3 The Linear Direct Current Motor Under Load

In the previous sections the linear dc motor was analyzed under the assumption of zero mechanical load. During startup we had to overcome only the inertia of the moving mass. Once the steady-state velocity was reached, the current was zero, and from that time on the voltage source supplied zero power to the motor.

In practice, for a motor to serve a useful function it must pull a load. In the case of the linear motor this load could consist of frictional drag on the rod and/or of a cart, representing a load force f_L. In the presence of this load force, the motor will no longer accelerate to the no-load speed given by equation (7.7). The speed will reach a lower value determined by the fact that the motor force f_m must be equal to the load force f_L. The former is obtained by the substitution of equation (7.5) into (7.2), which gives

$$f_m = BLi = BL\frac{V - BLs}{R} \qquad [\text{N}]. \tag{7.11}$$

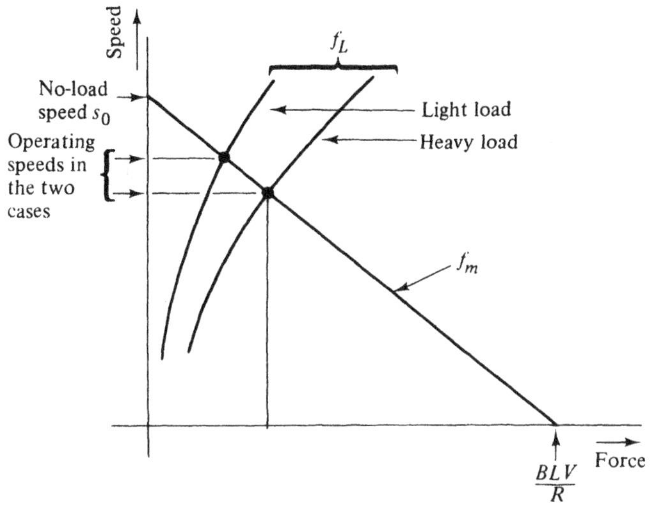

Figure 7.3

If we solve this equation for s, we obtain

$$s = s_0 - \frac{R}{B^2L^2} f_m \qquad \text{[m/s]}. \tag{7.12}$$

On a graph of speed against force, this is the equation for a straight line with a negative slope (see Figure 7.3.)

The load force f_L is also plotted on the same speed–force graph. Typically the load force increases with speed. (For example, wind drag increases approximately as the square of speed.) Actually, we have plotted two cases of loading, representing a "light load" and a "heavy load." The intersection of the f_L and f_m curves indicates that balance exists between the two forces. The intersection gives the operating speed and the corresponding force of the motor.

Note that the heavy load results in a lower speed. The reason for this is quite clear. The heavy load requires a higher value of motor force. The higher force can be attained only by a higher current. A higher current can occur only if the back emf of the motor is reduced. A reduced emf is obtained when the speed is reduced. (Note that the current–speed relationship (7.5) looks similar to the motor force–speed (7.12). The only difference is a scale factor. This means that the sloping f_m line in Figure 7.3 also represents the motor current, on a different scale.)

7.2.4 Motor Rating

When the motor is heavily loaded, it will draw such a high current that the ohmic losses in the circuit will become intolerable. Overheating of the motor sets a limit to its load-pulling capacity. If we were to overload the motor to the point when it *stalled*, the motor emf e would be zero. The motor current would now be limited only by the loop resistance, R. Not only would the current attain its maximum value (V/R), but the stalled motor would have lost all its cooling capacity, since the rod does not have the cooling effect of moving through the air and it would heat up very rapidly. It is logical, therefore, to rate the motor in terms of its maximum load force. We may also multiply this force by the speed and obtain a maximum power that the motor can deliver.

7.2.5 Turning the Motor into a Generator

Assume that the rod is traveling at the no-load velocity s_0. Its generated emf e is then equal to the source voltage, V, and the current, as we have noted, is zero.

Now assume that we reverse the load force f_L, that is, we make it act in the direction of the speed s, thus pulling the rod along at a speed in excess of s_0. The emf e will now exceed V. As a result the current I will change its direction, thus feeding current into the positive terminal of the source.

The linear motor has then turned into a *generator* feeding energy into the "source," which now acts as an energy "sink." A reversal of the load force f_L means that the "load" has turned into a "prime mover"—a source of the mechanical power required to pull the rod and supply energy to the sink.

7.2.6 Equivalent Circuits

As viewed from the source (or sink) the dc machine[2] behaves like a variable but unidirectional emf e in series with the resistance R. The latter consists of an "external" part (the current-limiting "starting" resistor) and an "internal" part (the contact resistance between fixed and moving parts plus conductor resistance).

In Figure 7.4 the "motor" and "generator" cases have been separated. Note the reversal of the current and power flow but the unchanged direction of the machine emf e.

7.3 Physical Motor Design

The linear motor shown in Figure 7.2 would clearly have very limited practical usefulness. Moreover, it would be very difficult to design it. How, for example, can we produce the strong magnetic field required? Let us now proceed to turn this impractical concept into a very practical rotating model.

[2] Calling the device "machine" rather than "motor" implies that we now realize its potential as a "generator."

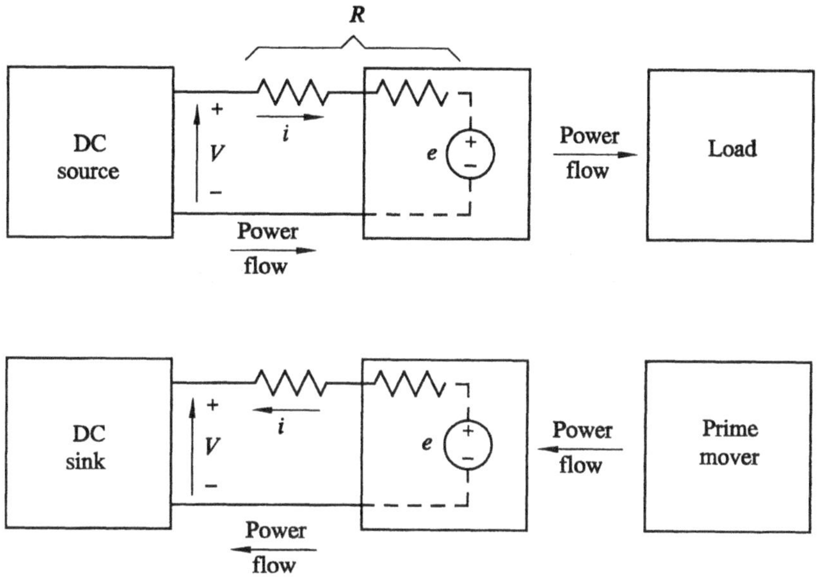

Figure 7.4

7.3.1 The Homopolar Machine

The first obvious improvement is to increase the motor force f_m by adding parallel conductors. We then obtain the ladderlike contraption depicted in Figure 7.5. If the "ladder" contains n steps (conductors) each carrying the current i, the total motor force will be equal to

$$f_m = nBiL \qquad [\text{N}]. \tag{7.13}$$

The next improvement is to bend the linear "racetrack" of Figures 7.2 and 7.5 into a circular shape. We can do this in several ways. By arranging the current-carrying conductors radially into a spokelike "rotor" as shown in Figure 7.6 the motor forces f_m acting on each spoke will result in a motor torque T_m. A practical variant of this design is the *homopolar* machine shown in Figure 7.7.

The rotor consists of a circular aluminum disc. Current is fed radially through the disc via the two ring-shaped carbon brushes. The magnetic flux is created by the *field current*, i_f, flowing in the field winding. The flux follows the path indicated through the stator iron. It penetrates the disc vertically, to give the flux geometry shown in Figure 7.7.

The homopolar machine is not very practical. As will be demonstrated in Example 7.1, this type of motor is a high-current/low-voltage device. This is not

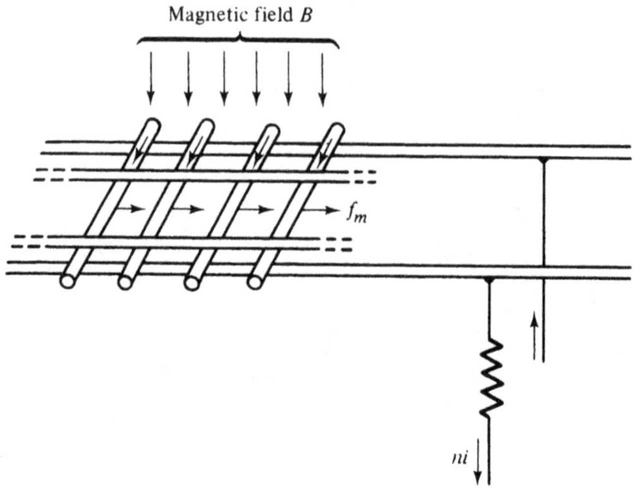

Figure 7.5

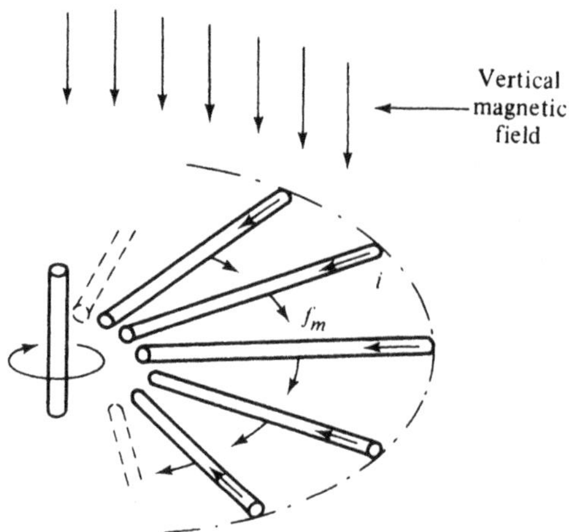

Figure 7.6

a good combination of characteristics. We remember, from earlier discussion of power transmission, that low-voltage levels result in high power losses.

Example 7.1

The aluminum disc in a homopolar motor runs at a speed of $n = 3000$ rpm. The disc (see Figure 7.8) has an outer and inner radii $R = 0.9$ m and $r = 0.05$ m, respectively. The flux density in the air gap is $B = 0.5$ T. A total current of i A

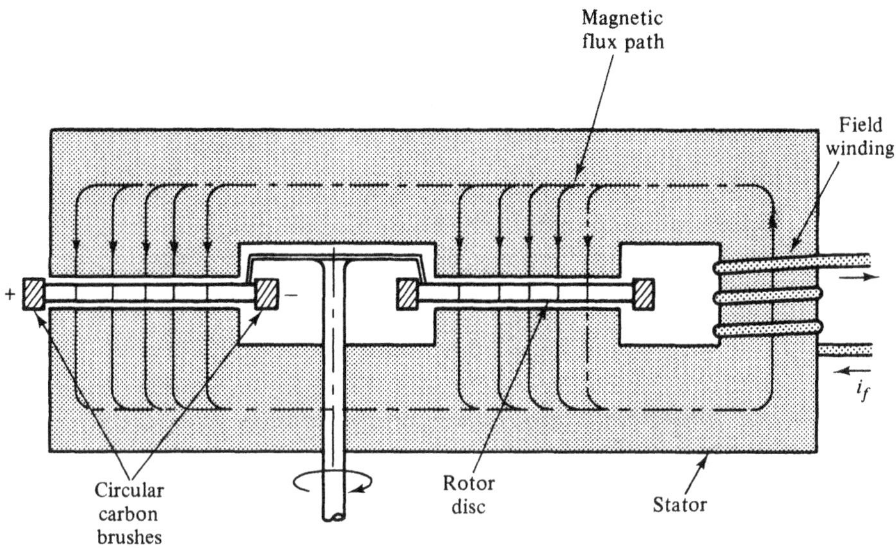

Figure 7.7

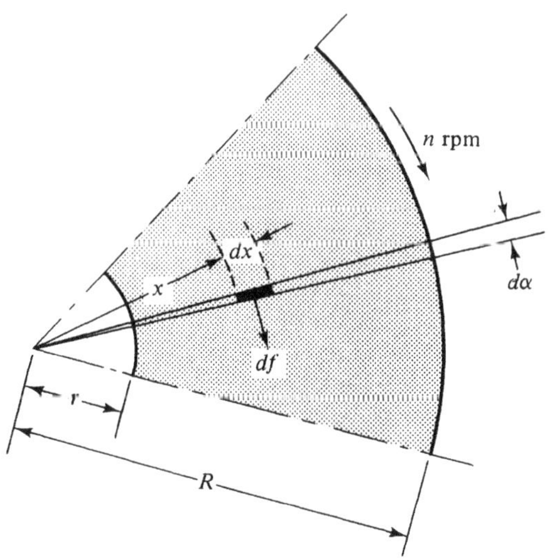

Figure 7.8

flows between the brushes. Find the magnitude of the emf e generated by the disc and the motor power p_m.

Solution: Consider the elemental emf, de, generated in the small shaded radial element of length, dx in Figure 7.8. According to equation (7.4) the differential emf will have the magnitude

$$de = Bs\,dx \qquad [\text{V}], \tag{7.1.1}$$

where

$$s = \omega x = \frac{n\pi}{30}x \qquad [\text{m/s}], \tag{7.1.2}$$

and ω is the rotating speed of the disc measured in radians per second.

Substituting for s and integrating from $x = r$ to $x = R$ we obtain the total emf e generated in the "spoke" between the two brushes. This gives

$$e = B\frac{n\pi}{30}\int_{x=r}^{x=R} x\,dx = \frac{Bn\pi}{60}(R^2 - r^2) \qquad [\text{V}]. \tag{7.1.3}$$

Substituting the given numerical values, we obtain

$$e = 2.945 \qquad [\text{V}]. \tag{7.1.4}$$

The current in the elemental "spoke" dx will interact with the magnetic flux to give the tangential elemental motor force df and a corresponding elemental motor torque dT_m. By a simple analysis (see Exercise 3.16) we find the total motor torque to be

$$T_m = \frac{Bi}{2}(R^2 - r^2) \qquad [\text{N}\cdot\text{m}], \tag{7.1.5}$$

where i is the total current fed radially into the disc.

For the motor power we get

$$p_m = \omega T_m = \frac{n\pi Bi}{60}(R^2 - r^2) \qquad [\text{W}]. \tag{7.1.6}$$

In view of equation (7.10) we can write:

$$p_m = ei \qquad [\text{W}]. \tag{7.1.7}$$

Assuming that $i = 500$ A, we get

$$p_m = 2.945 \cdot 500 = 1473\ [\text{W}] \approx 2\ [\text{hp}]. \tag{7.1.8}$$

A motor that supplies only 2 hp and needs a current of 500 A at a terminal voltage of about 3 V will not find many buyers. However, there are special applications where electric power of the low-voltage/high-current variety is available (electric solar panels and magnetohydrodynamic generators). Homopolar machines represent, at least, in principle, suitable electromechanical converters in such cases.

7.3.2 Cylindrical Conductor Direct Current Motor

A more practical motor design is obtained by arranging the current-carrying rotor conductors in a cylindrical geometry as shown in Figure 7.9. The rotor currents now have an axial orientation. The magnetic flux is radially directed at the location of all the rotor conductors. The electromechanical forces are therefore tangential, resulting in a net motor torque.

The prototype of an even more practical dc motor is shown in Figure 7.10. The flux is provided by a permanent magnet the pole shoes of which are shown. It is assumed that the air-gap flux is directed from the N pole to the S pole. A coil supported on a suitable axis is terminated in a two-segment commutator. Two spring-loaded carbon brushes make electrical contact with the commutator segments. Current i is supplied through the upper carbon brush and as it passes under the S pole, it will generate a force in the direction shown. On its return, the current

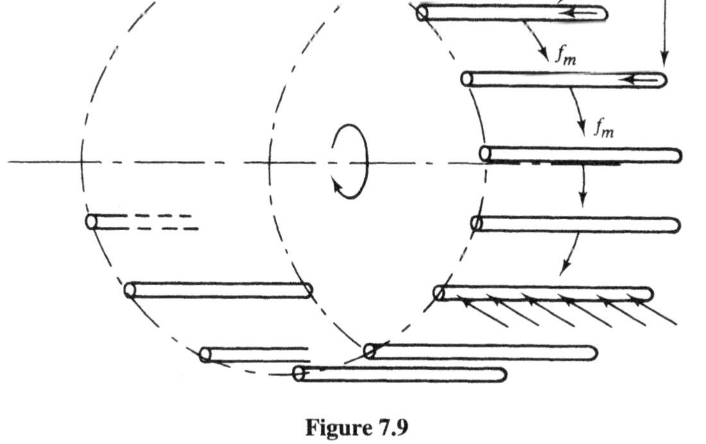

Figure 7.9

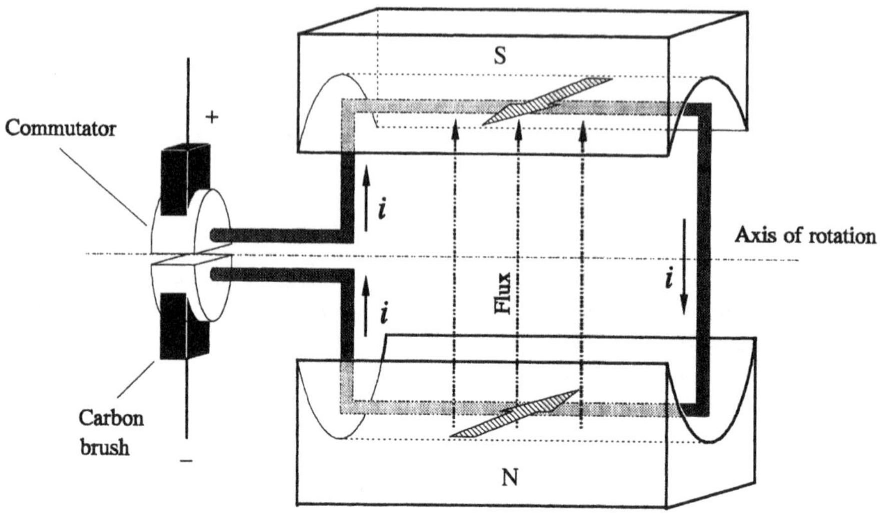

Figure 7.10

passes over the N pole of the permanent magnet, and it also generates a force in the direction shown. The two forces constitute a torque which tends to rotate the coil in a clockwise direction seen from the commutator end.

Figure 7.11 shows the coil in a series of positions and the direction of the torque generated by the coil. It can be seen that the magnitude of the torque is related to the position of the coil by a "rectified" sinusoidal function. When the coil is in the horizontal plane, (Figure 7.11d, the torque will be zero and at the same time the current i will have a value of zero, since the carbon brushes will not be in contact with the commutator segments. At this instant the motor should come to a halt, but if we assume that the coil has a finite moment of inertia, then it follows that it will rotate past the horizontal position. The carbon brushes then come into contact with the opposite commutator segments thereby reversing the current in the coil. The resulting torque is still in the clockwise direction and the motor will continue to rotate.

Several improvements can be made to the machine as follows:

1. Replace the permanent magnet with an electromagnet. This will enable us to control the flux density by changing the current in the magnetizing coil.
2. Decrease the length of the air gap by winding the coil in two suitable grooves on the outside of a cylinder made of a ferromagnetic material.
3. Increase the number of coils (parallel paths for the current) so as to increase the net torque generated. This will also help to "even out" the torques from the individual coils and avoid the "dead-zone" shown in Figure 7.11d.

When these improvements have been made, we have a cylindrical conductor dc motor.

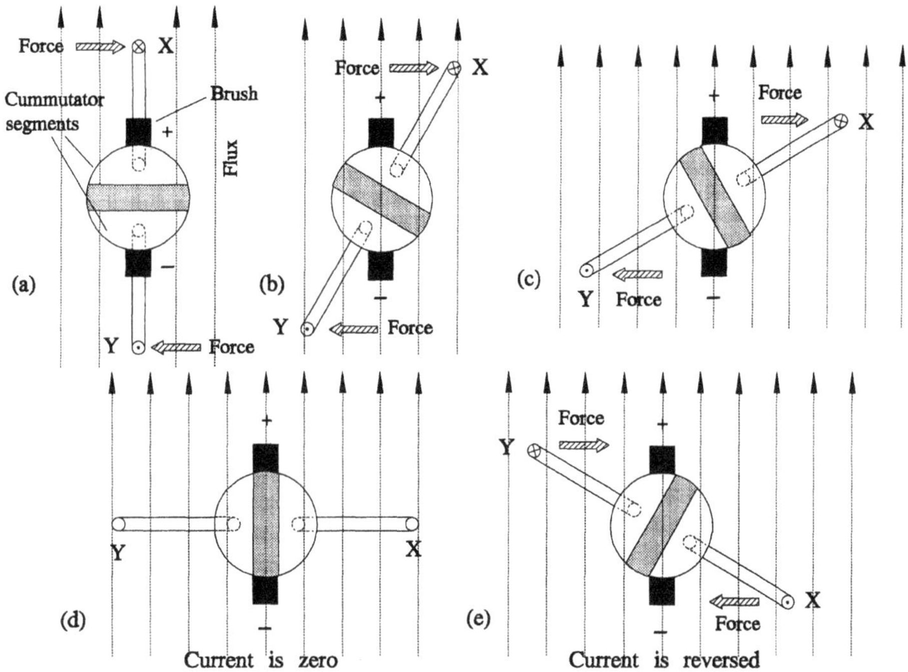

Figure 7.11

A simple but practical design incorporating the suggested improvements is shown in Figure 7.12. Practically all dc machines in use today are variants of this design. Figure 7.12 shows a two-pole machine with only 12 torque-creating rotor conductors (six coils). Normal dc machines are not quite that simple, but their operation can be fully explained in terms of this simple example. The magnetic flux is created by the dc field current i_f in the stator field winding. The flux passes through the air-gaps radially, going from the magnetic N pole to the S pole.

The rotor windings (coils) are placed in slots on the rotor surface. We refer to it as the *armature* winding. The armature current, i_a, is supplied by an external dc source via the *carbon brushes* and the *commutator*. It can be seen from Figure 7.12, that the conductor currents are in opposite directions under the two poles. The net torque T_m generated by the motor will have a clockwise direction, as shown in Figure 7.12.

Note that, although the stator iron experiences a constant or dc magnetic field, the rotor (when running) is subject to a changing or ac field.[3] In order to reduce the iron losses, the rotor is made of transversely laminated material.

[3] Remember, the opposite situation prevails in the synchronous machine.

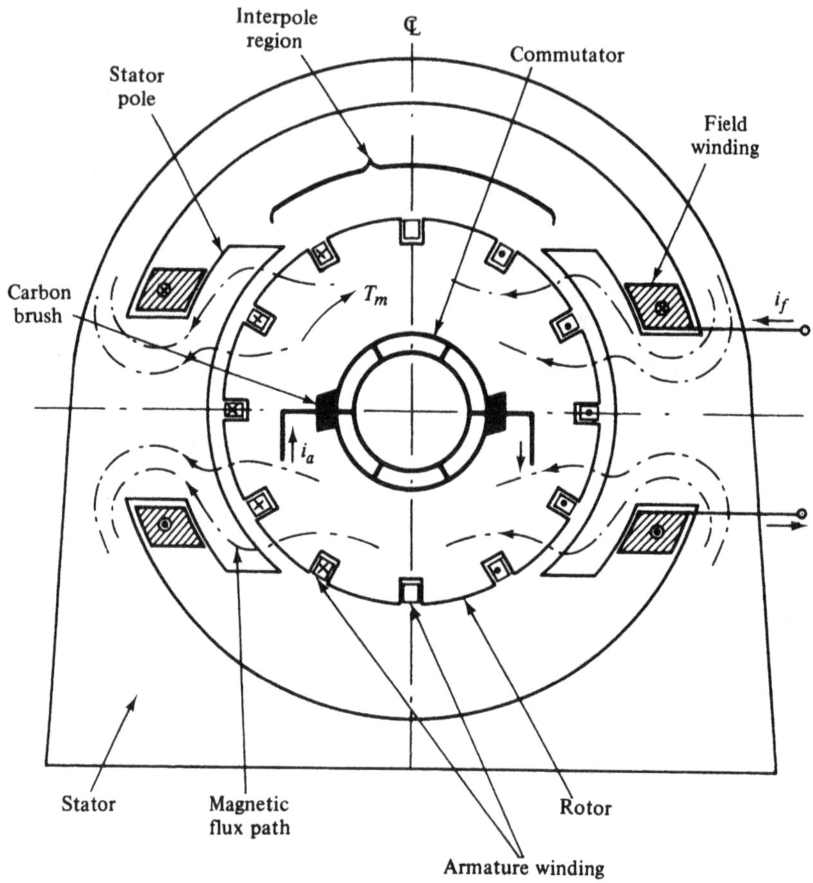

Figure 7.12

7.3.3 The Commutator Action

The action of the two-segment commutator of the prototype motor in reversing the direction of the current in the coil was explained earlier. It is most important that the current direction in all rotor conductors under each pole-shoe remains fixed as the rotor turns. Otherwise we could not preserve the constant direction of the electromechanical torque T_m. Clearly, to accomplish this, we must reverse the current direction in every conductor as it passes the interpole regions.

The continuous reversal of the current in the armature conductors is accomplished by the commutator. This device consists of a number (six in our case) of identical copper segments, insulated from each other and from the shaft, forming

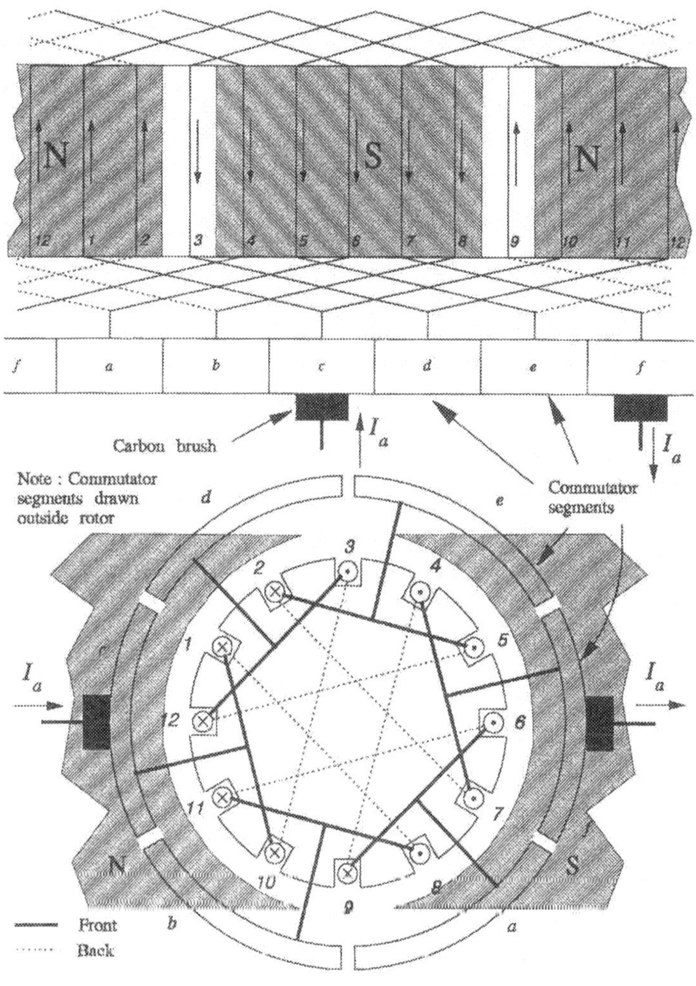

Figure 7.13

a ring fixed to the rotor shaft. Each segment is connected to the armature conductors in a symmetrical pattern as shown in Figure 7.13a. This connection can be made in many ways—the pattern of Figure 7.13a is called a *lap winding*. The reader can readily trace the current paths from the positive to the negative brush in Figure 7.13a, b. Figure 7.13b shows an axial view of the commutator. Note that the diameter of the commutator has been considerably enlarged to make it easy to trace the windings. Furthermore, only one conductor per slot is shown. With the brush position indicated (in contact with segments c and f) half of the armature conductors (3 through 8) carry currents in the same direction and they are under the S pole. The other half have the currents flowing in the opposite direction, and they are under the N pole. As the rotor turns and the brush position changes to the next segment pair (*b* and *e*), the currents in the conductors 1–2 and 7–8 will reverse direction. Note that as the rotor changes position, the commutator segment pairs *b*–*c* and *e*–*f* will be short-circuited for a short period. During this interval, the conductors 1–2 and 7–8 will be short-circuited. The current reversal or current *commutation* takes place during this period.

The current in the individual rotor conductors will look similar to the waveform shown in Figure 7.14. For one full turn of the rotor, the current waveform will be a highly nonsinusoidal ac. During the major portion of the cycle, the current will have a constant value. During the relatively short commutation interval, the current is reversed. Since the armature current i_a is split into two equal parallel paths (see Figure 7.13) the individual conductor currents will alternate between the peak values $\pm\frac{1}{2}i_a$.

7.3.4 The Motor Torque (T_m)

The electromechanical torque or motor torque, T_m is of fundamental importance in the study of motors. We proceed to derive an expression for this torque. We do so under the following general assumptions:

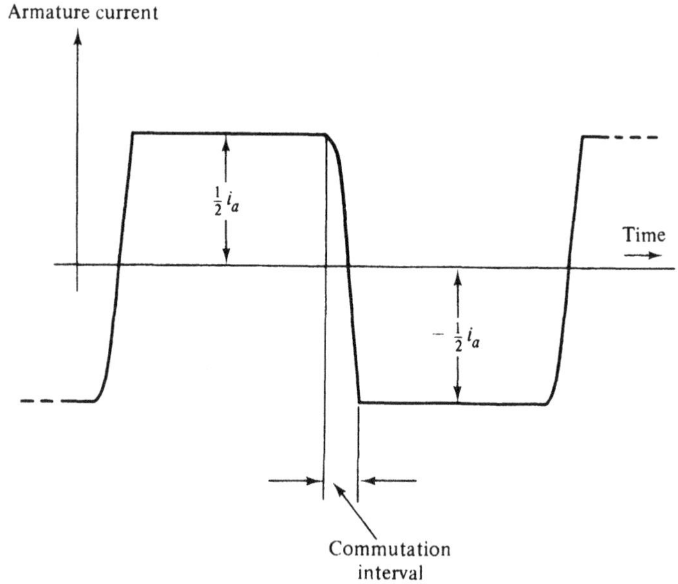

Figure 7.14

1. The armature winding consists of N conductors, each occupying, on the average, a rotor space $\pi D/N$ m, measured tangentially, as shown in Figure 7.15. (In the machine discussed previously, $N = 12$.)
2. The armature winding consists of a parallel paths. (The specific lap winding in Figure 7.13 was characterized by $a = 2$.)
3. The machine has p poles, where p is an even integer. (In the previous case, $p = 2$.)

Figure 7.15 shows the magnetic flux distribution and the rotor conductors under one adjoining pole pair of the machine. The individual forces on each conductor have been identified. The force f_ν on the individual conductor ν is given by

$$f_\nu = L\frac{i_a}{a}B_\nu \qquad [\text{N}]. \tag{7.14}$$

where B_ν is the magnetic flux density measured at the conductor in question, i_a/a is the conductor current, and L, the length of the conductor (the effective length that interacts with the magnetic field to produce the force).

The sum of the forces on all N/p conductors located under one pole will be equal to[4]

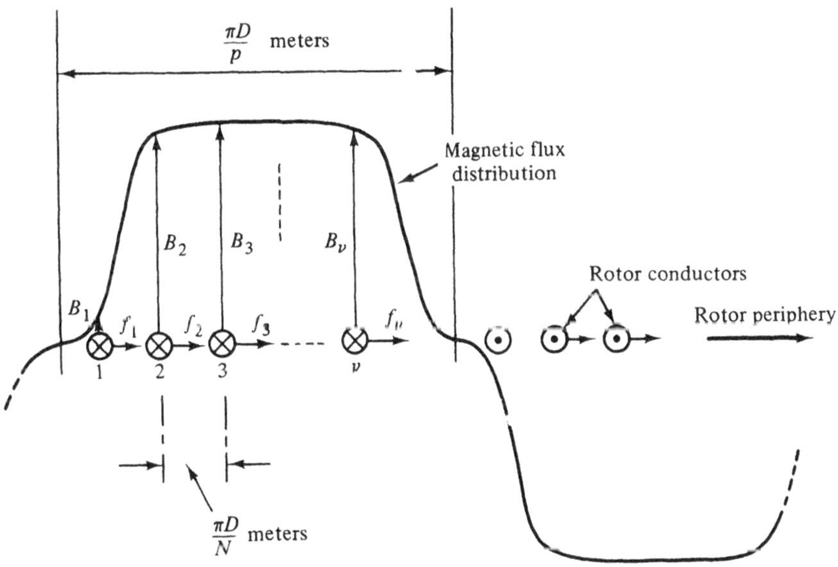

Figure 7.15

[4] Note that the expression for the force is slightly inaccurate because the conductors which are commutating do not carry the current, i_a/a. However, these conductors are located in the interpole regions where the flux density is very small. Therefore, the contribution which the commutating conductors make to the total force is negligible.

$$\sum_{\nu=1}^{N/p} f_\nu = L\frac{i_a}{a}\sum_{\nu=1}^{N/p} B_\nu \qquad [\text{N}]. \tag{7.15}$$

For the total motor force f_{m} acting on all armature conductors under the p poles we have

$$f_{\text{m}} = p\sum_{\nu=1}^{N/p} f_\nu = pL\frac{i_a}{a}\sum_{\nu=1}^{N/p} B_\nu \qquad [\text{N}]. \tag{7.16}$$

We had earlier concluded (see Figure 7.15) that each rotor conductor occupies on average a peripheral rotor space of $\pi D/N$ m. Through this rotor surface "window" passes a magnetic flux of magnitude:

$$\Phi_\nu = L\frac{\pi D}{N}B_\nu \qquad [\text{Wb}]. \tag{7.17}$$

The total pole flux, Φ_{p}, entering (or leaving) one pole will be equal to the sum

$$\Phi_{\text{p}} = L\frac{\pi D}{N}\sum_{\nu=1}^{N/p} B_\nu \qquad [\text{Wb}]. \tag{7.18}$$

Substitution of equation (7.18) into (7.16), gives

$$f_{\text{m}} = \frac{pN}{\pi a D}\Phi_{\text{p}} i_a \qquad [\text{N}]. \tag{7.19}$$

Finally, the motor torque T_{m} is obtained as the product of the force and the radius, that is,

$$T_{\text{m}} = f_{\text{m}}\frac{D}{2} = \frac{pN}{2\pi a}\Phi_{\text{p}} i_a = k_{\text{T}}\Phi_{\text{p}} i_a \qquad [\text{N}\cdot\text{m}], \tag{7.20}$$

where the constant,

$$k_{\text{T}} = \frac{pN}{2\pi a}, \tag{7.21}$$

is a machine design parameter.

7.3.5 The Induced Electromotive Force

As the rotor conductors move through the magnetic flux, emf's will be induced in each conductor. The magnitudes of these emf's follow from equation (7.4). If we assume that the rotor speed is constant (n rpm), then the tangential conductor speed s is likewise constant, and the instantaneous value of the emf will be proportional to the flux density, B. Consequently, as a conductor travels the total distance of one pole pair, the induced emf will complete a full ac cycle of the same waveform as the flux density in Figure 7.15. This waveform is highly nonsinusoidal. However, due to the action of the commutator, the emf, which appears between the brushes, will be a dc emf. Let us see why this must be so.

As we traced the path between the two brushes in Figure 7.13, we noted earlier that the 12 conductors constitute two parallel circuits, each containing 6 armature conductors. One such parallel circuit is shown schematically in Figure 7.16. At each instant the six emf's (identified in the figure) add up to the total emf, $e = V_{cf}$, which can be measured between the brushes c and f. (Note, however, that during the commutation interval some conductors are short-circuited. These "commutating conductors" are, however, located midway between the poles where the flux, and hence the emf, are negligible.)

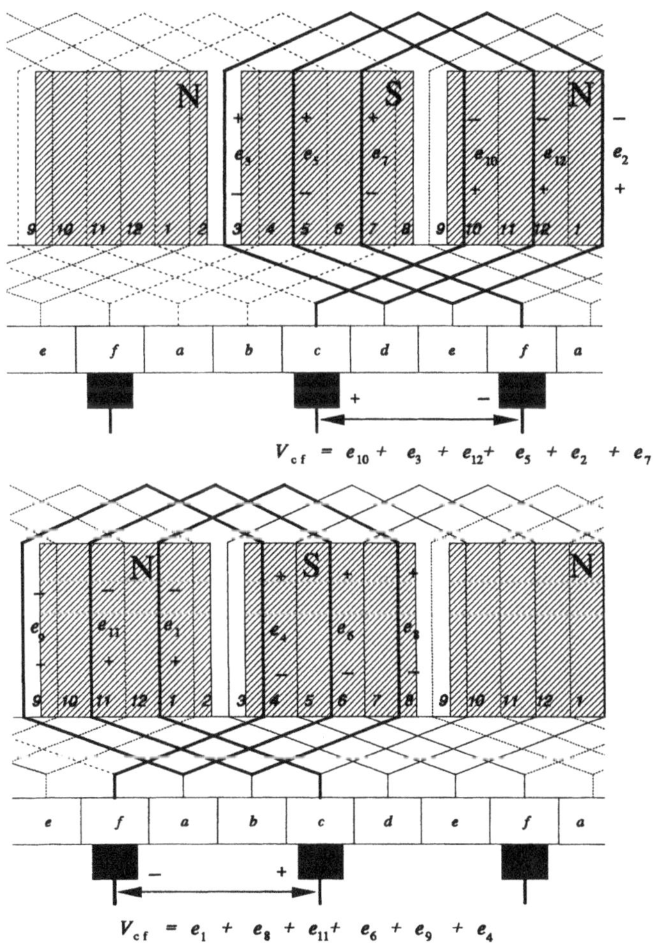

Figure 7.16

The total dc emf which is induced in the N/p conductors under one pole is the sum of the instantaneous[5] emf's existing in the conductors in question, and can be written as

$$\sum_{\nu=1}^{N/p} e_\nu = \sum_{\nu=1}^{N/p} sLB_\nu = sL \sum_{\nu=1}^{N/p} B_\nu \qquad \text{[V]}. \tag{7.22}$$

Since we have p poles and a parallel paths, the total emf appearing between the brushes will be

$$e = \frac{p}{a} \sum_{\nu=1}^{N/p} e_\nu = \frac{psL}{a} \sum_{\nu=1}^{N/p} B_\nu \qquad \text{[V]}. \tag{7.23}$$

In view of equation (7.18), e can be written in the form

$$e = \frac{pNs}{\pi Da} \Phi_\mathrm{p} \qquad \text{[V]}. \tag{7.24}$$

The tangential speed of the conductors, s is related to the angular speed of the motor, ω_m, by

$$s = \frac{D}{2} \omega_\mathrm{m} \qquad \text{[m/s]} \tag{7.25}$$

Substitution of this expression into equation (7.24) and making use of (7.21), we obtain the machine emf:

$$e = k_\mathrm{T} \omega_\mathrm{m} \Phi_\mathrm{p} \qquad \text{[V]}. \tag{7.26}$$

The emf e is not perfectly constant. It will contain a slight ripple, which is usually of negligible magnitude. This ripple is the result of the commutator action. Each commutator segment has a finite width and the brushes will therefore, in effect, pick off a finite "slice" of the armature coil emf. Because the coil emf varies with time, this "emf slice" will not be of constant magnitude. This is shown in Figure 7.17.

7.3.6 The Motor Power (p_m)

In view of equation (2.24) we can express the motor power p_m in terms of T_m and ω_m as

$$p_\mathrm{m} = T_\mathrm{m} \omega_\mathrm{m} \qquad \text{[W]}. \tag{7.27}$$

Making use of equation (7.20) we obtain

[5] This refers to the emf existing at the instant the brush makes contact with the commutator segment in question.

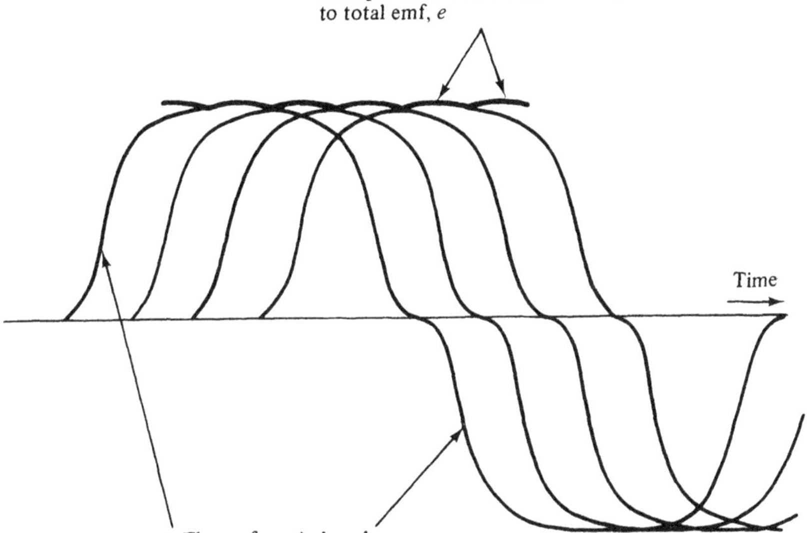

Figure 7.17

$$p_m = k_T \Phi_p i_a \omega_m \qquad [W]. \tag{7.28}$$

Using equation (7.26) we can also express the motor power in the alternate form:

$$p_m = e i_a \qquad [W]. \tag{7.29}$$

Note that equation (7.29) is identical to the linear dc motor equation (7.10).

7.3.7 The Equivalent Circuit of the Motor

Figure 7.18 shows the equivalent circuit of a dc motor. (cf. Figure 7.4.) The motor is fed from a voltage source supplying a terminal voltage V_a to the armature. The field coils are separately fed from a voltage source V_f. (This type of motor is often described as "separately excited." The field coils could, of course, be fed from the same source as the armature. We would then have a "shunt" motor. Shunt motors and separately excited motors have slightly different characteristics (see Section 7.4.4).) The total resistance of the armature winding, including the brush contact resistance is lumped into the *armature resistance*, R_a. The application of KVL around the armature loop gives

$$V_a - e = R_a i_a \qquad [V]. \tag{7.30}$$

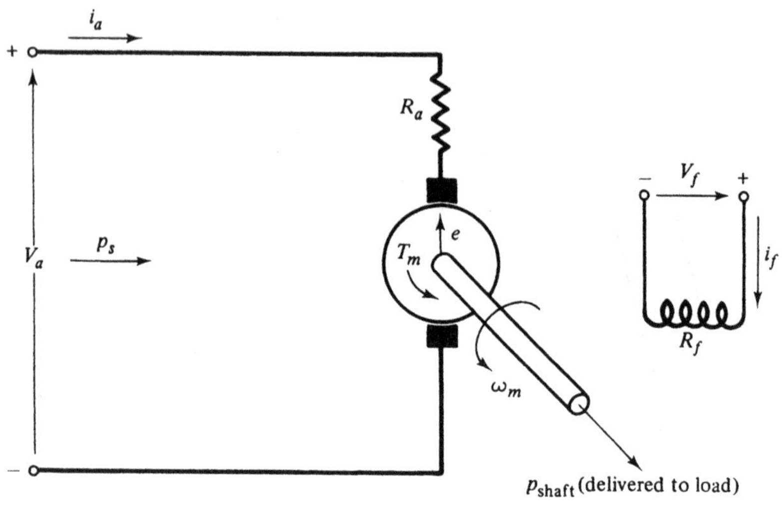

Figure 7.18

Example 7.2

A dc machine similar to the one shown in Figure 7.12 has the following parameters:

$$\text{flux per pole, } \Phi_{\text{p}} = 0.25 \text{ Wb;}$$
$$\text{armature current, } i_{\text{a}} = 25 \text{ A;}$$
$$\text{machine speed } n = 3000 \text{ rpm;}$$
$$\text{armature resistance, } R_{\text{a}} = 0.41 \ \Omega.$$

Calculate, the torque developed, the emf generated, and the power developed by the machine and the source voltage, V_{a} required to supply the 25-A armature current.

Solution: From equation (7.21)

$$k_{\text{T}} = \frac{2 \cdot 12}{2 \cdot \pi \cdot 2} = \frac{6}{\pi} = 1.910. \tag{7.2.1}$$

From equation (7.20) we then get

$$T_{\text{m}} = \frac{6}{\pi} \cdot 0.25 \cdot 25 = 11.94 \qquad [\text{N} \cdot \text{m}]. \tag{7.2.2}$$

We now compute ω_{m}:

$$\omega_{\text{m}} = \frac{n}{60} 2\pi = \frac{3000}{60} 2\pi = 314.2 \qquad [\text{rad/s}]. \tag{7.2.3}$$

From equation (7.26) we get

$$e = 1.910 \cdot 314.2 \cdot 0.25 = 150 \qquad [\text{V}]. \tag{7.2.4}$$

From equation (7.29) we get

$$p_{\text{m}} = 150 \cdot 25 = 3.75 \qquad [\text{kW}] \tag{7.2.5}$$

or

$$p_{\text{m}} = T_{\text{m}}\omega_{\text{m}} = 11.94 \times 314.2 = 3.75 \qquad [\text{kW}]. \tag{7.2.6}$$

Note that this is not the power delivered by the motor shaft. The shaft power is somewhat smaller due to rotational losses in the motor (see discussion in Section 7.3.8).

From equation (7.30) the required voltage is

$$V_{\text{a}} = 150 + 0.41 \cdot 25 = 160.25 \qquad [\text{V}]. \tag{7.2.7}$$

7.3.8 Additional Losses

Assume that the motor has reached a steady state, at a constant speed, ω_{m}. Its kinetic energy has then assumed a constant value. It follows that the power delivered by the source $V_{\text{a}} i_{\text{a}}$ is used to supply the power required to pull the load and to feed the ohmic losses in the armature.

7.3.8.1 Rotational Losses

There are additional losses that are not included in the above analysis. Consider first the load torque, T_{L}. A small portion of this torque consists of windage, brush and bearing friction torques. In addition, as the motor spins the rotor flux, as we noted earlier, will be of the ac type. Therefore there will be eddy current and hysteresis, that is, *core losses*, in the rotor. These must be supplied by the voltage source, V_{a}. The sum of windage, friction, and magnetic core losses are referred to as *rotational losses*, p_{rot}. The useful load power, also called *shaft power*, p_{shaft}, is obtained by deducting the rotational power loss from the motor power:

$$p_{\text{shaft}} = p_{\text{m}} - p_{\text{rot}} \qquad [\text{W}]. \tag{7.31}$$

7.3.8.2 Field Losses

The magnetic field coil (see Figure 7.18) in steady state consumes the power,

$$p_{\text{field}} = V_{\text{f}} i_{\text{f}} = i_{\text{f}}^2 R_{\text{f}} \qquad [\text{W}]. \tag{7.32}$$

Since this power does not reach the load, it must be considered a loss.

7.3.8.3 Stray Losses

A component of the lost power which is very hard to determine by either analysis or measurement, is the so-called stray loss, p_{stray}. It is caused by a nonuniform

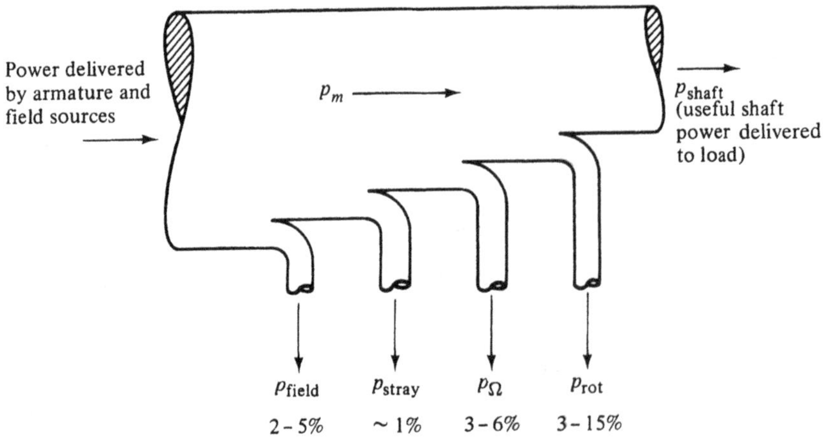

Figure 7.19

current distribution in the windings and a nonuniform flux density in the stator teeth (that is, in the spaces between the slots). It is usually estimated to be about one percent of the output power of the motor. This loss in effect, reduces the motor power, p_m. Figure 7.19 shows in schematic form the power flow within a dc motor. The figure also gives typical percentage ranges for the losses.

Example 7.3

Consider the dc motor discussed in Example 7.2. In order to measure the rotational losses of the motor, it is run at its rated speed (n = 3000 rpm) and hence unchanged emf (150 V) but without a mechanical load (no-load condition). The motor draws 2.21 A from the source.

Find the rotational losses and also the motor efficiency when operated at full-rated current (25 A) and full speed (3000 rpm). When operated under these conditions the field coil consumes 173 W and the stray loss is estimated to be 40 W.

Solution: Since the motor is running under no-load conditions, the shaft power, p_{shaft} is zero. The power taken from the voltage source V_a[6] must be equal to p_{rot}. We therefore have:

$$p_{rot} = V_a i_a = 150 \cdot 2.21 = 332 \qquad [W]. \tag{7.3.1}$$

[6] We know the p_{rot} varies with speed and rotor flux. If we wish the test to give the correct value of the loss, we must make sure that we measure it (as we did) at the proper speed and emf values. Note also that the ohmic armature loss during the no-load tests amounts to only:

$$p_\Omega = i_a^2 R_a = 2.21^2 \cdot 0.41 = 2\ W$$

Therefore, for all practical purposes, the power consumed during the no-load test goes into the rotational losses.

The ohmic loss at rated current is

$$p_{\Omega} = 25^2 \cdot 0.41 = 256 \qquad [\text{W}]. \tag{7.3.2}$$

In Example 7.2 we had computed the motor power at rated armature current to be 3750 W. We now adjust this value for stray losses:

$$p_{\text{m}} = 3750 - 40 = 3710 \qquad [\text{W}]. \tag{7.3.3}$$

From equation (7.31) we have for the shaft (or output) power,

$$p_{\text{shaft}} = 3710 - 332 = 3378 \qquad [\text{W}]. \tag{7.3.4}$$

The motor efficiency when operated at rated current and speed is

$$\begin{aligned} \eta &= \frac{p_{\text{shaft}}}{p_{\text{shaft}} + \text{total losses}}, \\ &= \frac{3378}{3378 + 332 + 256 + 173 + 40}, \\ &= 0.808. \end{aligned} \tag{7.3.5}$$

7.4 Operating Characteristics of the Direct Current Machine

We have now derived the most important expressions for motor torque, power, and emf. A more detailed analysis of the dc machine would include the study of phenomena such as:

1. Commutation voltages and the tendency for "sparkover" at the commutator. In an actual dc machine these voltages must be compensated for by means of *compensation windings.*
2. The *armature reaction*, a term that describes the magnetization effect of the armature current. The armature current represents an mmf that will be superimposed upon the mmf of the field winding, and will have a second order effect on both the emf and the torque. (It is also part of the cause of the stray losses discussed in the previous section.)
3. Nonlinear effects due to magnetic saturation.

These effects, although of great importance for the proper operation of the motor, are not of overriding significance to the motor user who wishes to learn about the basic operating features of the dc machine. They will therefore not be discussed further. Instead, we shall now proceed to explain the operating characteristics of the dc machine based upon the expressions derived earlier. We begin with the separately excited machine—the machine for which the pole flux is obtained from a separate dc source.

7.4.1 Starting the Direct Current Motor

At standstill, the dc motor generates zero emf. If we were to apply the full source voltage, V_a, to the armature winding, the current would be limited only by the armature resistance, R_a. The result would be a current that could damage the machine. The torque would likewise be very large and the corresponding sharp acceleration might damage the (mechanical) load. Actually, one of the very attractive features of the dc motor is its high starting torque (compared to the internal combustion engine). As the machine picks up speed, a back-emf, e, will be generated and the current will decrease.

Obviously, we need to control the magnitude of the starting current. The simplest way of doing this is to use a bank of starting resistors as shown in Figure 7.20. Initially, all three sections of the three-step starting resistor are in series with the armature. As the motor picks up speed, the sections are successively shorted out. As a result, the starting current will have a "sawtooth" shape as shown. Of course, the lower current obtained by the insertion of the starting resistor results in a longer startup period.

7.4.2 The Separately Excited Direct Current Machine Operated as a Generator

For the separately excited machine, the pole flux is constant if the field current, i_f (see Figure 7.18), is kept fixed. According to equation (7.26) the back-emf e will be proportional to the rotor speed.

Now assume that we reverse the "polarity" of the shaft power p_{shaft}. This requires that the mechanical load be replaced by a "prime mover," which drives the dc machine in the same direction as before. The prime-mover torque will tend to accelerate the rotor to a speed greater than it was before. Assume that the torque is of sufficient magnitude to speed up the rotor to the point where the emf e exceeds the source voltage V_a. The current i_a will now reverse direction, and equation (7.30) will change to

$$e - V_a = R_a i_a \qquad [\mathrm{V}]. \tag{7.33}$$

Since the current is fed into the voltage source at its positive terminal, the machine evidently delivers energy to the source—it operates as a generator. The power flow in the machine is shown in Figure 7.21. In comparing Figures 7.18 and 7.21, note that the current, torque, and power have reversed polarity, but the emf and speed have not.

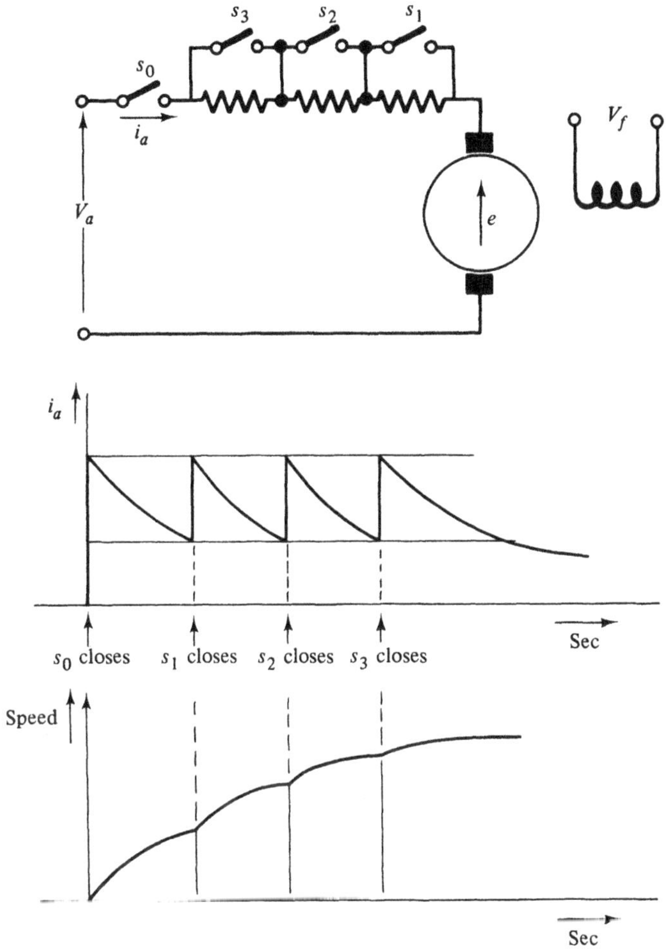

Figure 7.20

Example 7.4

The dc machine in Example 7.2 was fed from a dc source of voltage $V_a = 160.3$ V. The source voltages V_f and V_a are kept constant. The load is replaced by a prime mover that drives the rotor to $n - 3400$ rpm.

Find the emf, the armature current, and the power delivered by the machine.

Solution: With the motor running at 3000 rpm, the emf was 150 V (Example 7.2). At 3400 rpm the emf will be

$$e = \frac{3400}{3000} 150 = 170 \qquad [\text{V}]. \tag{7.4.1}$$

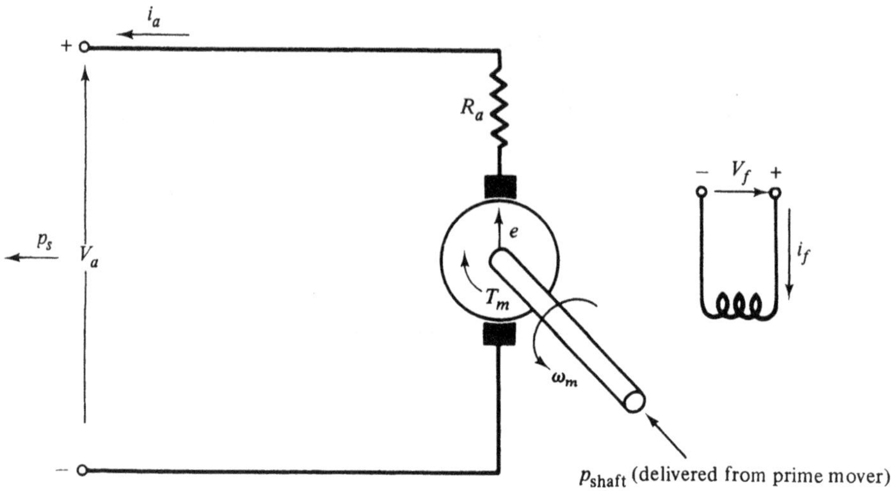

Figure 7.21

From equation (7.33) the current[7]

$$i_a = \frac{170 - 160.3}{0.41} = 23.7 \qquad [A]$$

(flowing into the voltage source, V_a at its positive terminal).

Finally, the power delivered to the voltage source is

$$p_s = 160.3 \cdot 23.7 = 3799 \qquad [W]. \tag{7.4.3}$$

The ability of a dc machine to change from a motor to a generator smoothly, and thus change the polarity of its motor torque T_m, equally smoothly is often described as *dynamic braking*. For example, decelerating trains and descending elevators require negative or braking torques. Instead of using mechanical brakes, which would dissipate the energy in heat, the dc motor may be operated during such periods as a generator, thus producing the negative torque needed and delivering the energy back into the source. In all such cases, the required current reversal is accomplished by field-current control of the dc machine. In other words, one accomplishes the required increase in the emf by an increase in the field current.

7.4.3 The Torque–Speed Characteristics of the Separately Excited Direct Current Machine

The torque–speed characteristics of the separately excited dc machine are very similar to the force–speed characteristics of the linear motor given by equation

[7] We have tacitly assumed that the source voltage V_a is "stiff," that is, it will not vary as the current changes polarity.

(7.12) and shown in Figure 7.3. Assuming a "ideal" motor, when it is running under no-load conditions, the speed will increase to its no-load value, at which speed the emf e is equal to the source voltage V_a. The motor torque and armature current are now both zero.[8] As the motor is loaded, the speed will decrease, resulting in a lower emf and hence a higher armature current. When the motor torque balances the load torque, the rotor speed settles down to its "operating" value (ω_m). We can obtain the speed-torque relationship by combining equations (7.20), (7.26), and (7.30) into the following:

$$\omega_m = \frac{V_a}{k_T \Phi_p} - \frac{R_a}{k_T^2 \Phi_p^2} T_m \qquad \text{[rad/s]}. \tag{7.34}$$

We introduce the no-load speed,

$$\omega_0 \equiv \frac{V_a}{k_T \Phi_p} \qquad \text{[rad/s]}, \tag{7.35}$$

and can now write equation (7.34) as follows:

$$\omega_m = \omega_0 - \frac{R_a}{k_T^2 \Phi_p^2} T_m \qquad \text{[rad/s]}. \tag{7.36}$$

Note the similarity of equations (7.12) and (7.36).

On a graph of ω_m against T_m (7.36) is the equation of a straight line with a negative slope. We have drawn several of these lines in Figure 7.22, each corresponding

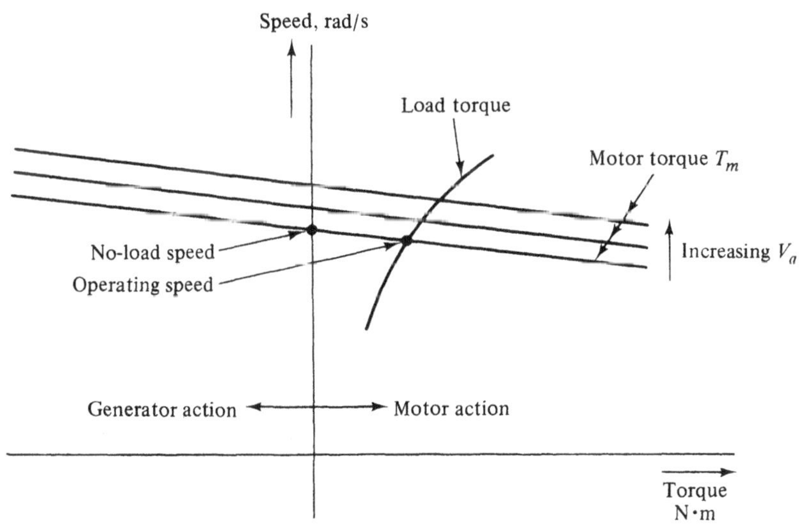

Figure 7.22

[8] In reality, as demonstrated in Example 7.3, both are nonzero but small because a small torque is needed to supply the no-load losses.

to a different value of V_a. Note that increased V_a results in an upward parallel shift of the torque–speed curve. If we extend these lines into the negative torque region we obtain the torque–speed characteristics of the machine when it is operated as a generator.

Also shown in Figure 7.22 is a *load–torque* curve. This line represents the torque characteristics of the load (including the rotational losses of the motor itself). The point of intersection of the load curve and the torque–speed curve of the motor indicates balance between the driving and load torques. This point gives the operating speed, and the torque generated for the corresponding applied voltage. Compare Figures 7.22 and 7.3.

Example 7.5

1. Find the no-load speed of the motor in Example 7.4.
2. Calculate the no-load speed when the field current is increased by 10%.

Solution:

1. We had earlier computed the following:

$$V_a = 160.25 \text{ V}$$

$$k_T = 1.910$$

$$\Phi_p = 0.25 \text{ Wb}$$

From equation (7.35) we get

$$\omega_0 = \frac{160.25}{1.910 \cdot 0.25} = 335.6 \qquad \text{[rad/s]}, \tag{7.5.1}$$

which corresponds to a no-load speed of $n_0 = 3205$ rpm.

In words, if the motor is running at its rated speed of 3000 rpm and it drops its load, it will accelerate to 3205 rpm, that is, a 6.83% increase in speed.

2. Let us assume that the pole flux, Φ_p is proportional to the field current i_f. (In reality this is not quite correct, as a more accurate analysis must consider the effects of magnetic saturation in the flux paths of the machine.)

A 10% increase in i_f will mean a 10% increase in Φ_p. The no-load speed (computed earlier) of 3205 rpm will then, according to (7.35), decrease to

$$n_0 = \frac{3205}{1.1} = 2914 \qquad \text{[rpm]}. \tag{7.5.2}$$

An increase in the armature voltage will increase the speed but an increase in the field current will decrease the speed [see equations (7.12) and (7.34)].

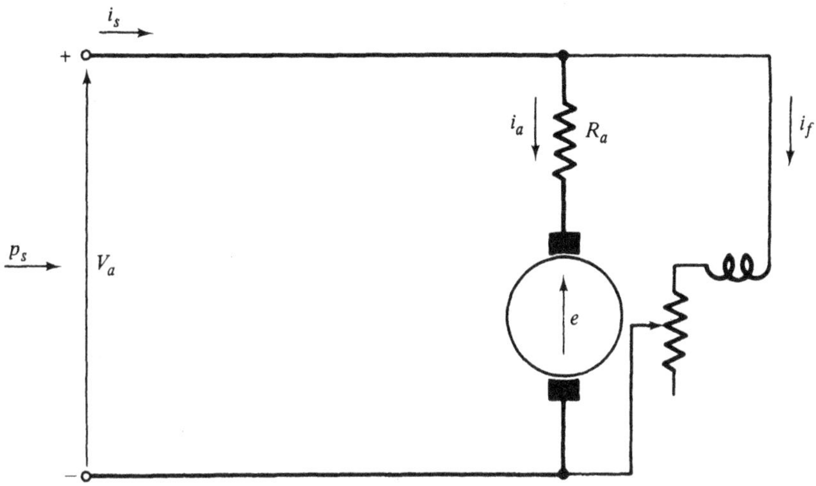

Figure 7.23

7.4.4 The Torque-Speed Characteristics of the Shunt Direct Current Motor

In a shunt-connected dc motor the field coil is fed from the same source voltage, V_a as the armature, usually in series with a rheostat, as shown in Figure 7.23. The rheostat is used to vary the field current i_f and hence the pole flux Φ_p and the emf e.

For a shunt motor, the source current is i_s, the sum of the armature and field currents:

$$i_s = i_a + i_f \qquad [A]. \tag{7.37}$$

Usually i_f is only a few percent of i_a.

With a fixed setting of the field rheostat we have

$$i_f = \frac{V_a}{R_f} \qquad [A], \tag{7.38}$$

where R_f is total field circuit resistance. If we make the assumption of linearity between the flux and the field current, then we have linearity between the pole flux, Φ_p and the supply voltage V_a, that is,

$$\Phi_p = k_1 V_a \qquad [Wb], \tag{7.39}$$

where k_1 is a parameter that varies with the rheostat setting.

Substituting equation (7.39) into (7.34) yields the following torque–speed relation:

$$\omega_m = \frac{1}{k_1 k_T} - \frac{R_a}{k_T^2 k_1^2 V_a^2} T_m \qquad [rad/s]. \tag{7.40}$$

We note the following dissimilarities between the shunt-connected and the separately excited motors:

1. The no-load speed, $\omega_0 = 1/k_1 k_T$, is a constant independent of the voltage V_a, but dependent upon the setting of the rheostat.
2. The slope of the torque speed curves decreases with increasing values of V_a.

Figure 7.24 shows a family of torque-speed curves for two different settings of the rheostat and three different values of V_a.

The magnitude of the motor torque T_m is proportional to the armature current i_a—from equation (7.20). As the current is a measure of the degree of loading on the motor, it is possible to plot the speed versus "percentage torque" as has been done in Figure 7.24. The "full-load torque" corresponds to the rated armature current.

7.4.5 Speed Control of a Direct Current Motor

The graphs in Figures 7.22 and 7.24 clearly indicate that for the dc motor, the change in speed is quite small for fairly large changes in torque. Typically, the speed will drop between 5% and 10% from zero to full-load torque for both the shunt-connected and the separately excited motors. (A series-excited motor behaves quite differently, as is shown in Section 7.4.7.) The dc motor is, how-

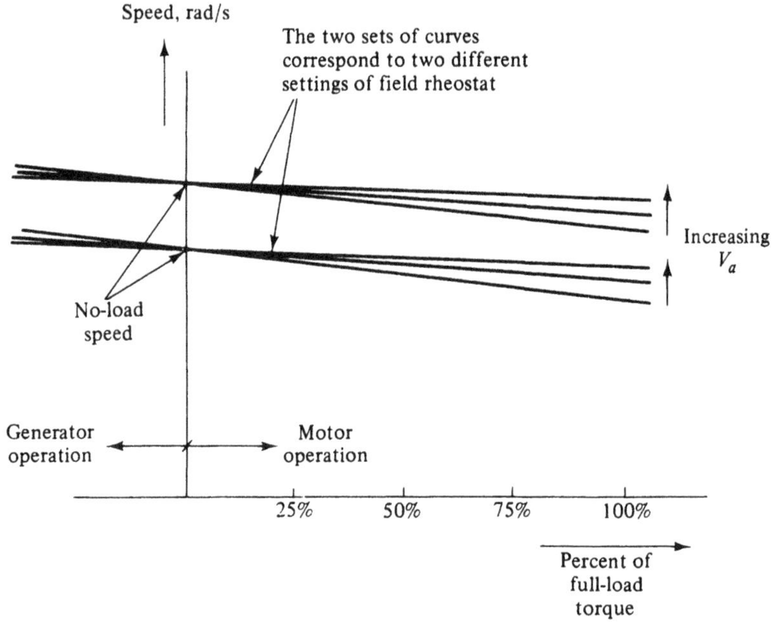

Figure 7.24

ever, not to be compared to the synchronous motor in terms of speed constancy. We remember from Chapter 4 that the latter will experience zero speed change as a result of changes in its load.

In many applications—for example, vehicular drive systems—a very important requirement is easy speed control. No electric motor surpasses the dc motor in this regard. There are two basic ways in which the speed of the dc motor can be controlled. Both of them are best understood from expression (7.35) for the no-load speed. We treat them separately.

7.4.5.1 Speed Control by Variation of Applied Voltage (V_a)

From equation (7.35) it is clear that the no-load speed is proportional to the applied voltage, V_a. A change in this voltage therefore results in a proportional parallel shift in the torque-speed curves as we already indicated in Figure 7.22. We offer the following comments on using the applied voltage to control the motor speed:

1. Although it works well for the separately excited machine, it does not work in the case of a shunt motor. In a shunt motor, the field current, and hence the pole flux, Φ_p is proportional to V_a [equation (7.39)]. Consequently, the ratio V_a/Φ_p will be unaffected by the change of the voltage. (Note that the no-load speed, $\omega_0 = 1/k_1 k_T$, is independent of V_a.)
2. By varying the armature voltage V_a throughout the range $\pm 100\%$, the speed will vary from full-forward to full-reverse (Figure 7.25a). Note that when V_a is zero, the speed is zero, a fact that permits smooth speed reversal. This type of speed control is very useful for many industrial and transportation applications where speed reversal is required.
3. This type of control (in contrast to the alternate method discussed in Section 7.4.5.2 below) will not reduce the pole flux Φ_p and hence the motor torque [equation (7.20)].

7.4.5.2 Speed Control by Variation of Field Resistance

The pole flux is approximately proportional to the field current i_f. From equation (7.35) it is clear that the no-load speed is inversely proportional to Φ_p. It follows that the speed can be conveniently controlled by variation of the field rheostat.

Because Φ_p appears in the denominator of equation (7.35), the graph of speed plotted against field current is the rectangular hyperbola shown in Figure 7.25b. We make the following observations about using the field current to control the dc machine:

1. It works for both the separately excited and the shunt motor.
2. Higher speeds are attained by a lowering the value of Φ_p. This reduces the magnitude of the torque [see equation (7.20)].

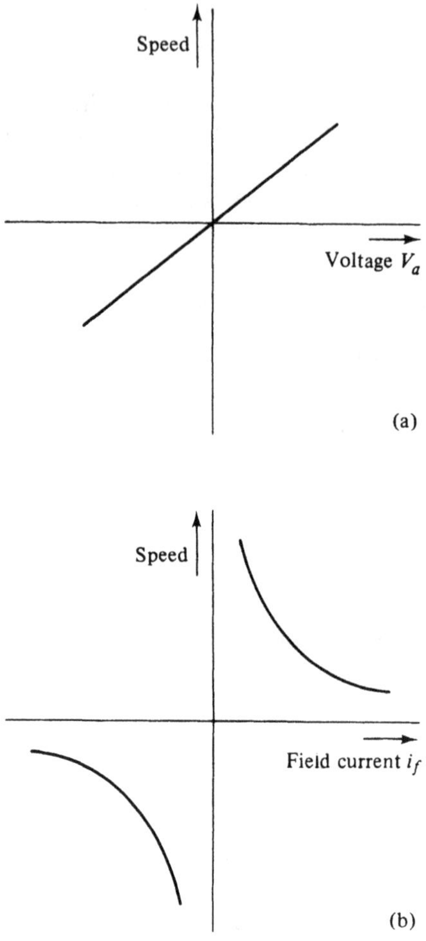

Figure 7.25

3. Equation (7.35) indicates that the speed approaches infinity when Φ_p goes to zero. Therefore care must be exercised not to run the motor without field current. (A classic danger is to "open-circuit" the field circuit accidentally when the motor is running; serious damage due to excessive speed is likely.)
4. Figure 7.25b shows clearly that there is no smooth way of going from positive to negative speeds by using field resistance variation—one must stop the motor, disconnect it from the source, and reverse the field current in order to reverse the speed. This control method is therefore not convenient when speed reversal is required.

Example 7.6

A separately excited dc motor is operated from a supply voltage of 300 V. Its no-load speed is 1200 rpm. When fully loaded, it delivers a motor torque of 400 N · m and the speed drops to 1100 rpm. Find the full-load motor torque, the power developed, and the speed if the supply voltage is changed to 600 V. The field excitation is assumed unchanged.

Solution: According to equation (7.35) the no-load speed will increase to 2400 rpm when the voltage changes from 300 to 600 V. Because Φ_p is unchanged the rated current[9] will give rise to the same torque (i.e., 400 N · m) at the higher speed.

At full-load, (at the lower speed, 1100 rpm) the motor power p_m is

$$p_m = \omega_m T_m = \frac{1100}{60} 2\pi \cdot 400 = 46.08 \qquad [\text{kW}]. \tag{7.6.1}$$

According to Figure 7.22, at the higher speed and full-load torque the speed will drop from 2400 to 2300 rpm. The motor power will therefore increase to

$$p_m = \frac{2300}{1100} 46.08 = 96.35 \qquad [\text{kW}]. \tag{7.6.2}$$

Example 7.7

Consider the motor in the Example 7.6. We now wish to increase the speed from 1200 to 2400 rpm by keeping the armature voltage, V_a constant at 300 V but decreasing the pole flux to half its original value. Find the full-load torque and power at the higher speed.

Solution: According to equation (7.20), the full-load torque will now be reduced to one-half the original value, that is, 200 N · m. (Note that we are not permitted to compensate for the reduced Φ_p by a higher armature current, i_a. This would cause excessive ohmic power loss in the armature.)

With the speed doubled but the torque at half the original value, the motor power will remain unchanged at 46 kW. (Compare this to the 96 kW obtained in the previous example.)

7.4.6 The Shunt Generator Feeding a Resistive Load

When a shunt machine is operated as a generator which feeds energy into the armature supply voltage source, V_a, its torque-speed characteristics are obtained from Figure 7.24 by extending them into the negative torque region. The situation

[9] Because of better cooling at the higher speed, the motor can actually accept even larger armature current without excessive heating.

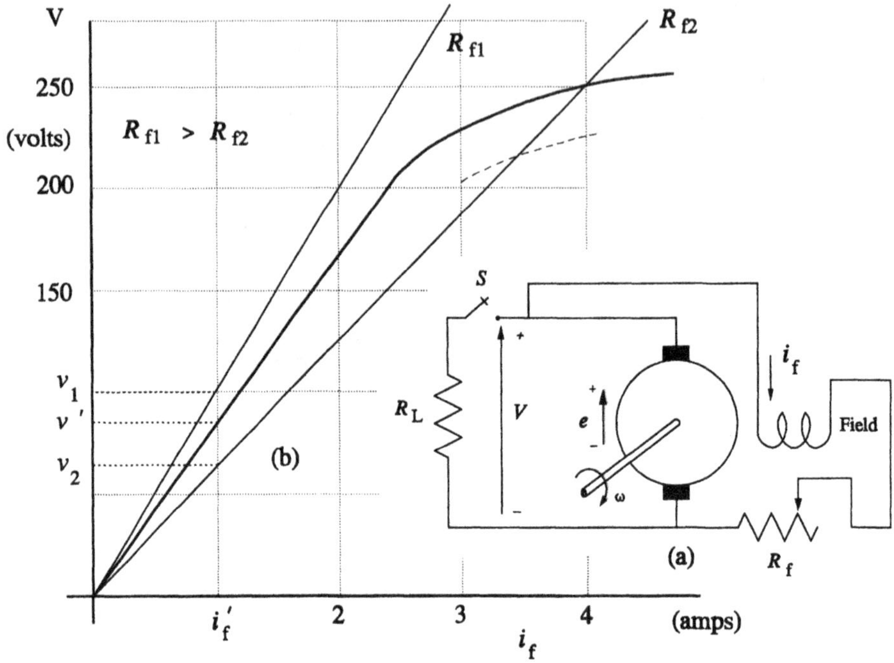

Figure 7.26

would be quite different if we feed the energy in to a "passive" load resistance R_L as shown in Figure 7.26a. The purpose of a dc generator is to generated dc power but to do this we have to supply a dc current to the field to create the flux. Is it possible to make the machine supply its own field excitation current?

Consider the graph of the generated emf, e plotted against i_f in Figure 7.26b.[10] We note that because of magnetic saturation the emf graph is linear only in its lower range. We also note that the graph is valid for a constant speed; a different constant speed will produce a similar but different graph [see equations (7.4) and (7.26)].

On the graph (Figure 7.26b) where a voltage is plotted against a current, a constant resistance is represented by a straight line through the origin with a slope proportional to the value of the resistance. Two possible values of resistance R_{f1} and R_{f2} are shown. Note that $R_{f1} > R_{f2}$.

Suppose that the field resistance of the machine is R_{f1} and it is connected across the armature as shown in Figure 7.26a, then, for the field current i'_f to flow, the machine must produce a terminal voltage v' but as the field resistance is R_{f1}, it

[10] The emf is proportional to the rotor speed. Figure 7.26b therefore refers to one particular speed—for example, the rated speed of the machine.

requires a voltage v_1 to supply the current i'_f to the field. Since the voltage $v_1 > v'$ the machine cannot supply its own excitation current.

If we now reduce the field resistance to R_{f2}. The voltage required to drive the field current i'_f is v_2 and since the emf generated is v' and $v' > v_2$, it follows that the machine can provide its own field excitation current.

In the discussion of the separately excited machine, the field current i_f was supplied from an external source. When the external source was disconnected, there must have been a finite residual flux in the magnetic path of the machine so that with i_f equal to zero, the machine will produce a small but finite emf corresponding to point a in Figure 7.27. When the machine is reconnected as a generator as shown in Figure 7.26a, this emf will produce in R_{f2} a current corresponding to point b, which in turn will cause the machine to produce an emf corresponding to point c, and so on. At point q, the system will reach a steady state and this is the voltage that the machine will produce at that speed. If the machine were driven at a different speed, the emf curve will change proportionately.

Example 7.8

Consider the generator in Figure 7.26 with R_f set at 63.5 Ω. What will be the terminal voltage V_a before and after closing the switch S to the load resistance $R_L = 5.0$ Ω? It is assumed that the prime mover maintains a constant speed. The armature resistance R_a of the generator is 0.5 Ω.

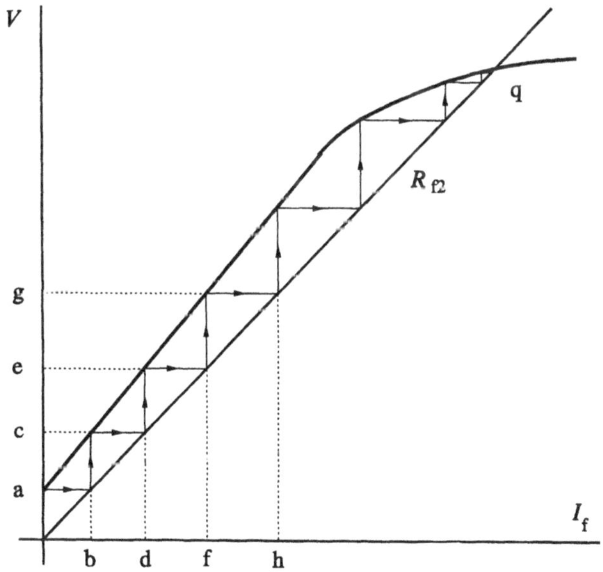

Figure 7.27

Solution:

Switch S is open: We have $i_a = i_f$. If we neglect the voltage drop $R_a i_a$, voltage equilibrium would occur according to the graph at $e = V_a = 250$ V and $i_f = 4.0$ A. The voltage drop $R_a i_a$ will be about 2.0 V so if we compensate for it the correct answer would be

$$V_a \approx 248 \qquad [\text{V}]. \tag{7.8.1}$$

Switch S is closed: By making the very rough and unjustified assumption that the insertion of the 5-Ω load will not affect the terminal voltage, we get for the load current:

$$i_L \approx \frac{248}{5.0} = 49.60 \qquad [\text{A}]. \tag{7.8.2}$$

The armature current would therefore be

$$i_a = 49.60 + 4.0 = 53.60 \qquad [\text{A}]. \tag{7.8.3}$$

The armature voltage drop can then be computed as

$$R_a i_a = 0.5 \cdot 53.60 = 26.8 \qquad [\text{V}]. \tag{7.8.4}$$

This voltage drop is too large to be neglected, so we deduct it from the emf curve as shown by the dashed line in Figure 7.26b. The intersection of this dashed line and the 63.5-Ω resistance line corresponds to $i_f = 3.4$ [A], and $V_a = 212$ [V]. With this corrected value for the terminal voltage we recompute a corrected value for the load current:

$$i_L = \frac{212}{5.0} = 42.40 \qquad [\text{A}]. \tag{7.8.5}$$

The corrected value for the armature current is

$$i_a = 42.40 + 3.4 = 45.8 \qquad [\text{A}], \tag{7.8.6}$$

and the corrected value for the armature voltage drop becomes

$$i_a R_a = 45.80 \cdot 0.5 = 22.9 \qquad [\text{V}]. \tag{7.8.7}$$

We can now go back and readjust all values once more, following the above procedure. In summary, the adjusted current and voltage values are

$$\begin{aligned} i_f &= 3.5 && [\text{A}], \\ V_a &= 220 && [\text{V}], \\ i_L &= 44.0 && [\text{A}], \\ i_a &= 47.5 && [\text{A}], \\ e &= 244 && [\text{V}]. \end{aligned} \tag{7.8.8}$$

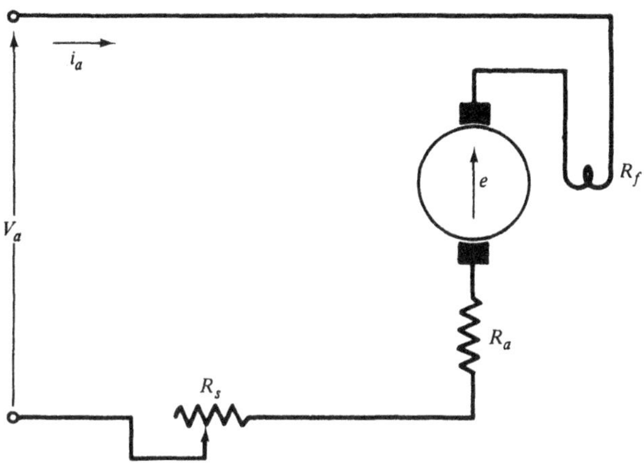

Figure 7.28

Power delivered,

$$p_{\text{out}} = V_{\text{a}} i_{\text{L}} = 10.45 \qquad [\text{kW}]. \tag{7.8.9}$$

7.4.7 The Series-Excited Direct Current Machine

Very special torque-speed characteristics are obtained if the armature of the dc motor is connected in series with its field winding as shown in Figure 7.28. We have also included a series control rheostat, R_{s}.

The pole flux, Φ_{p} is now controlled by the armature current rather than by the armature voltage as is the case in a shunt motor. Since the magnitudes of both the emf and torque are directly proportional to Φ_{p}, both can be expected to vary substantially with armature current. If we disregard magnetic saturation, we can assume the pole flux to be proportional to the current that produces it, that is, i_{a}, and we can therefore write:

$$\Phi_{\text{p}} = k_{\text{I}} i_{\text{a}} \qquad [\text{Wb}], \tag{7.41}$$

where k_{I}, is a machine constant.

Using equations (7.20) and (7.26) we then obtain for the motor torque and emf:

$$T_{\text{m}} = k_{\text{T}} k_{\text{I}} i_{\text{a}}^2 \qquad [\text{N} \cdot \text{m}], \tag{7.42}$$

$$e = k_{\text{T}} k_{\text{I}} \omega_{\text{m}} i_{\text{a}} \qquad [\text{V}]. \tag{7.43}$$

If the field winding has the resistance,[11] R_{f} the armature voltage equation (from KVL) reads:

[11] A shunt-connected field winding consists of many turns of light wire—its resistance is high and it carries a current, i_{f}. A series-connected winding consists of a few turns of heavy wire—its resistance is low. The armature current, i_{a} ($i_{\text{a}} \gg i_{\text{f}}$) must flow through the series winding and produce a comparable value of Φ_{p}.

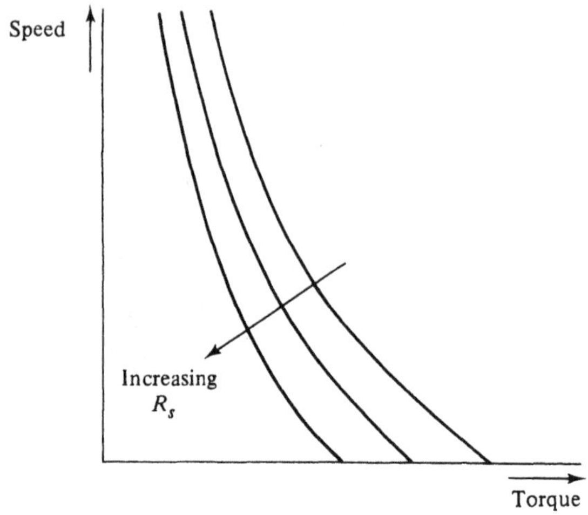

Figure 7.29

$$V_a = e + i_a(R_a + R_f + R_s) \qquad [\text{V}]. \tag{7.44}$$

Elimination of e between equations (7.43) and (7.44) gives

$$i_a = \frac{V_a}{(R_a + R_f + R_s + k_T k_I \omega_m)^2} \qquad [\text{A}]. \tag{7.45}$$

Finally, substituting this expression in (7.42) we get for the motor torque,

$$T_m = k_I k_T \frac{V_a^2}{(R_a + R_f + R_s + k_T k_I \omega_m)^2} \qquad [\text{N} \cdot \text{m}]. \tag{7.46}$$

We can plot the torque–speed curves as shown in Figure 7.29. The most prominent features of the torque–speed characteristics of the series motor are as follows:

1. For zero load the motor has a tendency to "run away." (Compare this to the other types of motors discussed earlier where the no-load speed is just a few percent above the full-load speed.)
2. The speed drops sharply with increasing torque. This means that a sharp increase in the load torque results in a sharp drop of the speed and only a modest change in the mechanical power produced. (A shunt motor which is "speed stiff" would respond with an equally sharp power increase.) The speed of the series motor may drop all the way to zero (stall) and the motor may not be damaged (if R_s is not too small).
3. Good starting torque.

We conclude that the series motor will "cushion" the power source against power peaks during severe torque overloads. It can also withstand severe starting duties. For these reasons the series motor is used extensively in hoists and cranes, and in traction applications. Its most common use is as a starter motor for automobile engines. In this application it is kind and gentle to the battery!

7.4.8 The "Universal" Motor

The torque of the series-motor will be positive for both positive and negative currents [equation (7.42)]. (The reason is, of course, that reversal of the current also reverses the flux, thus preserving the direction of the torque.) The series dc motor will preserve the direction of the torque if the source voltage changes polarity—it will operate on ac.[12] Because of the unique ability of the series dc motor to run on both ac and dc, it is referred to as a *universal* motor. When it is intended for ac operation, however, it must have both its rotor and stator laminated, to avoid excessive core losses.

When operating on ac the motor torque will pulsate as a function of a *rectified sinusoid* at twice the supply frequency. (Why?) The average value of the torque is used to specify the torque-speed characteristics as shown in Figure 7.29.

Universal motors can be designed for very high-speed operation, often in the range 5000 to 15,000 rpm. We know from earlier discussions that high-speed motors have a high power-to-weight ratio. This type of motor is therefore often found in portable equipment like vacuum cleaners and electric hand drills.

7.5 Direct Current Power Supply Systems

Essentially all electric power is distributed to the consumer in the form of ac. Before the power can be utilized for driving dc motors (or any other device requiring dc) it must be transformed—*rectified*. In this section, we shall briefly discuss devices and circuits that can be used to transform ac to dc.

There are many ways in which the transformation (or the opposite, *inversion* from dc to ac) can be made. One of the simplest systems consists of an ac motor driving a dc generator. This type of arrangement is finding increasingly less use, as it is comparatively expensive, it involves rotating equipment, and it is characterized by a relatively poor efficiency (the loss of two rotating machines). *Solid-state* devices such as *diodes, thyristors*, and *transistors* are dominant in dc motor power supply and control systems. They are increasingly finding uses in power inverters and the control of ac motors as well.

[12] The shunt motor will also change both flux and current directions thus preserving the unidirectional nature of the torque. Why will the shunt motor *not* operate on ac?

7.5.1 Basic Rectifier Elements

Diodes and thyristors are the basic elements used in the *rectifier circuits* to be discussed. Space and the scope of this book do not permit us to give a detailed explanation of the physics of these devices. For more details see Mazda, 1973. We limit ourselves to a description of the input–output characteristics of the diode and the thyristor.

7.5.1.1 The Diode

The diode constitutes the basic rectifier element. The circuit symbol for a diode is shown in Figure 7.30 along with its voltage–current characteristics. When the voltage, v across the diode assumes positive values it permits current to flow essentially without a voltage drop (short circuit). When the voltage, v changes polarity the resistance of the diode takes on very high values and the current flow through it is essentially blocked (open circuit).

An "ideal diode" could be described as having zero resistance from anode to cathode for $v > 0$ and infinite resistance for $v < 0$. These extreme values are not, or course, realized in practice. The diode has a small but nonzero resistance when

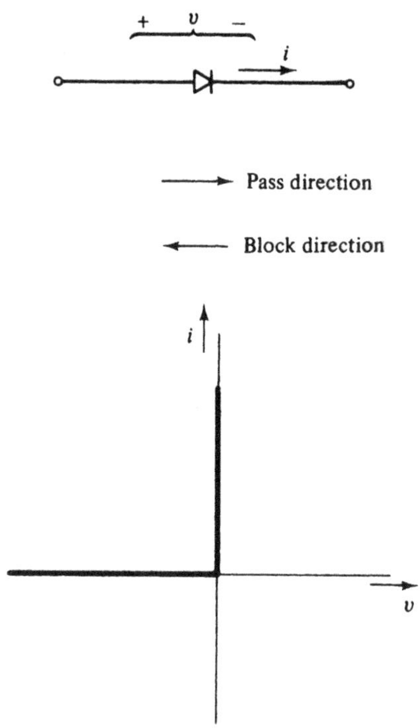

Figure 7.30

$v > 0$. This resistance results in a small ohmic loss which in effect sets the current limit for the device. When $v < 0$, the resistance is large but not infinite. A small leakage current flows.

The diode characteristics are evidently highly nonlinear, a fact that introduces considerable complexity into the analysis of circuits containing diodes. It is helpful to think of the diode as a switch, which, depending on the polarity of the voltage v, alternately and without inertia opens and closes.

7.5.1.2 The Thyristor

While the polarity of the voltage v is the sole determinant of the "open" or "closed" state of a diode, a thyristor is a *controlled-rectifier* element. Its "open" and "closed" states are controlled by a third terminal—the *gate*. *Silicon-controlled rectifier* (SCR) is an alternate name of the device. The symbol of the thyristor is shown in Figure 7.31, which also shows its voltage–current characteristics.

With no gate signal ($v_g = 0$) applied, the thyristor blocks current in both directions. When $v > 0$ and a gate signal is applied (usually a pulse of a few volts amplitude) the thyristor "fires," that is, it becomes conductive like the diode. It remains conductive for as long as $v > 0$.

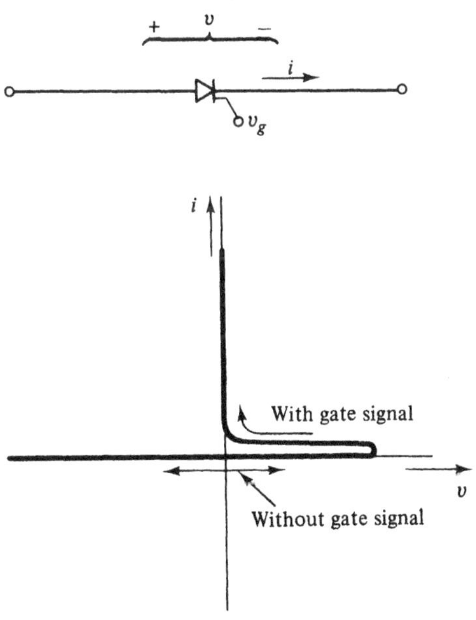

Figure 7.31

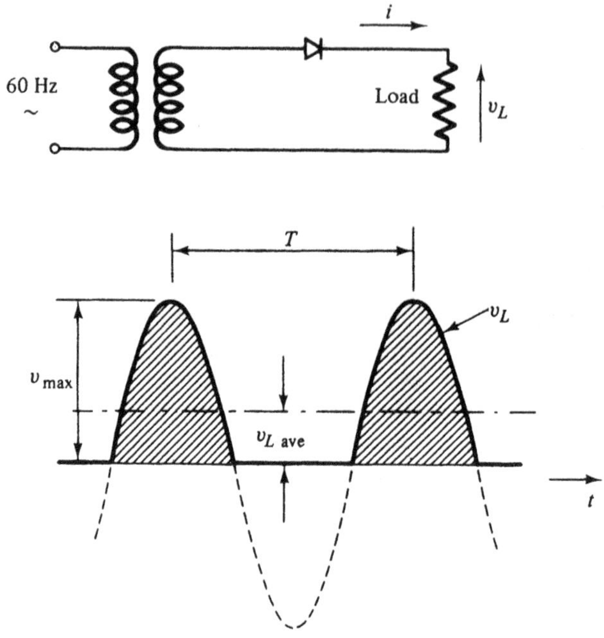

Figure 7.32

7.5.2 Single-Phase, Half-Wave Rectifier Circuits

The simplest type of dc supply circuit is obtained by using a diode in series with the load.

7.5.2.1 Diode Circuit

The circuit is shown in Figure 7.32. As only the positive parts of the ac voltage waves cause the diode to conduct current, the arrangement is referred to as a *half-wave* rectifier. The load[13] voltage v_L will be pulsating at the rate of 60 Hz, with an average value $v_{L\,ave}$ of magnitude:

$$v_{L\,ave} = \frac{1}{T}\int_0^{T/2} v_{max} \sin \omega t \, dt = \frac{v_{max}}{\pi} \quad [V]. \tag{7.47}$$

(We assume zero voltage drop across the diode.)

If the load cannot tolerate the pulsating unidirectional voltage, a *filter* can be placed between the diode and the load as shown in Figure 7.33. The series-inductor sections of the filter represent high impedance for the ac harmonics in the

[13] For simplicity, we assume that the load is purely resistive.

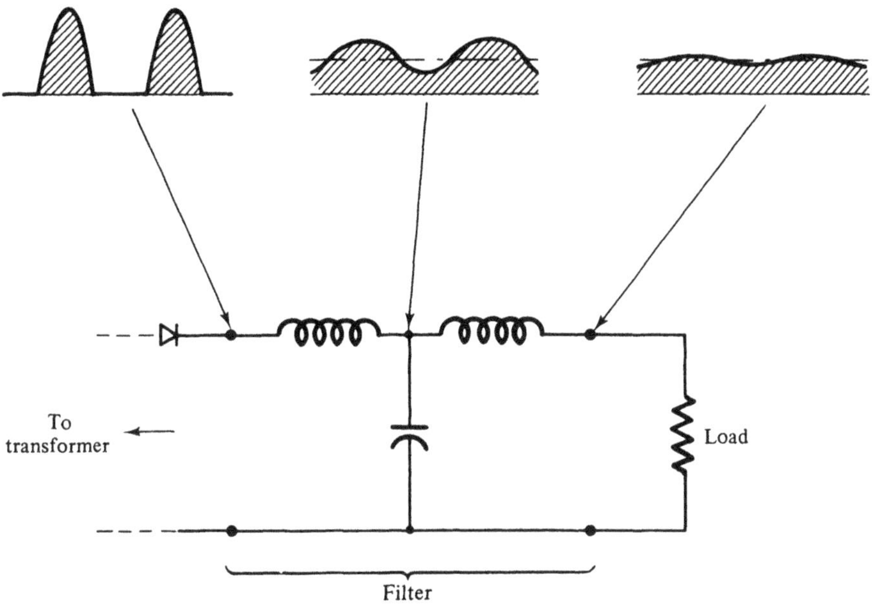

Figure 7.33

diode voltage. The shunt capacitor represents a low impedance. Consequently, only a slight ac ripple remains at the output terminals of the filter.

7.5.2.2 Thyristor Circuit

A single-phase, half-wave thyristor rectifier circuit is shown in Figure 7.34. Using a *control circuit* (not shown) the gate pulse is delayed by a period T_d. Clearly, this results in a reduction of the dc voltage, $v_{L\,ave}$ applied to the load. By changing T_d, we can control the dc load voltage, the magnitude of which is computed from

$$\begin{aligned} v_{L\,ave} &= \frac{1}{T}\int_{T_d}^{T/2} v_{max} \sin \omega t \, dt \qquad [V]. \\ &= \frac{v_{max}}{2\pi}\left[1 + \cos\left(2\pi \frac{T_d}{T}\right)\right] \qquad [V]. \end{aligned} \tag{7.48}$$

Note that the (dc) average value becomes zero for $T_d = T/2$. For $T_d = 0$, the average value is equal to the value given in equation (7.47)—that of the diode circuit.

7.5.3 Single-Phase, Full-Wave Rectifier Circuits

Using additional diodes, one can rectify both the positive and negative halves of the voltage and obtain full-wave rectification. A typical rectifier bridge circuit is

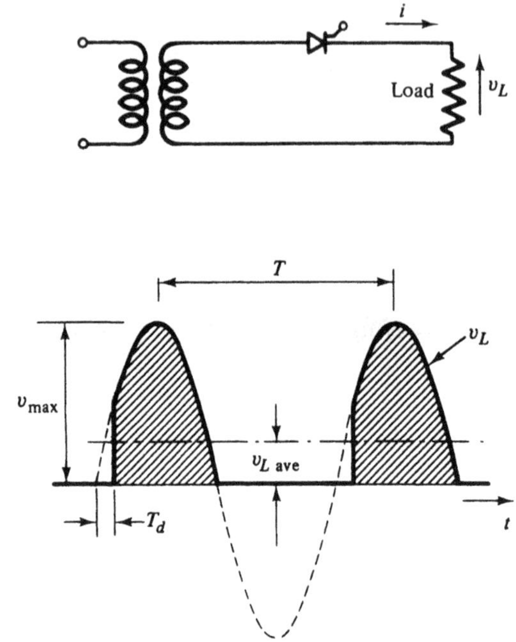

Figure 7.34

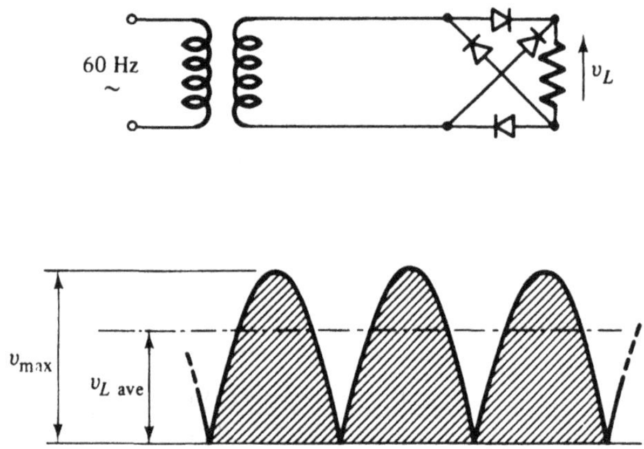

Figure 7.35

shown in Figure 7.35. As both half-waves are now rectified, the (dc) average value will have a value twice the magnitude given by equation (7.47); that is, we have for the dc component shown in Figure 7.35:

$$v_{\mathrm{L\,ave}} = \frac{2}{\pi} v_{\max} \qquad [\mathrm{V}]. \tag{7.49}$$

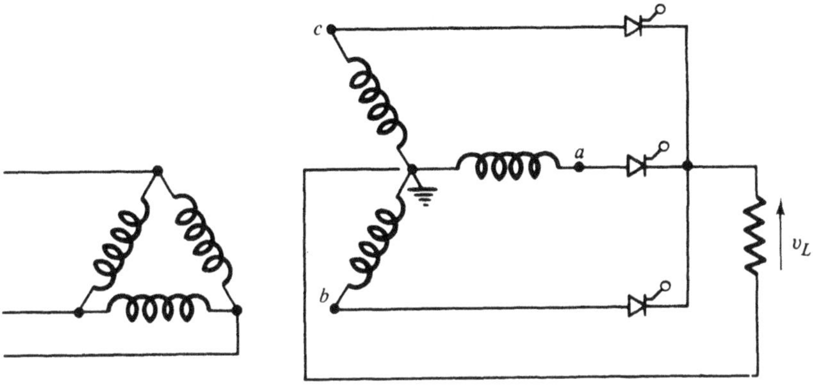

Figure 7.36

Note that we have also changed the lowest harmonic of the ripple frequency from 60 to 120 Hz.

7.5.4 Three-Phase Rectifier Circuits

If dc power in excess of about 5 kW is required, one must usually employ a three-phase source. Figure 7.36 shows one of the simplest three-phase rectifier circuits. Figure 7.37a shows the rectified voltage wave assuming the diodes are ideal. If thyristors are employed, the magnitude of the dc voltage may be controlled by changing the delay, T_d as shown in Figure 7.37b.

7.6 Summary

In this chapter we have discussed the dc machine in its three major configurations. We have also discussed very briefly, some methods for transforming ac to dc power. The dc motor (including its supply circuitry) is a comparatively expensive motor. Its use is therefore limited to those applications where the torque–speed requirements are so severe and exacting that no other motor is capable of satisfying them.

The principles of operation and characteristics of the dc motor were introduced through a prototype *linear* motor. The forces acting in a single rod and its behavior in a magnetic field were used to characterize the simple but impractical device. Working on the basis of step-by-step improvements to the linear motor, we developed a rotating and practical motor design.

The need for current commutation was explained and the principles of operation of the commutator were presented. The most important aspects of the different types of motors are their torque–speed characteristics. Expressions were derived for the torque, the emf and the torque–speed characteristics.

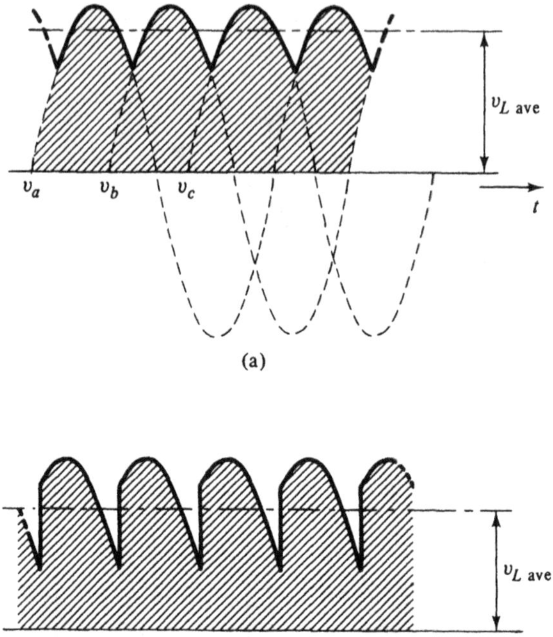

Figure 7.37

Various methods for speed control were discussed and compared in terms of torque and power ratings.

EXERCISES

7.1 The *linear* motor, discussed in Section 7.2, develops a motor force the magnitude of which depends upon the speed. What is the maximum value of the force? At what speed is this maximum force developed?

7.2 What is the maximum value of the motor power that the linear motor in Exercise 7.1 can deliver? At what speed (in terms of the no-load speed s_0) is this maximum power delivered? What is the power delivered by the source? What is the ohmic loss power? What is the motor efficiency?

7.3 What is the no-load speed s_0 of the linear motor in Example 7.1?

7.4 If the linear motor in Example 7.1 were to pull a load representing a constant load force of 0.3 N what would be its steady-state speed? What power would the motor develop? What would be the ohmic power loss?

7.5 Consider a separately excited dc motor. Show that it will deliver its maximum power at a speed half its no-load value. (Compare this to Exercise 7.2.) Would a motor ever be operated under this condition? Explain why not.

7.6 In this and several of the following exercises, we shall study the operating characteristics of a 6-pole dc machine that is characterized by the following design data:

a) It is to be operated from a 500-V dc armature supply voltage source.

b) It must tolerate a maximum armature current of 200 A.

c) The no-load speed is 2500 rpm (when operated from the 500-V source).

d) The field winding is separately excited from a 500-V dc source. When the machine is fed from the 500-V armature source and running at a no-load speed of 2500 rpm, the field current measures 5.05 A. We shall call this the *nominal excitation level.*

Laboratory tests gave the following data:

$$R_a = 0.211\ \Omega$$

$$p_{stray} = 0.9 \text{ kW at full-load}$$

$$p_{rot} = 4.56 \text{ kW, when the motor is running at the no-load speed of 2500 rpm and it is fed from a 500-V voltage source.}$$

To determine how the rotational losses vary with speed the no-load test was performed at three different armature supply voltage levels. For all three tests the excitation current is of nominal value. Test results are tabulated below:

Armature supply voltage [V]	Speed [rpm]	Rotational losses [kW]
450	2250	3.69
500	2500	4.65
550	2750	5.73

If we assume that the rotational losses increase as the xth power of the speed, n, we can express them by the equation:

$$p_{rot} = 4.65\left(\frac{n}{2500}\right)^x. \qquad (7.50)$$

Find x from the above test data. Why do the rotational losses increase with speed? Why will the speed increase as the armature supply voltage is raised?

7.7 Compute the emf and armature current for the machine in Exercise 7.6 if it is run at no-load from an armature supply voltage source,

a) $V_a = 500$ V,

b) $V_a = 525$ V.

Assume the nominal excitation level.

7.8 Assume the machine described in Exercise 7.6 to be running at no-load at the nominal excitation level. The armature supply source is 500 V. A load is now applied and increased slowly until the motor draws 200 A from the armature supply source. The motor is now considered to be fully loaded. What is the change in speed? How much power is drained from the armature supply source? How much power is drained from the field excitation source?

7.9 From Exercise 7.8 you should find that the fully loaded motor will draw a total of 102.53 kW from the armature and field excitation sources. How much power will it deliver to the load, in kW and in hp? What will be its operating efficiency? What is the magnitude of the shaft torque?

7.10 Consider the fully loaded machine in Exercises 7.8 and 7.9. The armature supply voltage is decreased by 7%. What is the percentage change in speed?

Assumption: The load torque varies as the square of the speed.

7.11 Consider the fully loaded machine in Exercises 7.8 and 7.9. Now we decrease the field excitation voltage by 7%. What is the percentage change in speed?

Assumptions:

a) The pole flux is proportional to the field current.

b) The load torque has the same speed characteristics as assumed in Exercises 7.8 and 7.9.

7.12 The dc machine in Exercise 7.6 is operated as a generator. It is run at 2500 rpm driven by a diesel motor. The field excitation voltage is adjusted until the voltage across the armature terminals on open-circuit is 500 V.

A load resistance R is connected across the armature terminals. What is the minimum value of R if the rated armature current of 200 A is not exceeded? When the armature current is 200 A, compute the power generated (that is, the power dissipated in the load) and the power delivered by the diesel engine.

Assumptions:

a) The armature current will give rise to an electromechanical torque that will tend to decrease the speed.

b) The diesel motor can maintain its speed constant

7.13 Repeat Exercise 7.12 but now assume that the diesel motor cannot maintain a constant speed. Assume a decrease of 1% in speed for each 35-A increase in the armature current.

7.14 As the load current in Exercise 7.12 is increased the terminal voltage decreases from its initial value (500 V) due to the voltage drop across R_a. For example, an armature current of 200 A will result in a voltage drop of 42.2 V, or 8.4%. This is a distinct drawback.

By adding a "compound" field winding in series with the armature (see Figure 7.38) the above voltage drop can be compensated for. The compound winding hav-

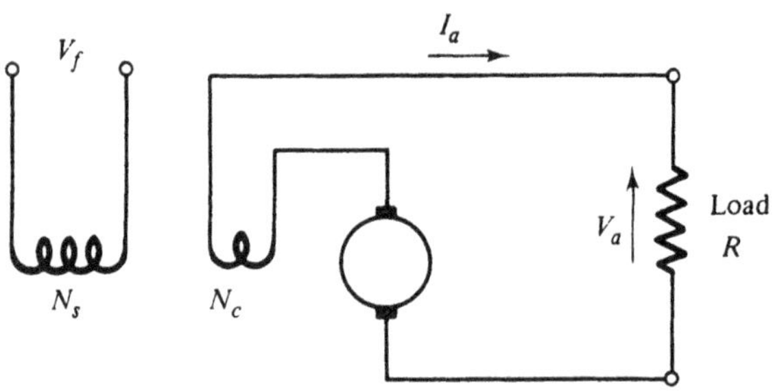

Figure 7.38

ing N_c turns with current I_a flowing in it adds an mmf to that caused by the separate excitation winding $N_s i_f$.

Explain how this added field winding works and also compute the turns ratio N_c:N_s if we want to maintain a terminal voltage V_a that is totally independent of load current I_a.

References

McPherson, G., Laramore, R.D. *An Introduction to Electrical Machines and Transformers*, 2nd ed. New York: John Wiley & Sons, 1990.

Nasar, S.A., Boldea, I. *Electric Machines: Steady-State Operation*. New York: Hemisphere Publishing, 1990.

Rashid, M.H. *Power Electronics: Circuits, Devices and Applications*, 2nd ed. Englewood Cliffs, NJ: Prentice Hall, 1988.

8

Induction Machines

8.1 Why Induction Motors?

There are many applications for which the dc motor is "overqualified." Its outstanding torque–speed characteristics and unmatched controllability come at a price. A dc motor always requires a dc supply, which entails extra costs. Numerous industrial, domestic, and commercial uses of motors require fairly simple torque–speed characteristics. For example, an air compressor motor is required to operate at a nominally constant speed and deliver a nominally constant torque. In such applications, a rugged design and the ability to operate directly off the ac network are very attractive.

The ac induction motor fills this niche very well and it is therefore the most widely used electric motor. It comes in sizes ranging from a fraction to several thousand horsepower. Large ac induction motors (>5 hp) are usually designed for three-phase operations because a constant torque and symmetrical network loading are desirable. Small fractional horsepower motors are often of single-phase design. This chapter begins with a treatment of three-phase motors. The final section is devoted to single-phase induction motors.

8.2 Basic Design Features

The essential parts of a three-phase induction motor are shown in Figure 8.1. The stator has a distributed three-phase winding essentially identical to that found in the stator of a synchronous machine, as described in Chapter 4 (Figure 4.14). The rotor winding design varies, depending on the need for torque control. The "squirrel-cage" winding is the most common design (Figure 8.1), consisting of solid copper or aluminum bars embedded in the rotor slots, with each bar short-circuited by two end-rings.

The electromechanical torque of the induction motor is obtained by the interaction between a stator-bound rotating magnetic flux and a rotor-bound current.

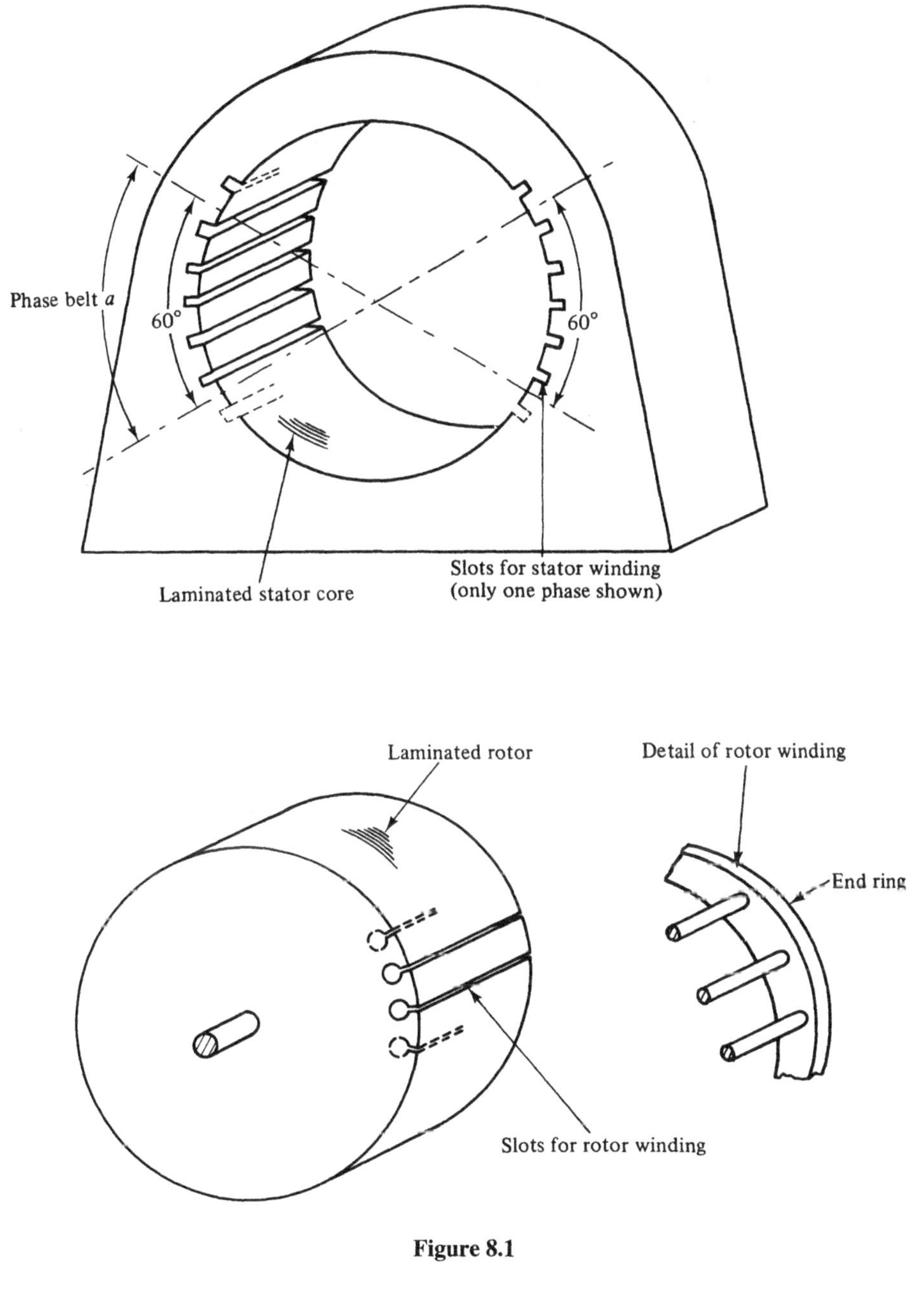
Phase belt a
60°
60°
Laminated stator core
Slots for stator winding
(only one phase shown)
Laminated rotor
Detail of rotor winding
End ring
Slots for rotor winding

Figure 8.1

The rotor-bound current is induced by the rotating stator flux through magnetic induction—thus giving the motor its name.

8.3 The Rotating Stator Flux Wave

In Section 4.7.1 we discussed the creation of a rotating magnetic flux of constant magnitude when a balanced three-phase current is applied to three coils spatially displaced by 120° from each other. We proceeded to make an analog of this phenomenon. It consisted of a horseshoe-shaped, permanent magnet, which is driven about its axis at a constant angular velocity by a prime mover (Figure 4.27)—reproduced here in Figure 8.2.

8.3.1 Harmonics of the Flux

We have assumed in the above analysis that because of the constant magnitude of the rotating flux all voltage and current waveforms are purely sinusoidal. This is not quite correct. Like an ocean wave surrounded by ripples of all sizes, our fundamental flux waveform is accompanied by a large number of small ripples of many different frequencies. These ripples can cause undesirable phenomena in the form of "cogging," "crawling," and magnetic noise and vibrations. These second-order side effects are of practical significance and cannot be ignored. However, by various design techniques they can be minimized to tolerable levels. The scope of our basic treatment does not permit us to enter into a discussion of these matters. The interested reader can find a good treatment in Alger, 1970.

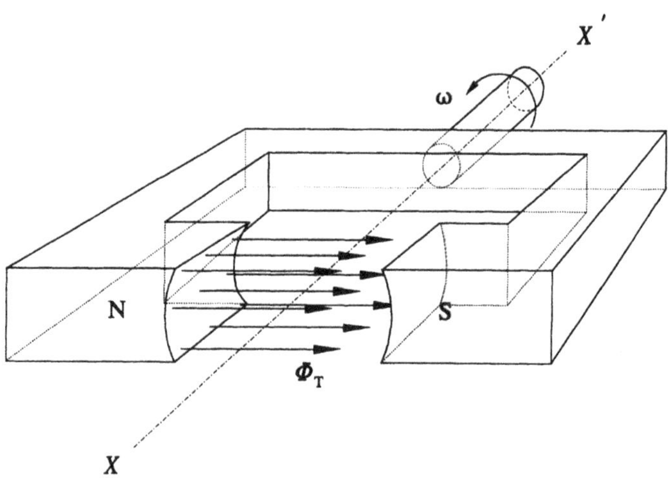

Figure 8.2

8.4 The Torque-Creating Mechanism

Returning to Figure 8.2, if we now place a short-circuited coil, supported on the axis X–X', between the pole-shoes of the permanent magnet, as shown in Figure 8.3, then as the permanent magnet rotates, its flux, Φ_T, rotates as well, and a voltage will be induced in the coil. As the coil is short-circuited, a current will flow. The direction of this current, according to Lenz's law, will be such as to oppose the rotating flux. The only way the coil flux (produced by induction from the permanent magnet) can oppose the rotating flux is for the coil itself to rotate in the same direction as the permanent magnet. If the coil is free to rotate, it follows that it will experience a torque that will accelerate it until it is rotating at the same angular velocity as the rotating horseshoe permanent magnet. The induced voltage will then be zero. This would be the case if the system were lossless. Due to the inevitable losses (bearings, windage, etc.) the coil will always rotate at a speed below that of the permanent magnet. The coil can be said to "slip" with respect to the permanent magnet.

The situation in the induction machine is an exact analog of the rotating horseshoe permanent magnet system shown in Figure 8.3. As the revolving flux (wave) Φ_T sweeps across the rotor and the stator windings, it will induce emf's in both. In the short-circuited rotor cage winding these emf's will cause currents to flow. The currents will interact with the stator flux to create an electromechanical torque. Under the influence of the torque the motor will accelerate to its full operating speed.

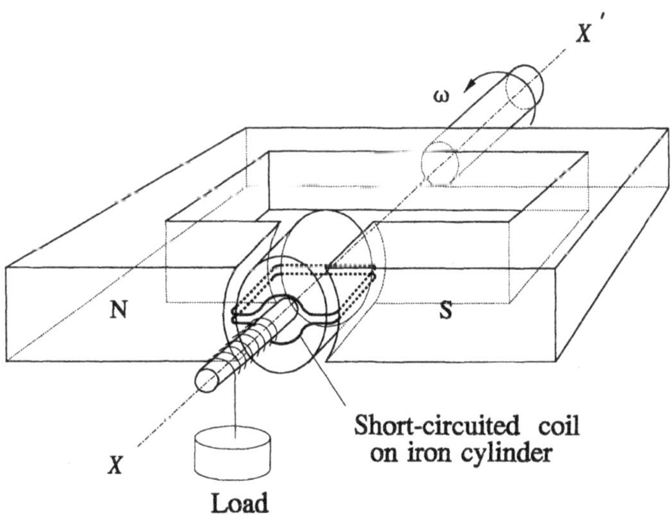

Figure 8.3

The following questions are important:

- What type of torque is obtained—constant or pulsating?
- What will be the magnitude of the starting torque?
- As the motor speeds up, how will the torque change?
- What will be the operating speed of the motor?
- What power will the motor absorb from the network to which it is connected?

We can address these questions by first investigating the nature of the induced rotor currents.

8.4.1 The Rotor Currents at Standstill

Consider the induction motor at standstill, with the stator winding energized. This condition occurs at the start or stall of the motor. The speed of the flux wave relative to the stator and rotor windings is now the same, that is, n_s rpm. The frequencies of the induced emf's will therefore be the same in both windings, that is, the frequency, f Hz, of the energizing source.

The instantaneous value of the emf induced in a particular conductor will be proportional to the rate of change of the flux at the conductor in question [see equation (3.52)]. The stator and rotor windings will experience "emf waves" that will accompany the (rotating) flux wave.

Consider the rotor emf wave. In each rotor copper bar the emf will drive a current. If the impedance of the rotor bars were purely resistive, the instantaneous magnitude of the current would be directly proportional to the instantaneous magnitude of the emf and the result would be a current wave in phase with the emf wave. However, the impedance of the rotor bars has a significant reactive component,[1] (see Figure 8.4) and the current wave will therefore lag the emf (and flux) wave by an angle, γ.

Figure 8.5 shows one full cycle of the system of waves. We make the following observations from the diagram:

1. As the flux wave is essentially sinusoidal, both the emf and current waves will also be sinusoidal.
2. The emf and current waves will have as many maxima and minima (poles) as does the stator flux wave. This number is determined solely by the number of poles of the stator winding (two poles in Figures 8.1, 8.2, and 8.3).
3. The rotor end-rings serve as short-circuit paths for the copper bars. The currents in the rings vary sinusoidally around the periphery as shown in Figure 8.5, having their

[1] The reactance is due to the self-inductance of the rotor winding. Each rotor bar is surronded by a magnetic field (Figure 8.4) which is induced by its own current.

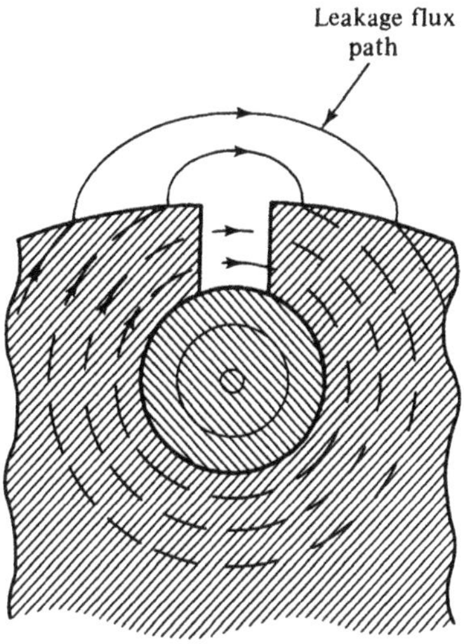

Figure 8.4

instantaneous maximum values opposite to those bars that have the instantaneous current zeros. The current waves in the rings and bars travel with equal speed.

8.4.2 The Motor Torque

Returning to the synchronous machine we remind ourselves that its electromechanical air-gap torque resulted from the interaction between the following:

1. A rotor-bound flux wave of amplitude B_{max} revolving at the synchronous speed n_s rpm.
2. A stator-bound current wave revolving at the same speed but lagging the flux wave by the angle γ. The current wave, although existing only in the discrete stator slots, could be considered in a macrosense as "smeared out" over the total stator surface to form a surface or sheet current wave of amplitude A_{max}.

We derived equation (4.88) for the torque and restate it here for easy reference:

$$T_{em} = V_{rot} B_{max} A_{max} \cos \gamma \qquad [\text{N} \cdot \text{m}], \tag{8.1}$$

where V_{rot} is the rotor volume.

The three-phase induction motor has a stator-bound revolving flux wave of amplitude B_{max}. The current wave is rotor-bound, lagging the flux wave by γ degrees and existing in the discrete rotor slots. If, as before, we "smear" the current over the rotor surface[2] we obtain a continuous surface or sheet current wave

[2] The cage winding shown in Figure 8.1 contains a discrete *finite* number of rotor bars. By allowing the number of bars to approach infinity, the cage approaches a cylindrical shell. Some induction motors have this type of rotor-winding design ("drag cup" motors). When the rotor winding consists of a continuous cylindrical shell, the rotor currents appear as a continuous sheet current.

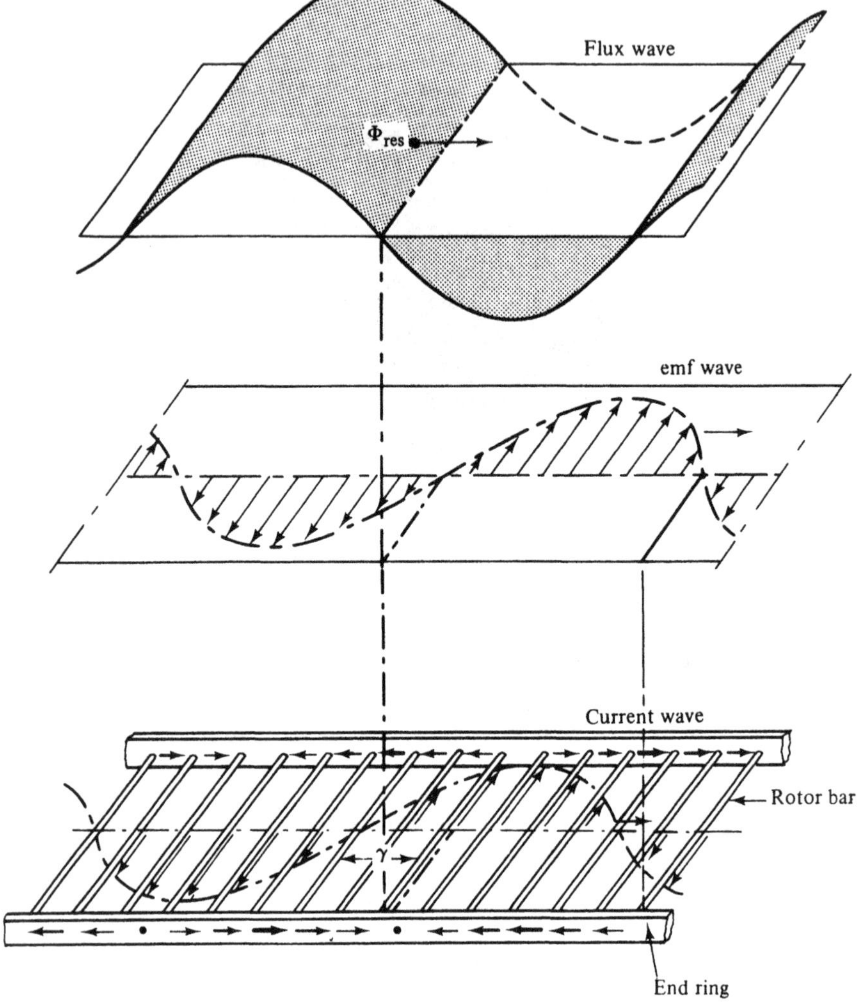

Figure 8.5

of amplitude A_{max} [A/m] revolving at the speed n_s rpm relative to the stator and lagging the B wave by γ degrees.

The only difference between the synchronous and induction machines is the reversed "home bases" of the two waves. Since this should not affect[3] the torque significantly, we conclude that equation (8.1) applies to induction motors as well. In particular, we are reminded that the torque is constant.

[3] You can pull a boat standing either in the boat or on the shore.

8.4.3 Current Wave in the Running Rotor: Concept of Slip

Under the influence of the torque the rotor will (unless constrained) accelerate and reach an operating speed, n rpm. How will the torque change as the rotor speeds up?

The relative speed, n_{rel}, between the rotor and the flux wave is

$$n_{rel} = n_s - n \qquad \text{[rpm]}. \tag{8.2}$$

As the frequency f_r of the rotor emf and current will be proportional to this relative speed we can write:

$$\frac{f_r}{f} = \frac{n_{rel}}{n_s} = \frac{n_s - n}{n_s} = 1 - \frac{n}{n_s}, \tag{8.3}$$

where f, as before, represents the stator or power source frequency. The rotor frequency can be computed from

$$f_r = \frac{n_s - n}{n_s} f = sf \qquad \text{[Hz]}. \tag{8.4}$$

We introduce the *slip*, s, defined by

$$s \equiv \frac{n_s - n}{n_s}. \tag{8.5}$$

The slip is a measure of the relative speed between the rotor and the stator flux wave. At standstill, n is zero, and the slip is then 1 (or 100%). If the rotor were to run at synchronous speed the slip would be zero. Figure 8.6 shows the relationship between rotor speed and slip. Note in particular that the slip is negative for speeds in excess of the synchronous. Note also that negative speeds correspond to slips in excess of 100%.

As the motor speeds up, the relative speed decreases, thus reducing both the rotor frequency and the magnitude of the rotor emf. As a result, the magnitude of the rotor current will also decrease.

The rotor emf and current waves, in spite of their lower speed relative to the rotor, still have the same synchronous speed, n_s, relative to the stator. The flux, emf, and current waves, as before, constitute a wave system that revolves with the speed n_s rpm relative to the stator, independent of the rotor speed n.

8.4.4 Torque as a Function of Slip: A Qualitative Analysis

As the motor speeds up and approaches the speed n_s of the flux wave, both the frequency and magnitude of the rotor currents will diminish. Our first guess would be that the torque would also decrease. Decreased rotor current undoubtedly means a decrease in A_{max}, the amplitude of the current wave. According to equation (8.1) this tends to decrease the torque. However, the reduced rotor frequency,

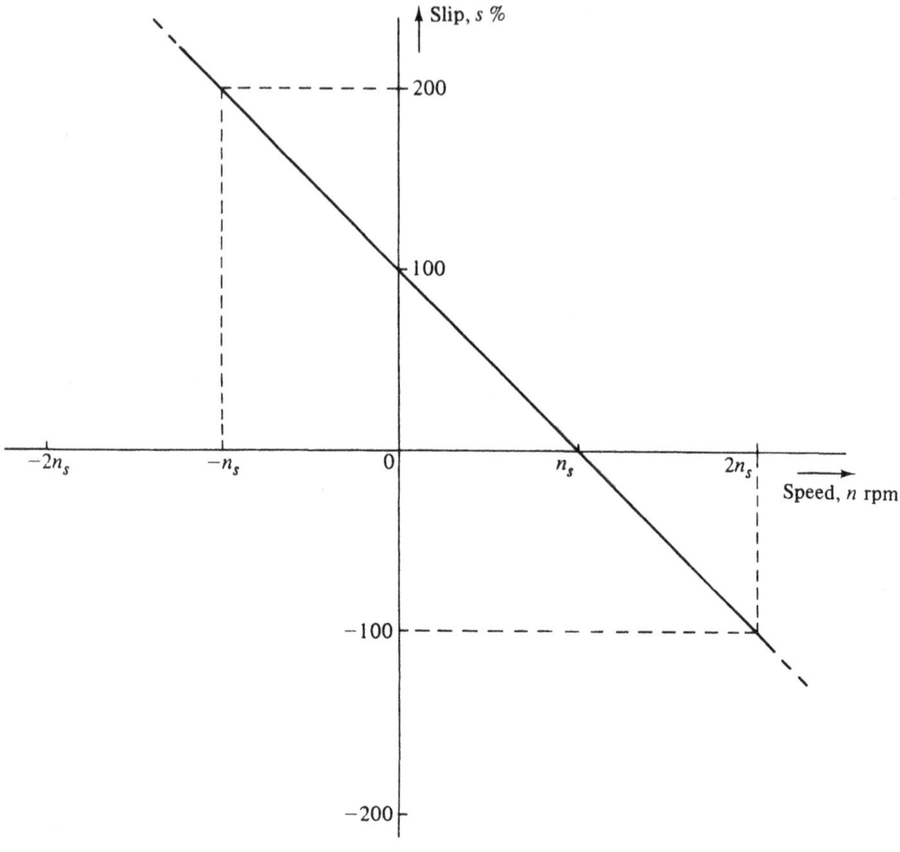

Figure 8.6

f_r, will result in a lower rotor reactance and a corresponding decrease in the phase angle, γ. The resultant increase in $\cos\gamma$ will tend to increase the torque. As the torque [equation (8.1)] depends on the product $A_{max} \cdot \cos\gamma$, and A_{max} decreases, but $\cos\gamma$ increases, with rotor speed, it is not immediately clear whether the result will be an increase or decrease of the torque.

For a typical motor, as the rotor speed increases from standstill, the increase in $\cos\gamma$ has a greater effect on the torque than the decrease in A_{max}. The torque, which at zero speed has a *standstill* or *starting* value, T_{st}, will at first increase with speed as shown in Figure 8.7. When the rotor speed has reached the value n_{max}, the rate of increase of $\cos\gamma$ is equal to the rate of decrease of A_{max}. The growth of the torque will be zero—the torque has reached its highest value, T_{max}. As the speed increases beyond n_{max} the decrease in A_{max} will have a greater effect on the torque than the increase in $\cos\gamma$, resulting in a lower torque. Finally, when the

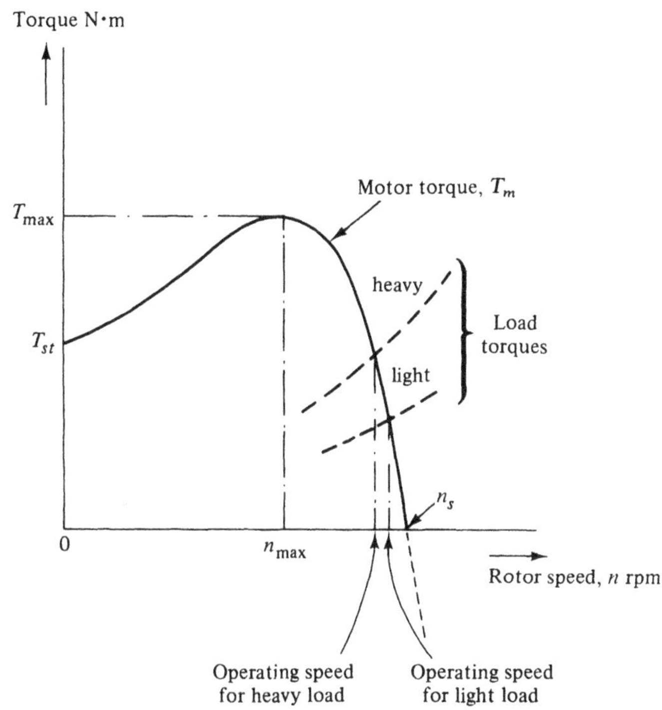

Figure 8.7

rotor speed reaches the value n_s, there is no longer a relative speed between the flux wave and the rotor. No emf is induced in the rotor and therefore no current flow, and the torque will be zero.

It is clear that an induction motor can never pull a load at synchronous speed n_s. Its speed, if run as a motor, will always be less than n_s. For this reason the induction motor is sometimes referred to as an *asynchronous* machine.

8.4.5 Determination of Motor Operating Speed

Assume that we have found either by measurement or by analysis the motor torque curve shown in Figure 8.7. It is then very easy to determine the operating speed, n, of the motor. As the speed will depend on the load being pulled, it is necessary to have a knowledge of the torque–speed characteristics of the load in question. Remember that the same was true of dc motors.

In Figure 8.7 two different load curves are plotted, labeled "light" and "heavy." The intersections between these curves and the motor torque curve indicate torque balance. These points give the operating speeds and the corresponding torques of the motor under the load conditions in question.

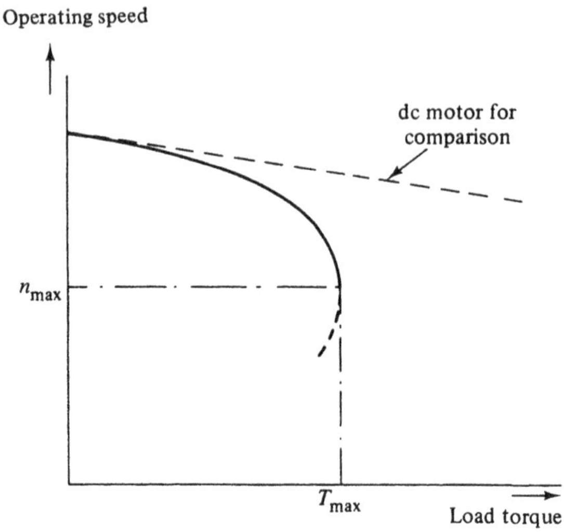

Figure 8.8

As the load torque increases, the induction motor responds by reducing its speed. We can actually plot the operating speed versus load torque and obtain the speed–torque curve shown in Figure 8.8. In this respect the induction motor resembles the shunt dc motor (the torque–speed curve of which has been included in Figure 8.8 for comparison).

Figure 8.8 also demonstrates a distinct difference in the behavior of the two types of motor. The induction motor cannot deliver a torque in excess of T_{max} shown in Figure 8.7. If we continue to increase the load the speed would decrease to the value n_{max}. A further increase in the load torque would result in a speed collapse (or "stall"). The shunt dc motor would not stall (but the high armature current would cause the motor to overheat).

Typically, an induction motor will normally operate at speeds of 95% to 99% of synchronous speed, corresponding to a slip range of 5% to 1%, respectively.

8.4.6 The Induction Generator

Under no-load conditions, the induction motor will run at a speed slightly less than synchronous speed. The presence of inevitable losses will require a small torque, and to deliver this torque the motor must run at a slight slip. Now assume that, by means of a prime mover, we accelerate the motor beyond the synchronous speed. How will the machine behave?

As the rotor now runs faster than the stator flux, the relative speed, n_{rel} [equation (8.2)], becomes negative. Consequently, both the induced rotor emf and the rotor current will reverse direction. As a result, the electromechanical torque will reverse direction—indicating, as in the case of dc motors, *generator action.* (See the extrapolation (dashed line) of the motor torque curve beyond synchronous speed in Figure 8.7.)

The induction machine now acts as a generator receiving mechanical energy from the prime mover and transforming it into electrical energy, which is supplied to the electrical grid. For example, if we used an induction motor to drive a mine elevator it could serve as a "dynamic brake" during the descent phase of the work cycle. In this respect it would match the dc motor as long as it was running above synchronous speed. In practice, elevators are never driven by induction motors because of the possibility of stalling when overloaded. Note that a crucial assumption in the above discussion was the presence of a flux wave in the stator. This means that the induction motor can be turned into a generator only if its stator winding is connected to a three-phase source that can hold up its voltage. An induction machine cannot be turned into a generator and used to feed energy into a set of (passive) impedances.

8.4.7 "Wound-Rotor" Induction Motors

The qualitative discussion above indicates clearly that the magnitude of the motor torque depends to a great extent on the magnitude of the induced rotor currents. As the magnitudes of these currents in turn depend primarily on the impedance of the rotor bars and end-rings, it follows that this impedance will, to a great extent, affect the magnitude of the torque. (The specific relationship between torque and rotor impedance is discussed in Section 8.5.8.)

Variation of the rotor impedance seems an obvious way to vary the torque. However, a squirrel-cage rotor winding has a built-in impedance and therefore does not lend itself to this option for control of the torque. In cases where torque control is required, a *wound-rotor* design is preferred to the cage structure.

In a wound-rotor induction motor the rotor winding consists of a symmetrical three-phase winding, of the same type found in the stator. However, the rotor winding does not need to be identical to the stator winding. The only important restriction is that the two must have the same number of poles.

Of the six rotor-winding terminals, three are connected together to form an "internal" neutral. The remaining three terminals are connected to three slip rings. Three external variable resistors are connected to the three slip rings by carbon brushes. The rotor circuit is closed by means of an "external" neutral, usually grounded. The arrangement is shown schematically in Figure 8.9. Figure 8.10 shows details of the slip rings and the carbon brushes. It is important to note that in order to maintain the symmetry of the three rotor phases, the three variable external resistors must be mechanically interlocked to keep them equal at all times.

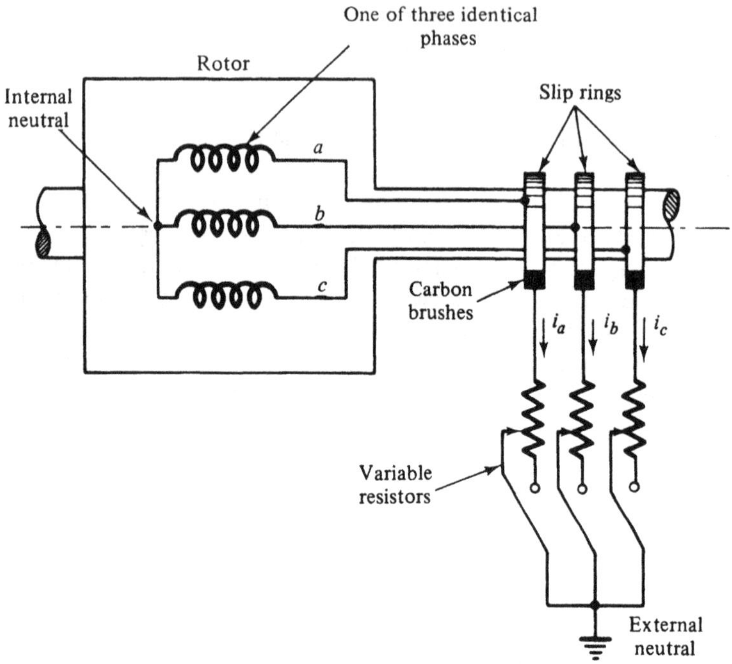

Figure 8.9

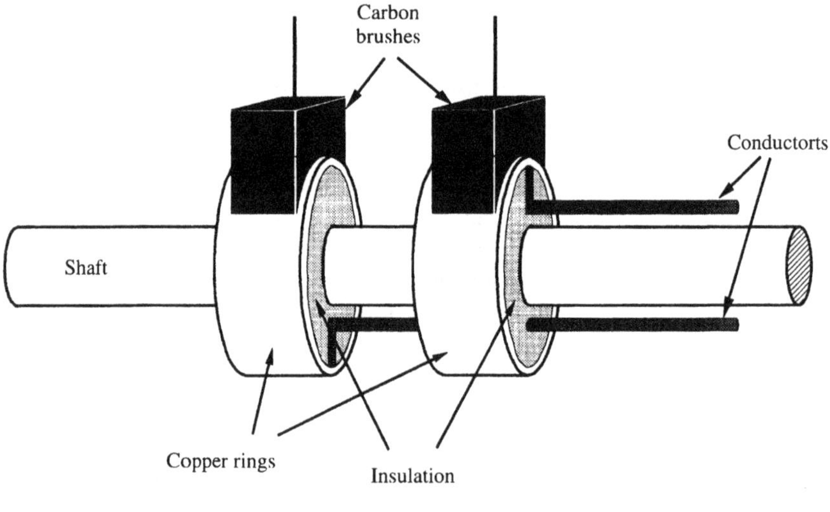

Figure 8.10

8.5 Three-Phase Induction Motor Performance Analysis

The expression for the torque, (8.1), proved extremely valuable in conveying a qualitative impression of the performance of the three-phase induction motor. However, it is not very practical for the purpose of obtaining quantitative performance data, because the variables B_{max}, A_{max}, and $\cos\gamma$ are very difficult to measure. We need to develop an expression that is based on variables that are easier to measure.[4]

The first step in deriving quantitative performance criteria is to develop an *equivalent electric circuit* for the induction motor. This circuit will permit us to compute the more relevant electric variables that determine the performance of the motor.

8.5.1 The Transformer as an Analog of the Induction Motor

Before we attempt to find an equivalent circuit for the induction motor we should point out some far-reaching similarities between this type of motor and the transformer.

For simplicity, we consider the ideal transformer. When the primary winding of a transformer is energized from an ac source, a core flux is generated that induces emf's in both the primary and secondary windings. The primary current is zero. When the secondary winding, N_2, is connected to a load, a secondary current, I_2, will flow, the magnitude and phase of which depend on the load impedance. In response to the change of the mmf equal to N_2I_2, a primary current will arise that will be of opposite polarity and have a magnitude proportional to the secondary current so as to restore magnetic mmf balance in the core ($N_1I_1 = N_2I_2$). The primary current drawn from the ac source accounts for the power that, after transformation, is supplied to the secondary load.

For simplicity, we consider an "ideal" induction motor. When the "primary" (stator) winding of an induction motor is energized from an ac source a magnetic flux is generated that will induce emf's in both the primary and the "secondary" (rotor) winding. A secondary current will flow in the rotor, the magnitude and phase of which will depend on the speed (which is a function of the load). A primary current will arise that will be of opposite polarity and have a magnitude proportional to the secondary current so as to preserve magnetic mmf balance in the iron core. This primary current drain from the ac source accounts for the power supplied, which, upon transformation, is used to pull the mechanical load.

We take advantage of the similarities of an induction motor and a transformer to develop an equivalent circuit for the induction motor.

[4] The reader will remember that in a discussion of synchronous machines we also found it necessary to develop a more practical expression, (Section 4.7.4), for obtaining quantitative data.

8.5.2 The Concept of an "Ideal Motor"

In deriving the equivalent circuit of the transformer in Chapter 5, we started with assumptions that idealized the transformer. We introduced the concept of the "ideal transformer" (IT), a device characterized by

1. Zero winding resistances (and hence zero copper loss)
2. Zero core reluctance (and as a consequence zero leakage fluxes and reactances)
3. Zero core losses

After developing the simple mathematical model for the IT (Figure 5.5), we removed the assumptions one by one, and arrived eventually at a model that represented the physical transformer accurately (Figure 5.10).

We shall follow the same procedure here, but we must exercise special care in doing so. For example, we cannot use the above assumptions in defining an *ideal motor* (IM). If we make the assumptions of zero winding resistances and reactances, then we would be saying that the impedance of the rotor of the squirrel-cage motor is zero. This would be an absurd assumption, as it would lead to infinite current[5] in the rotor. As we have already suggested, and as we shall further confirm, the resistance of the rotor of an induction motor plays an important role in its theory, and we must retain it in our model.

Considerable simplicity and clarity in our analysis can be obtained without compromising the accuracy of our model if we define an "ideal" induction motor that is characterized by the following features:

1. A magnetic path whose reluctance is zero (that is, infinite permeability), and also the core losses are zero
2. An air-gap width that approaches zero
3. Zero friction and windage losses

The air gap of a practical motor is made as narrow as is practically possible. Figure 8.11 shows that the magnetic flux path crosses the air gap twice. By reducing

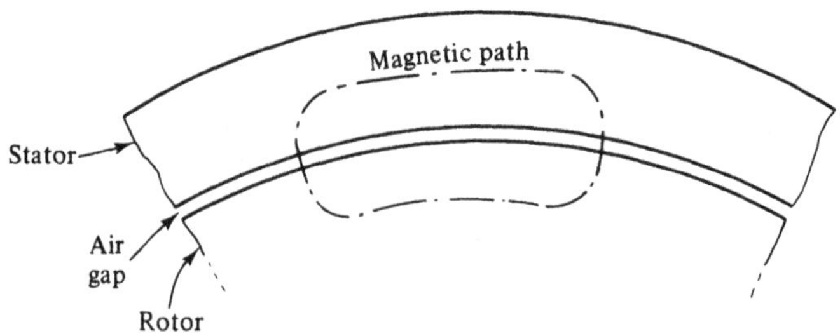

Figure 8.11

[5] In the case of the transformer, this caused little difficulty as the load impedance is in series with the secondary.

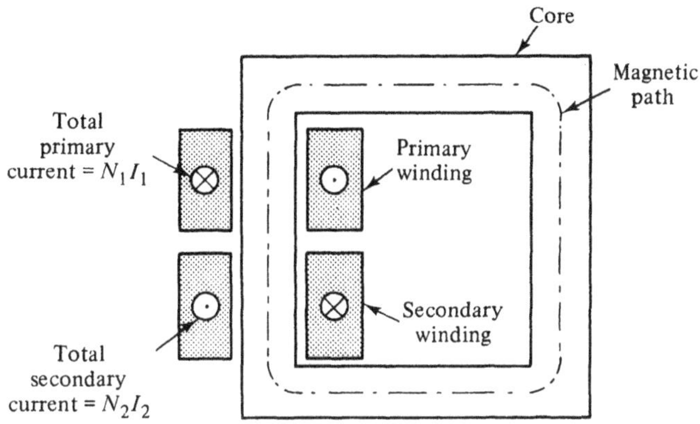

Figure 8.12

the width of the air gap, we increase the air-gap flux and the torque of the motor. Ideally, the width of the air gap should approach zero, which means that the reluctance of the air gap would vanish. If, in addition, the permeability of the iron is infinite, that part of the magnetic path will have zero reluctance.

We remember that zero reluctance in the case of the IT core resulted in the requirement that its magnetic path must encircle zero total current (Figure 8.12). From this we can write the equation for "mmf balance" as

$$N_1 I_1 = N_2 I_2 \qquad [\text{A} \cdot \text{t}]. \tag{8.6}$$

Similarly, for the ideal motor the need for mmf balance requires that the magnetic path must encircle zero total current. We have already concluded that the rotor surface contains a sinusoidally varying surface sheet current. We now conclude that the stator surface must contain a matching surface sheet current but of the opposite polarity if mmf balance is to be maintained for any arbitrary position along the magnetic path. Figure 8.13 shows a segment of the surface currents in

Figure 8.13

the rotor and stator, and it is not very different from the actual situation in a physical motor. In practice, the stator surface current flows in the discrete stator conductors in the slots.

The rotor sheet current revolves at a speed n_s rpm relative to the stator and so must the stator sheet current. In order to generate a stator current wave traveling at synchronous speed the stator currents must constitute a symmetrical three-phase set. As these currents are drawn from the ac supply source, we conclude that the three-phase induction motor constitutes a balanced three-phase load on the network.

8.5.3 Circuit Equations for the Ideal Motor

The analysis in this section is based on the following assumptions:

1. The motor has IM magnetic characteristics, that is, the stator and rotor current waves are equal in magnitude but of opposite sign (see Figure 8.13).
2. Both the stator and rotor have three-phase, Y-connected windings.[6] There are an equal number of slots on the stator and rotor.[7] The number of conductors per stator slot is a times the number of conductors per rotor slot.
3. The stator winding is fed from a three-phase source with $|V|$ volts (rms) between each phase and ground. Phase a of the stator is chosen as a reference.
4. In the analysis to follow the subscript 1 refers to the primary side (the stator); the subscript 2 refers to the secondary side (the rotor).
5. All reactances used in the analysis are assumed to have been computed at 60 Hz. As we have different frequencies in the rotor and stator, it is important to agree on the frequency at which the reactances are computed.

Application of KVL to the stator circuit (Figure 8.14) gives

$$V = E_1 + I_1(R_1 + jX_1) \qquad [\text{V}], \tag{8.7}$$

where

E_1 = induced emf in stator winding, volts per phase;

R_1 = stator resistance, ohms per phase;

X_1 = stator reactance, ohms per phase;

I_1 = stator current, amperes per phase.

Application of KVL to the rotor circuit gives

$$E_2 = I_2(R_2 + jsX_2) \qquad [\text{V}], \tag{8.8}$$

[6] The analysis is simplified by assuming that both the rotor and stator have similar types of winding.

[7] This assumption is really not necessary. We make it to ensure that the stator and rotor windings have equal distribution factors.

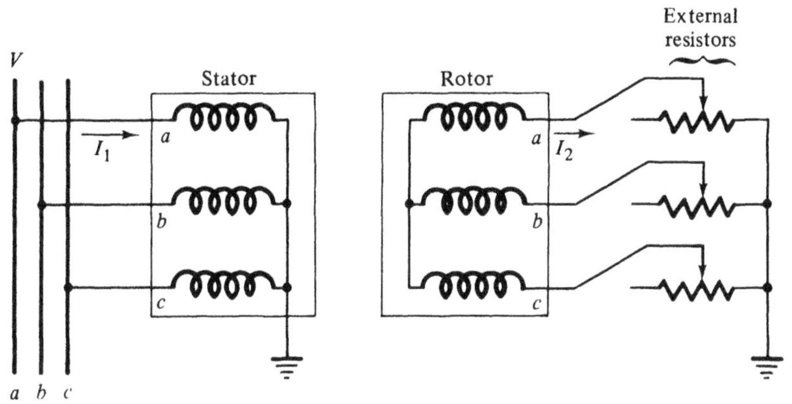

Figure 8.14

where

E_2 = induced emf in rotor winding, volts per phase;

R_2 = rotor resistance, ohms per phase (including the external resistance if any);

X_2 = rotor reactance, ohms per phase;

I_2 = rotor current, amperes per phase.

Note that the reactance term in equation (8.8) is multiplied by s because the induced voltage in the rotor is at the slip frequency, sf. In addition to the two voltage equations we have these relations between primary and secondary emf's and currents:

$$E_2 = s\frac{E_1}{a} \qquad [\mathrm{V}], \tag{8.9}$$

$$I_2 = aI_1 \qquad [\mathrm{A}], \tag{8.10}$$

where

$$a = \frac{\text{number of conductors per stator slot}}{\text{number of conductors per rotor slot}}.$$

From equation (8.9) it is clear that the rotor emf is considerably reduced due to both the slip (s) and the winding ratio (a).

Equation (8.10) is identical to (8.6); they express mathematically what is depicted in Figure 8.11. We may rewrite (8.10) as follows:

$$I_1 = \frac{I_2}{a} \equiv I_2' \qquad [\mathrm{A}]. \tag{8.11}$$

The current I_2' is defined as "the rotor current referred to the stator" [cf. equation (5.36)].

Substitution of the expressions for E_2 and I_2 into equation (8.8) and subsequent elimination of E_1 between equations (8.7) and (8.8) gives the following:

$$V = I_1\left[R_1 + \frac{a^2R_2}{s} + j(X_1 + a^2X_2)\right] \quad [\text{V}]. \tag{8.12}$$

In order to establish the analogy with equation (5.34) we define

$$\begin{aligned} R_2' &\equiv a^2R_2 \quad [\Omega], \\ X_2' &\equiv a^2X_2 \quad [\Omega]. \end{aligned} \tag{8.13}$$

These are the "rotor impedances referred to the stator." In terms of these impedances equation (8.12) becomes

$$V = I_1\left[R_1 + \frac{R_2'}{s} + j(X_1 + X_2')\right] = I_2'\left[R_1 + \frac{R_2'}{s} + j(X_1 + X_2')\right] \quad [\text{V}] \tag{8.14}$$

8.5.4 Equivalent Circuit of the Ideal Motor

Using equation (8.14) we can draw the circuit shown in Figure 8.15a for the ideal induction motor. In a practical case all of the four impedance elements can be

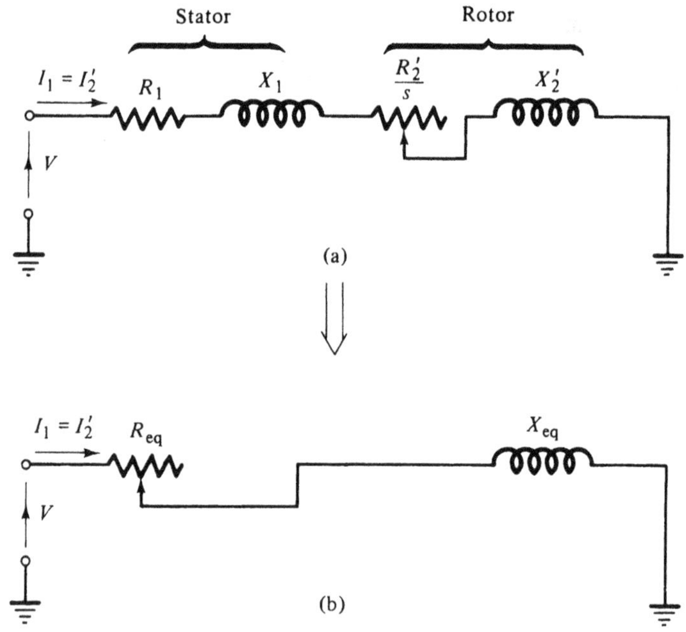

Figure 8.15

found either by tests or by computation. From the equivalent circuit we can obtain the value of the current I_1 for any speed (expressed in terms of slip s) of the motor. From equation (8.10) we can obtain I_2. The torque and power can then be computed. The equivalent circuit provides a practical means for extracting quantitative performance data in a convenient way.

8.5.5 The Circle Diagram for the Ideal Motor

Using the circuit in Figure 8.15a, we can develop some interesting characteristics of the induction motor. In particular let us study the relationship between the current taken by the motor and the speed. In Figure 8.15b we have simplified the circuit by lumping the resistive elements into an equivalent resistance:

$$R_{eq} \equiv R_1 + \frac{R_2'}{s} \qquad [\Omega]. \tag{8.15}$$

Similarly, we combine the reactive elements into an equivalent reactance:

$$X_{eq} \equiv X_1 + X_2 \qquad [\Omega]. \tag{8.16}$$

We make the important observation that the speed affects R_{eq} but not X_{eq}. From Figure 8.15b we note that the current is

$$I_2' = I_1 = \frac{V}{R_{eq} + jX_{eq}} \qquad [A]. \tag{8.17}$$

The magnitude (rms) of the current is

$$|I_2'| = |I_1| = \frac{|V|}{\sqrt{R_{eq}^2 + X_{eq}^2}} = \frac{|V|}{X_{eq}\sqrt{1 + (R_{eq}/X_{eq})^2}} \qquad [A]. \tag{8.18}$$

Its phase angle ϕ (relative to V) can be computed from

$$\phi = \cos^{-1}\left[\frac{R_{eq}}{\sqrt{R_{eq}^2 + X_{eq}^2}}\right] = \cos^{-1}\left[\frac{1}{\sqrt{1 + (X_{eq}/R_{eq})^2}}\right] \tag{8.19}$$

Equations (8.18) and (8.19) confirm the following:

> Increasing the speed (that is, decreasing slip) has the effect of decreasing the values of both $|I_1|$ and ϕ.

We obtain a very interesting interpretation of the variation of the current, I_1 by the following geometric process.

Decompose the current $|I_1|$ into the "in-phase" component, I_p and the "out-of-phase" component, I_q as shown in Figure 8.16. For these components we have

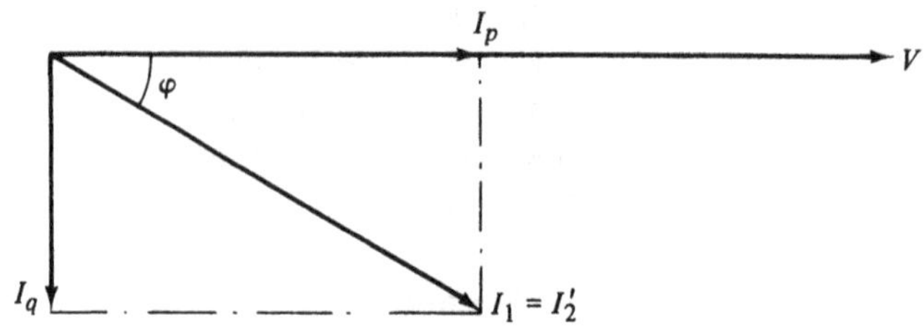

Figure 8.16

$$I_{\mathrm{p}} = |I_1| \cos \phi = \frac{|V|}{R_{\mathrm{eq}}^2 + X_{\mathrm{eq}}^2} \cdot R_{\mathrm{eq}} \qquad [\mathrm{A}]; \tag{8.20}$$

$$I_{\mathrm{q}} = |I_1| \sin \phi = \frac{|V|}{R_{\mathrm{eq}}^2 + X_{\mathrm{eq}}^2} \cdot X_{\mathrm{eq}} \qquad [\mathrm{A}]. \tag{8.21}$$

Squaring equations (8.20) and (8.21) and adding them we obtain

$$I_{\mathrm{p}}^2 + I_{\mathrm{q}}^2 = \frac{|V|^2}{R_{\mathrm{eq}}^2 + X_{\mathrm{eq}}^2} = \frac{|V|}{X_{\mathrm{eq}}} \cdot I_{\mathrm{q}}. \tag{8.22}$$

The last step follows directly from equation (8.21).

Equation (8.22) can be rewritten as

$$I_{\mathrm{p}}^2 + \left(I_{\mathrm{q}} - \frac{|V|}{2X_{\mathrm{eq}}}\right)^2 = \left(\frac{|V|}{2X_{\mathrm{eq}}}\right)^2. \tag{8.23}$$

This is the equation for a circle in the I_{p}–I_{q} plane. The circle has the radius $|V|/2X_{\mathrm{eq}}$. Its center is located in the point $[0, |V|/2X_{\mathrm{eq}}]$. As the slip s changes, causing a change in R_{eq}, the tip of the current phasor, I_1, will move along a circular locus (Figure 8.17). This is called an "induction motor circle diagram."

Example 8.1

A three-phase, 6-pole induction motor rated at 10 hp, 220 V, 60 Hz has the following impedance parameters referred to the stator:

$$R_1 = 0.295\ \Omega \text{ per phase,}$$

$$R_2 = 0.150\ \Omega \text{ per phase,}$$

$$X_1 = 0.510\ \Omega \text{ per phase,}$$

$$X_2 = 0.210\ \Omega \text{ per phase.}$$

1. Draw the circle diagram for this motor if it is operated from a 220-V, three-phase network. Model the motor as an IM.

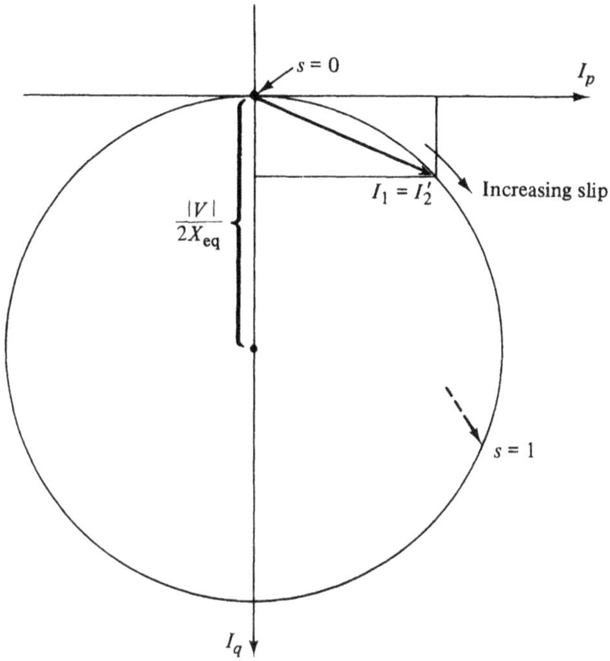

Figure 8.17

2. Find the stator current if the motor runs at synchronous speed.
3. Find the stator current at standstill. Also compute the ohmic losses and the power drained from the network.
4. Compute the stator current and ohmic losses if the motor is running at 97% of synchronous speed.
5. Compute the power drained from the ac source in step 4.

Solution: With the numerical data given we compute:

$$|V| = \frac{220}{\sqrt{3}} = 127.0 \qquad \text{[volts/phase]} \tag{8.1.1}$$

$$X_{eq} = 0.510 + 0.210 = 0.720 \qquad [\Omega/\text{phase}] \tag{8.1.2}$$

1. The radius of the circle will be:

$$\frac{|V|}{2X_{eq}} = \frac{127.0}{2 \times 0.720} = 88.2 \qquad \text{[A/phase]} \tag{8.1.3}$$

2. At synchronous speed we have $s = 0$. Thus,

$$R_{eq} = 0.295 + \frac{0.150}{0} = \infty \tag{8.1.4}$$

From equations (8.18) and (8.19) we obtain

$$|I_1| = 0; \qquad \cos\phi = 1. \tag{8.1.5}$$

3. At standstill we have $s = 1$. Thus,

$$R_{eq} = 0.295 + \frac{0.150}{1} = 0.445 \qquad [\Omega \text{ per phase}]. \tag{8.1.6}$$

From equations (8.18) and (8.19) we obtain:

$$|I_1| = \frac{127.0}{\sqrt{0.445^2 + 0.720^2}} = 150.0 \qquad [\text{A per phase}]; \tag{8.1.7}$$

$$\phi = \cos^{-1}\left(\frac{0.445}{\sqrt{0.445^2 + 0.720^2}}\right) = 58.28°. \tag{8.1.8}$$

We have marked the position of the current phasor, I_1, for these two cases in Figure 8.17.

According to the equivalent circuit, the real power drained from the ac network will be

$$P_1 = R_{eq}|I_1|^2 = \left(R_1 + \frac{R_2'}{s}\right)|I_1|^2 \qquad [\text{W per phase}]. \tag{8.1.9}$$

In this case we have, at standstill,

$$P_1 = \left(0.295 + \frac{0.150}{1}\right) \times (150.0)^2 = 10.01 \qquad [\text{kW per phase}] \tag{8.1.10}$$

or

$$3 \cdot 10.01 = 30.03 \qquad [\text{kW}] \qquad (\text{3-phase}). \tag{8.1.11}$$

The power dissipated in the stator and rotor windings is

$$P_\Omega = R_1|I_1|^2 + R_2|I_2|^2 = R_1|I_1|^2 + R_2|aI_1|^2 \qquad [\text{W per phase}]; \tag{8.1.12}$$

$$P_\Omega = (R_1 + a^2R_2)|I_1|^2 = (R_1 + R_2')|I_1|^2 \qquad [\text{W per phase}]. \tag{8.1.13}$$

Thus,

$$P_\Omega = P_1 \qquad [\text{W per phase}]. \tag{8.1.14}$$

In other words, all the power drained from the network goes into losses. This is, however, only true at standstill because

$$R_{eq} = R_1 + R_2' \qquad [\Omega \text{ per phase}] \tag{8.1.15}$$

only for $s = 1$.

Caution: With a total power loss of 30 kW this motor would rapidly overheat. A stalled motor must be quickly disconnected from the source.

4. The slip is now 3%. From equation (8.15)

$$R_{eq} = 0.295 + \frac{0.150}{0.03} = 5.295 \qquad [\Omega \text{ per phase}]. \tag{8.1.16}$$

From equations (8.18) and (8.19) we get

$$|I_2'| = |I_1| = \frac{127.0}{\sqrt{5.295^2 + 0.720^2}} = 23.77 \qquad \text{[A per phase]} \tag{8.1.17}$$

and

$$\phi = \cos^{-1}\left(\frac{5.295}{\sqrt{5.295^2 + 0.720^2}}\right) = 7.74°. \tag{8.1.18}$$

Where will this current phasor be positioned in the circle diagram of Figure 8.17?

The ohmic losses accordingly will be

$$P_\Omega = 0.445 \cdot 23.77^2 = 251 \qquad \text{[W per phase]}. \tag{8.1.19}$$

The total power loss in all three phases is 753 W. Compare this value with the 30-kW loss at standstill. Note, in addition, that the running motor has a better self-cooling capability than the stalled motor.

5. The power drained from the ac source is

$$P_1 = R_{eq}|I_1|^2 = 5.295 \cdot 23.77^2 = 2.992 \qquad \text{[kW per phase]}, \tag{8.1.20}$$

or

$$3 \cdot 2.992 = 8.975 \qquad \text{[kW]} \qquad \text{(3-phase)}. \tag{8.1.21}$$

8.5.6 Motor Power and Torque in the Ideal Motor

In Example 8.1, with the IM running at 3% slip it drained 8.975 kW from the ac source. Only 0.753 kW (or 8.39%) was actually dissipated in the resistors as ohmic heat. Where did the remaining 8.222 kW (or 91.61%) go? (Note that in the stalled motor in Example 8.1 **(3)**, 100% of the received power went into losses.)

As the IM has no other "power sinks" to account for the 8.222 kW we can only conclude that it passes through the motor to the load being pulled. In other words, it constitutes the motor power P_m, which corresponds to the electromechanical torque, T_m of the motor.[8] In general, we can write:

$$P_m = P_1 - P_\Omega \qquad \text{[W per phase]}, \tag{8.24}$$

where

$$P_1 = R_{eq}|I_1|^2 \qquad \text{[W per phase]} \tag{8.25}$$

and

$$P_\Omega = (R_1 + R_2')|I_1|^2 \qquad \text{[W per phase]}. \tag{8.26}$$

Substituting equations (8.25) and (8.26) into (8.24) we get

[8] Note that 8.222 kW corresponds to an output of 11.0 hp. This motor is rated at 10 hp; we conclude that when running at 3% slip this motor is slightly overloaded.

$$P_m = \frac{1-s}{s} R_2' |I_1|^2 \qquad \text{[W per phase]} \tag{8.27}$$

or

$$P_m = 3\frac{1-s}{s} R_2' |I_1|^2 \qquad \text{[W]} \qquad \text{(3-phase)}. \tag{8.28}$$

The motor power,

$$P_m = \omega_m T_m \qquad \text{[W]}, \tag{8.29}$$

where

$$\omega_m = 2\pi \frac{n}{60} = \frac{\pi}{30}(1-s)n_s \qquad \text{[rad/s]}. \tag{8.30}$$

By combining equations (8.28), (8.29), and (8.30) we obtain

$$T_m = \frac{90}{\pi n_s} \frac{R_2'}{s} |I_1|^2 \qquad [\text{N} \cdot \text{m}]. \tag{8.31}$$

Equations (8.28) and (8.31) give two of the most important criteria of performance. Equation (8.31) gives the same information as (8.1). The important difference, however, is that (8.31) expresses the motor torque in measurable quantities.

8.5.7 Maximum Torque of the Ideal Motor

From equation (8.31) it is clear that the motor torque is a function of speed (or slip). We had earlier (see Figure 8.7) alluded to the fact that somewhere in the speed range, $0 - n_s$ rpm, (or slip range, $1 - 0$) one can expect to find a maximum value, T_{max}, for the torque. It is important to know this value in order to avoid the possibility of stall.

Before we proceed to find T_{max} we substitute the expression for $|I_1|$ in (8.18) into (8.31), which gives

$$T_m = \frac{90}{\pi n_s} \frac{R_2'}{s} \frac{|V|^2}{(R_1 + R_2'/s)^2 + (X_1 + X_2')^2} \qquad [\text{N} \cdot \text{m}]. \tag{8.32}$$

The torque is now expressed as a function of slip.

Several observations can be made about equation (8.32):

1. Although in the range of slip $0 < s < 1$ is of particular importance in normal motor operation, equation (8.32) gives the torque in the complete range of slip $-\infty > s > +\infty$.

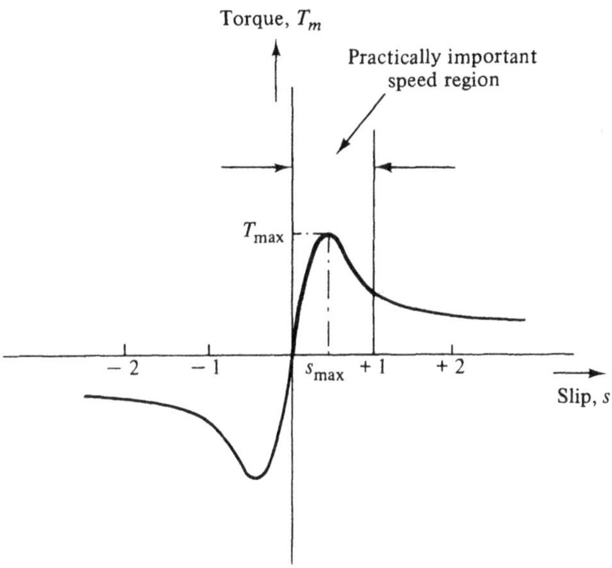

Figure 8.18

2. The torque is positive for all $s > 0$. This is the region of speed for motor operation.
3. The torque is negative for all $s < 0$. This is the region of speed for generator operation.
4. The magnitude of the torque approaches zero for $s = 0$ and for $s = \pm\infty$.
5. For the same magnitude of slip, s the numerical value of the torque is larger when s is negative than when s is positive. (This follows because the denominator of equation (8.32) is numerically smaller for negative values of s.)

Taking all of the above into account, we obtain the curve for torque against slip as shown in Figure 8.18. We have identified the important speed[9] region, $0 < s < 1$.

We can expect to find the maximum positive torque, T_{max} occurring at $s = s_{max}$. We can also expect to find a negative maximum. The positive maximum located in the important speed region is the object for our search. One seeks the value s_{max} that satisfies the equation[10]

$$\frac{dT_m}{ds} = 0, \tag{8.33}$$

[9] Note that the torque in Figure 8.7 appears "inverted" as compared to Figure 8.18 because it is plotted against n rather than s.

[10] If one realizes that the torque (8.32) is a function of R_2'/s one can simplify the analysis by solving the simpler equation

$$\frac{dT_m}{d(R_2'/s)} = 0.$$

which gives

$$s_{max} = \pm \frac{R_2'}{\sqrt{R_1^2 + (X_1 + X_2')^2}}. \tag{8.34}$$

If we limit our attention to the positive torque and substitute the positive value of s_{max} into equation (8.32), we obtain

$$T_{max} = \frac{45}{\pi n_s} \frac{|V|^2}{R_1 + \sqrt{R_1^2 + (X_1 + X_2')^2}} \quad [\text{N} \cdot \text{m}] \tag{8.35}$$

Example 8.2

Consider the motor whose impedance data is given in Example 8.1.

1. Compute the torque
   (a) at standstill
   (b) at a speed corresponding to $s = 0.03$.
2. Compute the slip at maximum torque, s_{max} and the value of the maximum torque, T_{max}.

Solution: As the motor is a 6-pole machine, the synchronous speed $n_s = 1200$ rpm.

1. From Example 8.1
   (a) we have $s = 1$, when

$$|I_1| = 150.0 \quad [\text{A}]. \tag{8.2.1}$$

From equation (8.31) we get for the start or standstill torque,

$$T_{st} = \frac{90}{\pi \times 1200} \frac{0.150}{1} (150.0)^2 = 80.6 \quad [\text{N} \cdot \text{m}]. \tag{8.2.2}$$

   (b) From Example 8.1 we have $s = 0.03$, when

$$|I_1| = 23.77 \quad [\text{A}]. \tag{8.2.3}$$

From equation (8.31), the running torque is

$$T_m = \frac{90}{\pi \times 1200} \frac{0.150}{0.03} (23.77)^2 = 67.4 \quad [\text{N} \cdot \text{m}]. \tag{8.2.4}$$

2. From equation (8.34) we get

$$s_{max} = \frac{0.150}{\sqrt{0.295^2 + 0.720^2}} = 0.193. \tag{8.2.5}$$

From equation (8.35) we get

$$T_{max} = \frac{45}{\pi \cdot 1200} \cdot \frac{127.0^2}{0.295 + \sqrt{0.295^2 + 0.720^2}} = 179.4 \quad [\text{N} \cdot \text{m}]. \tag{8.2.6}$$

Compare T_{max} to the start and running torques.

8.5.8 Torque Control by Variation of Rotor Resistance

Earlier we surmised that by inserting external resistances into the rotor circuit we could exert some control over the magnitude of the motor torque. The theory developed above permits us to determine the effect of such resistors.

From equation (8.34) we find that the value of s_{max} is directly proportional to the rotor resistance. Therefore by adding resistance to the rotor circuit we can move the point at which maximum torque occurs toward higher values of s, that is, in the direction of lower speeds.

From equation (8.35) we find that T_{max} is independent of the rotor resistance.

By combining these two observations we realize that insertion of extra resistance in the rotor circuit shifts the torque curves in a manner indicated in Figure 8.19.

Example 8.3

By shifting the torque curve for the machine in Example 8.2 so that the maximum torque moves from $s_{max} = 0.193$ to $s_{max} = 1.00$ we have arranged to have the maximum torque occur at $n = 0$, that is, at the start. This will result in a fast starting motor.[11] How much resistance must be inserted in the rotor circuit?

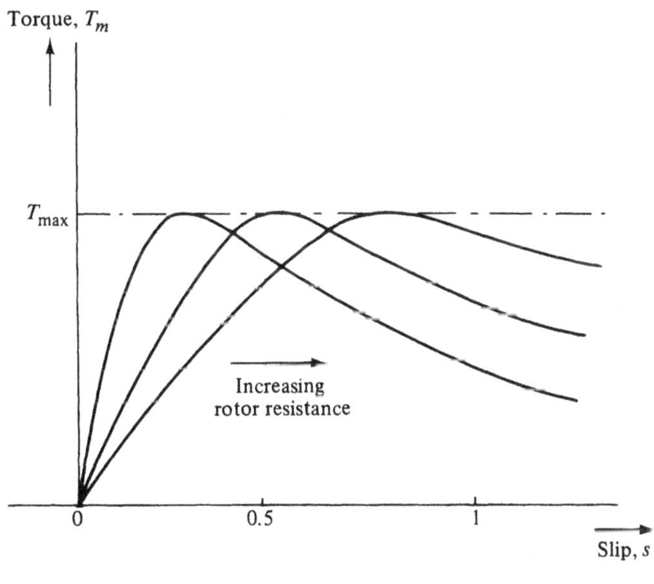

Figure 8.19

[11] As the motor accelerates one can shift s_{max} toward lower s values (by reducing the external rotor resistance) and seeking to match s_{max} with the actual s. In this manner the motor torque is at its peak value during the total startup time.

Solution: From equation (8.34) we get

$$s_{max} = 1 = \frac{R_2'}{\sqrt{0.295^2 + 0.720^2}}. \tag{8.3.1}$$

This equation yields

$$R_2' = 0.778 \qquad [\Omega \text{ per phase}]. \tag{8.3.2}$$

If we deduct the rotor winding resistance (0.150 Ω) we obtain the required external resistance:

$$R_{2ext}' = 0.778 - 0.150 = 0.628 \qquad [\Omega \text{ per phase}]. \tag{8.3.3}$$

We should remember that this resistance is referred to the stator side. To obtain the actual value we must divide it by a^2.

8.6 Modification of the Model for Nonideal Motor Characteristics

The previous analysis was based on the assumptions made for the "ideal motor." It is appropriate at this time to make modifications to the various mathematical models to account for the "nonideal" behavior of a real motor.

8.6.1 Inaccuracy of the Ideal Motor Model at Light Load

If the mechanical load of an ideal motor were removed, its speed should rise to n_s rpm. The slip would become zero and the impedance element R_2'/s in the equivalent circuit would approach infinity. The IM equivalent circuit would then predict zero current and power.

A practical motor would behave quite differently. Under no-load conditions, its slip would not be reduced to zero but to a value usually about 1%. Windage and friction losses will require a small but nonzero input power to overcome them, hence a nonzero slip can be expected. The impedance element R_2'/s would be large but not infinite. Should we use the IM equivalent circuit to model the actual motor at such low-load levels, it would produce current and power values which are in poor agreement with reality.

8.6.2 Existence of Excitation or Magnetization Current

In developing the theory that led to the "ideal" model for the induction motor we neglected the reluctance of both the iron core and the air gap. This assumption leads to the conclusion that it requires zero mmf and thus zero current to maintain the core flux. In reality, the iron core does not have infinite permeability and the air gap is not of zero width.[12] Thus, the core of a real machine requires a nonzero mmf to maintain the flux. Expressed differently, the core of a real machine does not represent an infinite reactance as viewed from the source but a finite reactance of value X_m Ω per phase.

In addition, the real core requires a finite amount of real power, to overcome hysteresis and eddy current losses. These losses as viewed from the source can be represented by a finite resistance of value R_m Ω per phase. Together, R_m and X_m constitute a *magnetization impedance*, which absorbs a magnetization or excitation current, I_m, which is essentially independent of the mechanical loading of the motor.

8.6.3 Modification of the Ideal Motor Equivalent Circuit

One logical way[13] of accounting for the presence of I_m is shown in Figure 8.20. The magnetization current I_m is drawn by the shunt elements R_m and X_m. As before, the rotor slip results in the torque-creating rotor current I_2, which is reflected to the stator side and appears in our circuit as I_2'. The stator current is the phasor sum of the two, hence

$$I_1 = I_m + I_2' \quad \text{[A per phase]}. \tag{8.36}$$

As long as the source voltage remains constant, the I_m component is unchanged. As before, the mechanical load and hence the slip will determine the value of I_2'.

8.6.4 The Effect of the Circuit Modification on the Circle Diagram

As before, the tip of the current phasor I_2' will follow a circle (see Figure 8.17). As the primary current I_1 is the phasor sum of I_2' and the constant phasor, I_m we obtain the modified circle diagram shown in Figure 8.21. The modified circle diagram reveals the following important facts:

[12] Example 3.22 shows how even a very small air gap in a magnetic circuit drastically increases the mmf required (that is, the current needed) to maintain the flux. Contemplate what factors determine the minimum air gap width for a real motor.

[13] Many will disagree and point out that a "better" model of the physical motor can be obtained by placing the shunt elements *after* the stator impedance, $R_1 + jX_1$, as shown in the dashed line in Figure 8.20. This is debatable for several reasons. There are *additional* losses beyond windage, friction, core, and copper losses. They are sometimes conveniently lumped into a group and called "stray" losses. These losses are due to harmonic effects (which we have neglected), rotor hysteresis, and other causes. Generally, they are difficult to model. Sometimes one simply lumps them together with the core losses, sometimes one divides them between the core and copper losses. The point is that the modification of the model under discussion represents a second-order effect. The question of whether the shunt impedance should be connected before or after the stator impedance therefore corresponds to a second-order effect of a second-order effect.

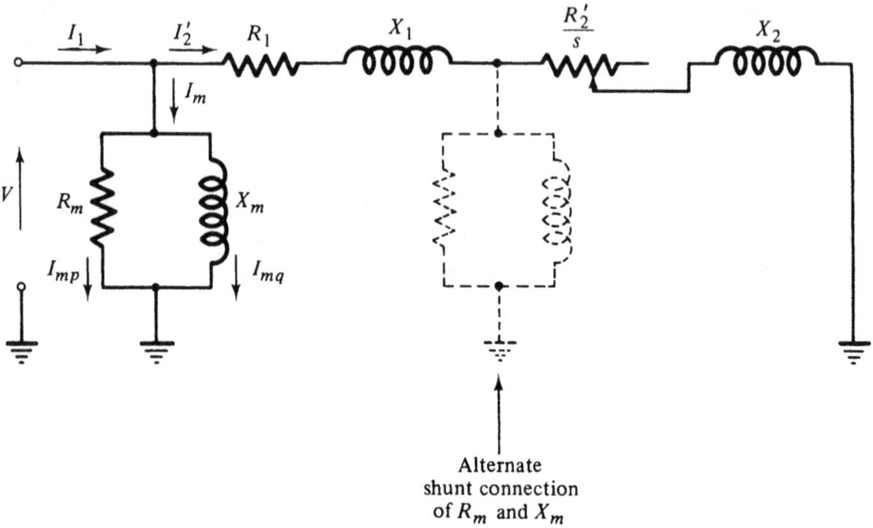

Figure 8.20

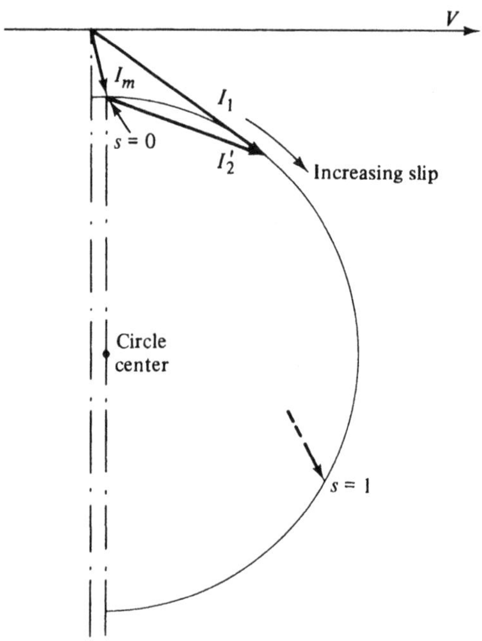

Figure 8.21

1. For a heavily loaded motor, that is, when the slip is relatively large, the currents I_2' and I_1 tend to approach each other in both magnitude and phase as measured in relative terms. For this mode of operation equation (8.11) gives the best approximation and the IM model is valid.
2. For light loads, that is, when the slip approaches zero, the disparity between I_1, and I_2' becomes quite large. For this mode of operation, the IM model will cause significant errors. We have to use the modified version of the circle diagram.

Example 8.4

Consider the 6-pole motor discussed in Example 8.1. The motor is made the subject of a no-load test (it carries no mechanical load other than its own friction and windage loads) at rated voltage. The speed, stator current, and power recorded are as follows:

$$\text{speed} = 1198.6 \text{ rpm},$$

$$\text{stator current} = 7.51 \text{ A per phase},$$

$$\text{power drain from network} = 0.503 \text{ kW (total, 3-phase)}.$$

1. Compare these test results with the power and current you can compute from the IM equivalent circuit.
2. Clear up the disagreements in the model in item 1. Also, from the test data compute the magnetization impedance and core losses.
3. Use the modified model to "improve" the current and power data computed in Example 8.1, items 4 and 5. We make the very reasonable assumption that the windage and friction losses require about the same input power as before and therefore the slip is 3%.

Solution:

1. The measured speed corresponds to a slip,

$$s = \frac{1200 - 1198.6}{1200} = 0.001167. \tag{8.4.1}$$

From the IM equivalent circuit diagram we obtain

$$R_{eq} = 0.295 + \frac{0.150}{0.001167} = 128.9 \qquad [\Omega \text{ per phase}] \tag{8.4.2}$$

and

$$X_{eq} = 0.510 + 0.210 = 0.720 \qquad [\Omega \text{ per phase}] \tag{8.4.3}$$

The current is

$$I_1 = I_2' = \frac{V}{R_{eq} + jX_{eq}} = \frac{127.0}{128.9 + j0.72} = 0.985\angle -0.32^\circ \qquad [\text{A per phase}]. \tag{8.4.4}$$

The real power drawn from the source is

$$P_1 = 127.0 \cdot 0.985 \cdot \cos 0.32° = 125 \qquad \text{[W per phase]} \tag{8.4.5}$$

or 375 W (total, 3-phase).

We note that the computed current, $|I_1| = 0.985$ A, compares poorly with the measured value, 7.51 A. There is likewise a considerable difference between the computed (375 W) and measured (503 W) power. An even more noticeable disagreement occurs in the phase angles. The computed value is, $\phi = 0.32°$ lagging. The power factor, according to the measurements, is

$$\cos\phi = \frac{503/3}{127.0 \cdot 7.51} = 0.176, \tag{8.4.6}$$

which corresponds to a phase angle between voltage and current of $\phi = 79.9°$.

Conclusion: In this case, the IM model, is not a good representation of the physical device. We shall identify the reasons for the disagreements later.

2. In view of the need to modify the IM model to account for the magnetization current we proceed as follows:

(a) The current computed in item 1 (0.985 A) is not I_1 but I_2'. Figure 8.22 shows an enlarged view of the circle diagram in the region of zero slip and also the relationship between I_1, I_2', and I_m.

(b) The power computed in item 1 (375 W) is the sum of the ohmic losses in the rotor and stator windings plus the motor power, that is, the power needed to overcome windage and friction losses.

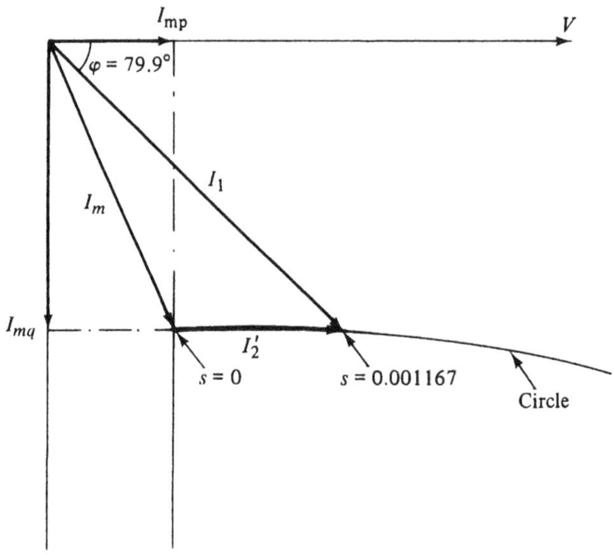

Figure 8.22

As the ohmic losses in the test was only 17 W (how does one get this figure?), we can for all practical purposes say that the windage and friction losses constitute the total computed power of 375 W.

(c) From Figure 8.22 we can compute the magnetization current I_m as follows:

Step I. I_m is first resolved into the in-phase component I_{mp} and the out-of-phase component I_{mq}.

Step II. Because I_2' for all practical purposes, is parallel to V, we obtain

$$I_{mp} \approx |I_1| \cos\phi - |I_2'| = 7.51 \cos 79.9° - 0.985 = 0.332 \quad [\text{A}]; \tag{8.4.7}$$

$$I_{mq} \approx |I_1| \sin\phi = 7.51 \sin 79.9° = 7.39 \quad [\text{A}]. \tag{8.4.8}$$

Step III.

$$|I_m| = \sqrt{I_{mp}^2 + I_{mq}^2} = \sqrt{0.332^2 + 7.39^2} = 7.40 \quad [\text{A}]. \tag{8.4.9}$$

Step IV.

$$\angle I_m = -\tan^{-1}\left(\frac{I_{mq}}{I_{mp}}\right) = -87.4°. \tag{8.4.10}$$

(d) The magnetization impedance elements shown in Figure 8.20 are obtained from

$$R_m = \frac{|V|}{I_{mp}} = \frac{127.0}{0.332} = 383 \quad [\Omega \text{ per phase}]; \tag{8.4.11}$$

$$X_m = \frac{|V|}{I_{mq}} = \frac{127.0}{7.39} = 17.2 \quad [\Omega \text{ per phase}]. \tag{8.4.12}$$

(e) Total core loss is equal to (503 − 375) = 128 W. (Note that you can also compute it using $3|V|^2/R_m$.)

3. In Example 8.1, item 4 we computed

$$I_2' = 23.77\underline{/-7.74°} \quad [\text{A per phase}]. \tag{8.4.13}$$

In item 2, equations (8.4.9) and (8.4.10), we found

$$I_m = 7.40\underline{/-87.4°} \quad [\text{A per phase}]. \tag{8.4.14}$$

From equation (8.36) we have

$$I_1 = 23.77\underline{/-7.74°} + 7.40\underline{/-87.4°} = 26.13\underline{/-23.92°} \quad [\text{A per phase}]. \tag{8.4.15}$$

The power drained from the ac source is

$$\begin{aligned} P_1 &= 3|V|\,|I_1| \cos\phi = 3 \cdot 127.0 \cdot 26.13 \cdot \cos 23.92° \\ &= 9.100 \quad [\text{kW}] \quad (\text{3-phase}) \end{aligned} \tag{8.4.16}$$

(as compared to 8.975 kW based on the IM model).

The output power, P_{out} is equal to the input power minus the losses:

$$P_{out} = P_1 - (P_{core} + P_{WF} + P_\Omega) \quad [\text{W}]. \tag{8.4.17}$$

But from the test results we have

$$P_{core} + P_{WF} = 0.503 \qquad [kW] \tag{8.4.18}$$

and

$$P_{\Omega} = 3R_1|I_1|^2 + 3R_2'|I_2|^2 \qquad [W] \tag{8.4.19}$$

$$P_{\Omega} = 3 \cdot 0.295 \cdot 26.13^2 + 3 \cdot 0.150 \cdot 23.77^2$$
$$= 0.857 \qquad [kW] \qquad (3\text{-phase}). \tag{8.4.20}$$

Therefore,

$$P_{out} = 9.100 - (0.503 + 0.857) = 7.740 \qquad [kW]. \tag{8.4.21}$$

(This value should be compared to $P_m = 8.222$ [kW] computed on the basis of the IM model.)

For the motor efficiency we have

$$\eta = \frac{P_{out}}{P_1} = \frac{7.740}{9.100} \times 100 = 85.1\%. \tag{8.4.22}$$

(The IM model gave the better value of 91.6%.)

8.7 Operational Considerations

The greatest advantages of the induction motor are its simple design, ruggedness, reliability and its ability to run directly off the ac power network. Its major disadvantages are as follows:

1. High starting current
2. Limited control of speed and torque
3. Low efficiency when operating at high slips (low speed)

8.7.1 High Starting Current During Direct Start

By "direct start" we mean that the motor is switched directly onto the ac power network. It can be seen from the equivalent circuit that the motor at standstill ($s = 1$) offers a very low[14] impedance to the network. The equivalent reactance X_m, amounts to the small leakage value. The equivalent resistance R_{eq} is equal to the sum of the rotor and stator winding resistances. This means that as the motor is connected to the network a high starting current will flow. As the motor speeds up, the equivalent resistance R_2'/s will increase with a resulting decrease in the current.

[14] The lowest impedance and hence the highest current, occurs when $R_{eq} = R_1 + R_2'/s = 0$; that is, for $s = -R_2'/R_1$. (Find the point representing this condition on the circle diagram)

The high starting current will cause undesirable voltage drops in the feeding transformer and/or feeder lines. These voltage drops can interfere with the operation of other load objects on the line. In practice, the problem can become acute if the load, driven by the motor, has high inertia and the starting torque is low. Such a combination would result in long starting times. Direct start can also damage the load if it is not designed to tolerate the sudden application of the torque.

We now discuss a few methods for alleviating the problem arising from the voltage drop due to starting the induction motor.

8.7.1.1 Insertion of External Rotor Resistance

If the motor is of the wound-rotor type (which would rarely be the case with a 10-hp motor because a squirrel-cage rotor is considerably cheaper) external rotor resistance would normally be inserted to shift the maximum torque to $s = 1$. (See Example 8.3.) The added resistance, will of course, help reduce the starting current and hence the voltage drop. The effect is, however, not too pronounced because the major portion of the voltage drop across the transformer and line impedances (which are predominantly reactive) is caused by the out-of-phase component of the current. Addition of rotor resistance will reduce the in-phase component of the current.

8.7.1.2 Using a Starting Compensator

The only possible means of softening the shock of the starting current in a squirrel-cage-rotor motor is to reduce the starting voltage. A common way of doing this is to use a starting autotransformer sometimes referred to as a "compensator" (see Figure 8.23). The motor is started with the circuit breaker in the position *S*. When the motor attains running speed, the breaker is thrown to position *R*. The

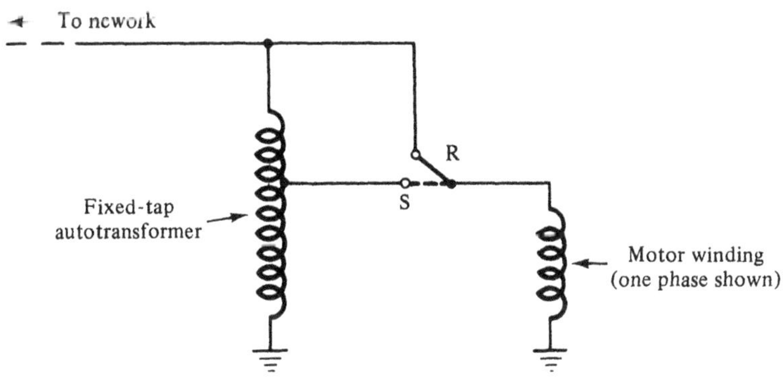

Figure 8.23

starting current and hence the voltage drop are reduced in the same ratio as the autotransformer tap.

The price to be paid for this type of starting current control is a greatly reduced starting torque and therefore prolonged starting periods. Note from equation (8.32) that the motor torque varies as the square of the voltage.[15] For example, if the transformer tap is set at 50%, the torque will be reduced by the ratio 1:4.

Note that the motor is disconnected from the network during the breaker transition from S to R. If the transition is slow then the flux may decay and/or fall back one pole. A very high but short reclosing current surge will then occur. To prevent this one can add in the S lead a transition impedance, which will preserve current continuity during the changeover.

8.7.1.3 The Y–Δ Starting Method

This starting method requires that the stator winding has all the six terminals accessible. By means of three double-throw switches (details not shown in Figure 8.24) the stator windings at start, are connected in Y. When the motor is running, the winding is reconnected in Δ.

In normal (run) operation each phase winding is connected to the line voltage. During start the same winding has across it only the phase voltage, that is, a voltage reduction in the ratio $1{:}\sqrt{3}$. The starting torque is therefore reduced in the ratio 1:3.

8.7.2 Efficiency at Low Speed

A distinct disadvantage of the induction motor is the reduced efficiency at low operating speeds. The major reason is the high ohmic losses in the rotor and stator resistances at high slip. From equation (8.26) we note that these losses, based on the IM model, are

$$P_\Omega = |I_1|^2(R_1 + R_2') \qquad \text{[W per phase]}. \tag{8.37}$$

The output or motor power, from equation (8.28) is

$$P_\text{m} = |I_1|^2 \frac{1 - s}{s} R_2' \qquad \text{[W per phase]}. \tag{8.38}$$

If we neglect the relatively small core, friction, and windage losses, the efficiency will be

$$\eta = \frac{P_\text{m}}{P_\text{m} + P_\Omega} = \frac{(1 - s/s)R_2'}{R_1 + (R_2'/s)} = \frac{(1 - s)R_2'}{sR_1 + R_2'}. \tag{8.39}$$

[15] The reason is the following: A reduction of the stator voltage in the ratio r results in a reduction of *both* the stator flux and rotor current. According to equation (8.1) the reduction of the torque has to be in the ratio r^2.

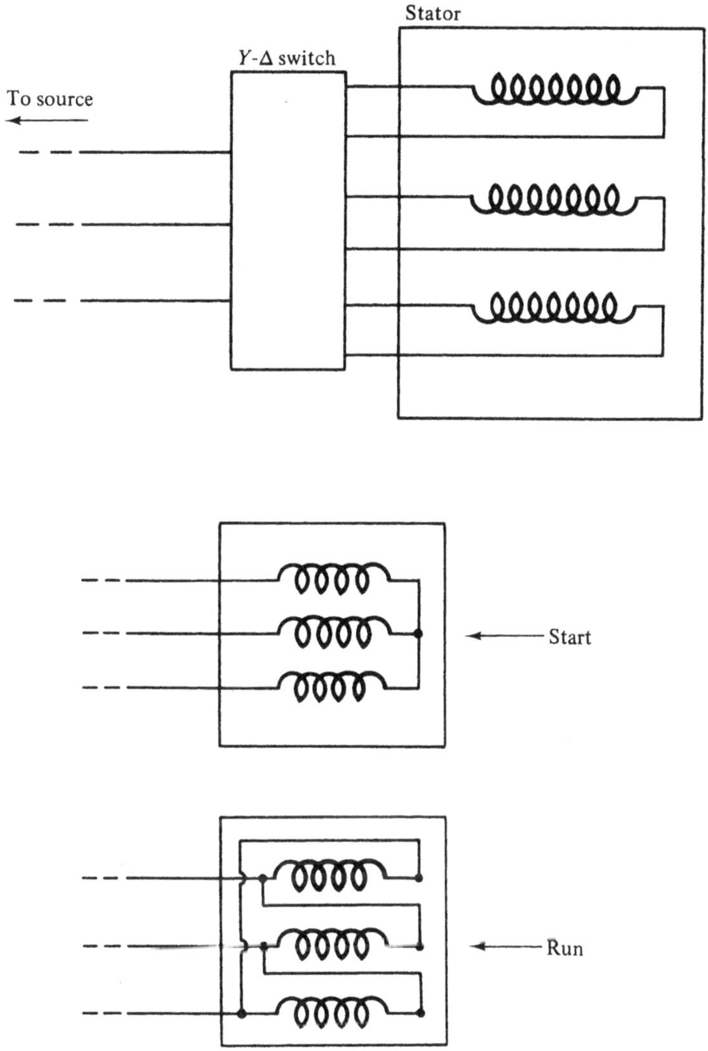

Figure 8.24

If, for simplicity, we set $R_1 \approx R_2'$, we get

$$\eta = \frac{1 - s}{1 + s}. \tag{8.40}$$

We note that the efficiency will be high at low slip, that is, at high speed. At half the synchronous speed, corresponding to $s = 0.5$, the efficiency will be 33%.

If the motor is of the squirrel-cage type, all of the ohmic losses are dissipated within the machine. If the motor is of the wound-rotor type with external resistors, at least part of the rotor losses occur outside the machine. These facts have important consequences:

1. The motor can be subjected to heavy heat stress when starting with loads of high inertia.
2. High sustained load torques will cause sustained low speeds. Careful attention must be paid to the heat dissipation within the motor in such situations. (The danger is increased by the fact that a slow rotor has decreased self-cooling capacity.)

8.7.3 Torque-Speed Control

One of the outstanding features of the separately excited dc machine is its ability to deliver a torque of any magnitude and direction at any speed.[16] The induction motor cannot claim such glowing attributes. In fact, as we have seen, one of the very basic features of the induction motor is the strong dependence of its torque on speed.

Furthermore, the speed of the induction motor is closely related to the frequency of its power supply (i.e., its synchronous speed) it becomes clear that we have very limited influence on the torque-speed characteristics. Our options are summarized in Figure 8.25. The figure shows the torque-speed curve, T_{m} of an induction motor. It also shows the torque speed curve, T_{L} of a given load that is

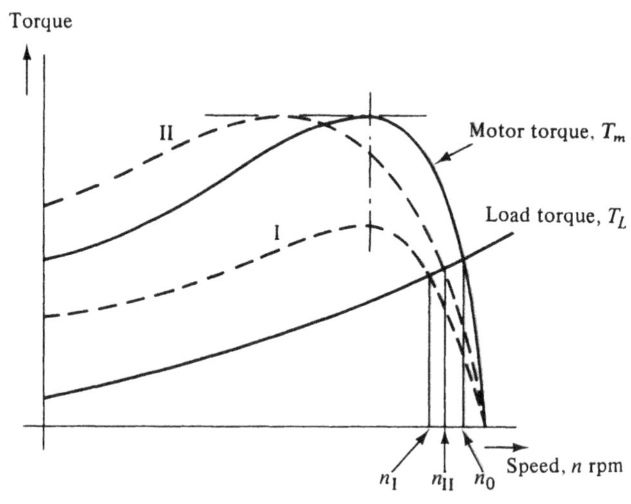

Figure 8.25

[16] Within the limitations set by the ratings, of course.

being pulled. The motor is running at the subsynchronous speed n_O rpm. Assuming that we want to lower the operating speed. Two possibilities exist.

8.7.3.1 Speed Control by Voltage Variation

The stator voltage V of the motor is lowered by some means available such as an autotransformer. As the magnitude of the torque is proportional to $|V|^2$, the reduced voltage gives the reduced torque shown in curve I. Torque equilibrium will now occur at the lower speed, n_1, which will be the new operating speed.

8.7.3.2 Speed Control by Rotor Rheostat Adjustment

By adding external resistance to the rotor circuit, the motor torque curve is shifted toward the left (curve II). Again torque equilibrium occurs at a lower speed, n_{II}. The three speeds, n_O, n_I, and n_{II}, all lie within a narrow subsynchronous range.

The speed of an induction motor may be varied over a wider range only if provisions are made to change its synchronous speed, n_s. The equation $n_s = 120f/p$ tells us that n_s can be varied by changing either p or f. In practice, both p and f can be changed. For example, it is not difficult to reconnect a 4-pole stator winding as an 8-pole stator winding. This would then change n_s from 1800 rpm to 900 rpm.

Neither is it difficult, by means of modern solid-state circuitry, to vary the frequency f within a wide range. Since the source voltage, V, is related to the frequency, f, and the flux, Φ, through the relationship

$$|V| \propto f \cdot \Phi, \tag{8.41}$$

it is necessary to vary $|V|$ in proportion to f in order to keep Φ, and hence the maximum torque, constant.

8.8 Single-Phase Induction Motors

For reasons of cost and simplicity one cannot always count on having a three-phase ac power supply. In a multitude of domestic, commercial, and sometimes even industrial cases only single-phase ac power is available. In applications where the power required is limited (usually below 1 hp), single-phase induction motors, sometimes collectively referred to as "fractional-horsepower motors," can fill the need. Together with the universal motor, the single-phase induction motor account for practically 100% of the low-power electric-motor market.

8.8.1 Stator Flux as a Function of Time and Distance

When dc current is applied to the coil shown in Figure 8.26, it is clear that the flux density, B_a in the air gap will be a function of distance, x, with B_{max} occurring at $x = 0$ and at $x = \pi D/2$, where D is the diameter of the rotor. When the current is ac, the flux density B_a will also be a function of time such that

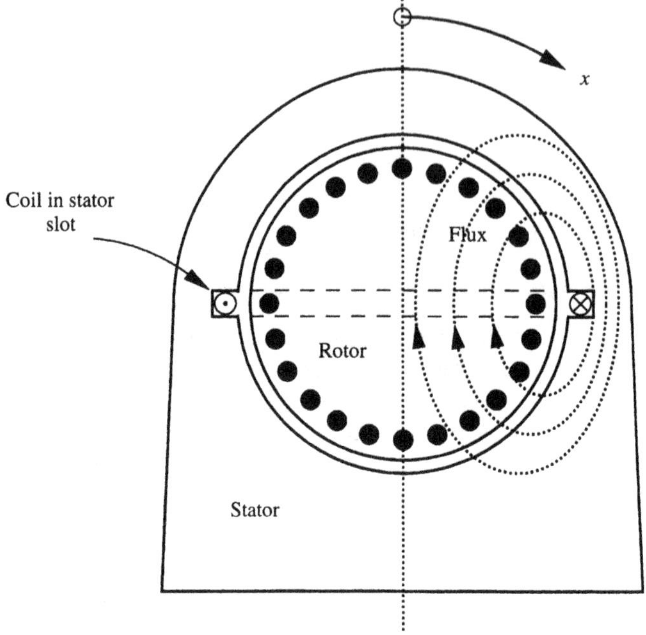

Figure 8.26

$$B_a = f(x,t). \tag{8.42}$$

Following the form of equation (4.34), we can write:

$$B_a(x,t) = B_{max} \cos(\beta x) \cos \omega t \qquad [\text{T}], \tag{8.43}$$

where $\beta = p/\text{D}$; p = number of poles; D = diameter of the rotor. It is clear that B_a will pulsate sinusoidally, in time, as a cosine function. B_a will have a maximum value B_{max} when $x = 0$ and when $x = \pi D/p$.

Figure 8.27 shows the flux density, B_a for a 2-pole machine as a function of x and for a few discrete values of t.

8.8.2 Equivalence of Pulsating and Revolving Fluxes

Using the trigonometric identity

$$\cos\alpha \cos\gamma = \frac{1}{2}\cos(\alpha - \gamma) + \frac{1}{2}\cos(\alpha + \gamma), \tag{8.44}$$

we can write:

$$B_a(x, t) = \frac{B_{max}}{2}\cos(\beta x - \omega t) + \frac{B_{max}}{2}\cos(\beta x + \omega t) \qquad [\text{T}]. \tag{8.45}$$

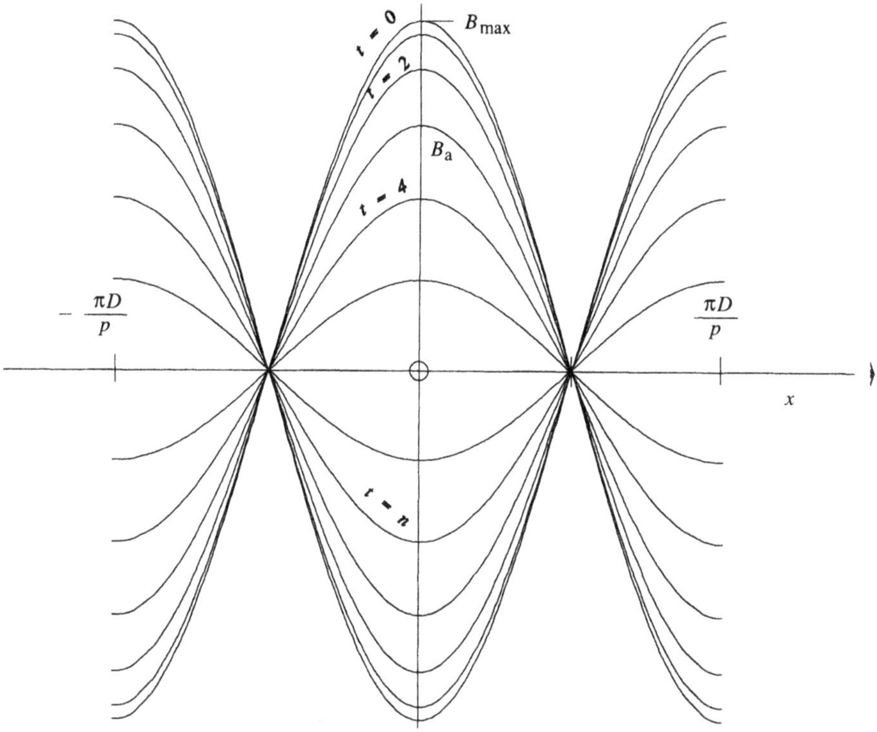

Figure 8.27

Let

$$B_a^+ = \frac{B_{max}}{2} \cos(\beta x - \omega t) \qquad [T] \tag{8.46}$$

and

$$B_a^- = \frac{B_{max}}{2} \cos(\beta x + \omega t) \qquad [T] \tag{8.47}$$

Figure 8.28 represents equation (8.45) in terms of rotating phasors. From Figure 8.28 it is clear that $B_a(x,t)$, the sinusoidally pulsating flux, can be resolved into two rotating fluxes B_{a+} and B_{a-}, rotating in opposite directions at angular velocities $+\omega$ and $-\omega$.

The pulsating flux produced by the single-phase ac current in the stator can be resolved into two equal rotating fluxes with directions opposite to each other. Each rotating flux will generate a torque in the rotor, but as they are in opposite directions, the resultant torque will be zero.

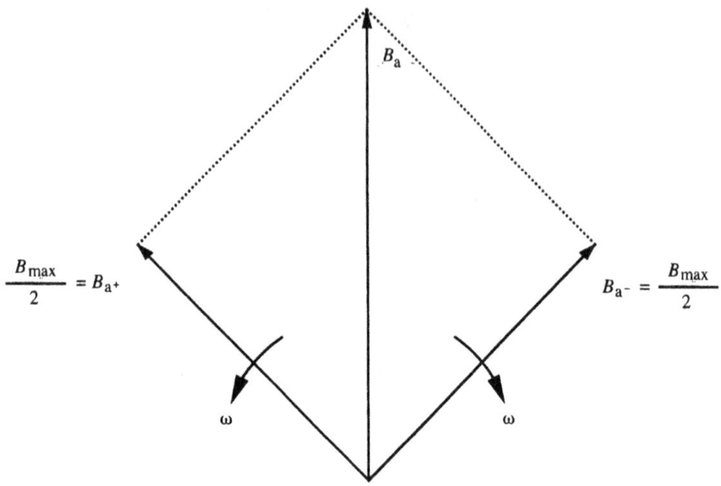

Figure 8.28

8.8.3 Representation of the Single-Phase Motor in Terms of the Three-Phase Motor

Since in the single-phase induction motor, we have two equal fluxes rotating in opposite directions, we can replace the single-phase induction motor by two identical three-phase machines connected so that they would rotate in opposite directions. We recall that the phase sequence, and hence the direction of rotation of a synchronous motor, can be reversed by interchanging any two of its phases (see Section 4.7.1). The two three-phase induction motors shown in Figure 8.29 are

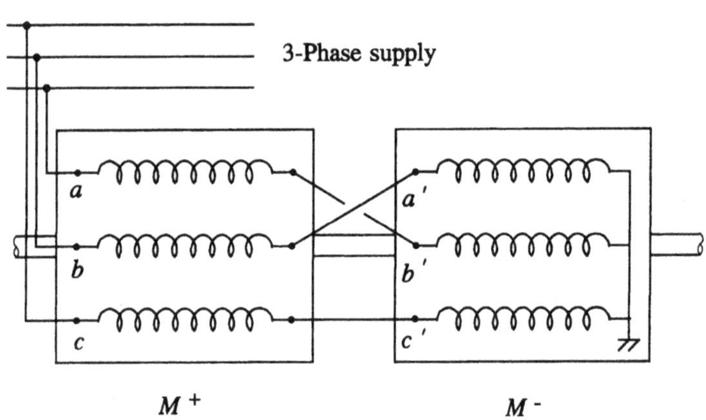

Figure 8.29

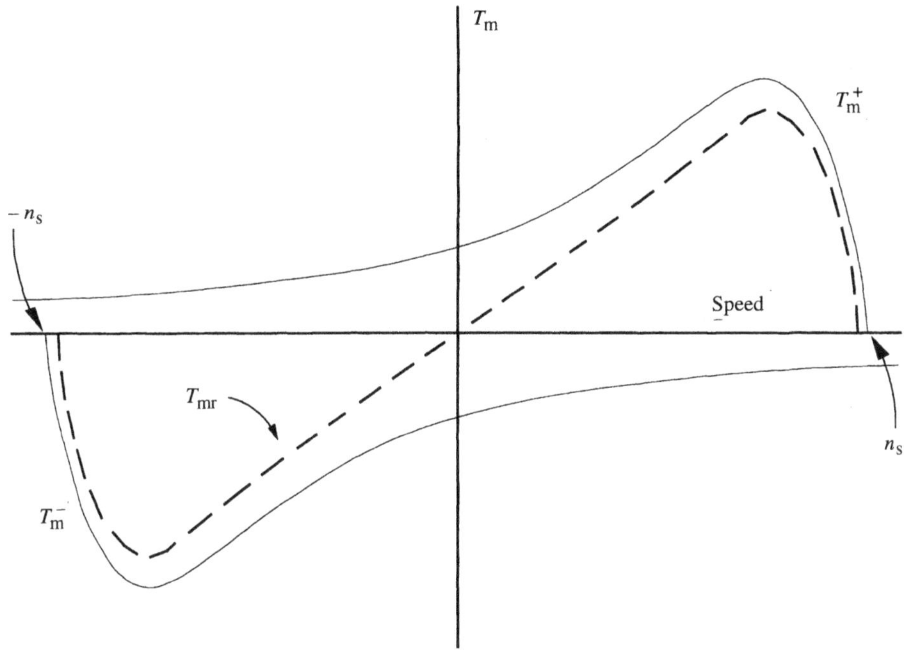

Figure 8.30

mechanically coupled to the same shaft and connected with phases a and b interchanged. Induction motor, M^+ generates a torque in the counterclockwise direction while M^- generates an equal torque but in the clockwise direction.

Figure 8.30 shows the torque-speed characteristics of the three phase induction motors, M^+, M^- and their combined characteristics. It can be seen from the diagram that at standstill, the resultant torque, T_{mr} is zero. The single-phase induction motor will therefore not start by itself. However, if it could (somehow) be persuaded to rotate in either direction then the torque is nonzero and the machine can do useful work.

We can write an expression for the torque using equation (8.31) where the current I_1 is the stator current for both motors:

$$T_m^+ = \frac{90}{\pi n_s} |I_1|^2 \frac{R_2'}{s_+} \qquad [\mathrm{N \cdot m}] \tag{8.48}$$

and

$$T_m^- = \frac{90}{\pi n_s} |I_1|^2 \frac{R_2'}{s_-} \qquad [\mathrm{N \cdot m}]. \tag{8.49}$$

The synchronous speed for the two machines are n_s and $-n_s$, respectively. The slips are, therefore,

$$s_+ = \frac{n_s - n}{n_s} = 1 - \frac{n}{n_s} \tag{8.50}$$

and

$$s_- = \frac{-n_s - n}{-n_s} = 1 + \frac{n}{n_s}. \tag{8.51}$$

From equations (8.50) and (8.51) we get

$$s_- = 2 - s_+. \tag{8.52}$$

From equations (8.48) and (8.49) we have

$$\frac{T_m^+}{T_m^-} = \frac{s_-}{s_+} = \frac{2 - s_+}{s_+}. \tag{8.53}$$

At standstill, $s_+ = 1$ and therefore

$$\frac{T_m^+}{T_m^-} = 1. \tag{8.54}$$

The dual drive system, similar to the single-phase motor, has zero starting torque.

8.8.4 The Torque: A Qualitative Assessment

Before we resume the discussion of the single-phase induction motor, it is useful to point out a number of its interesting characteristics.

Our *first observation* is that in the single-phase motor we have the following:

1. the stator flux B_{a+} revolving with speed $+\,n_s$ rpm,
2. the stator flux B_{a-} revolving with speed $-\,n_s$ rpm,
3. the rotor current A_+ revolving with speed $+\,n_s$ rpm,
4. the rotor current A_- revolving with speed $-\,n_s$ rpm.

Our *second observation* is that the fluxes and currents in a single-phase machine is considerably more complex than that in a three-phase machine. As a result the mathematical analysis of a single-phase induction motor is more complex than that of a three-phase motor.

Our *third observation* is that the single-phase motor lacks a starting torque.

Our *fourth observation* is that the single-phase induction motor will contain a pulsating torque component of frequency equal to twice that of the network frequency.

The flux wave, B_{a+} and current wave, A_- glide past each other with a relative speed of $2n_s$ rpm. The resulting torque between them will vary sinusoidally with the frequency, $2f$ Hz. (This can best be seen from equation (8.1) by setting the angle γ equal to $2\omega t$ where $\omega = 2\pi f$.) The same applies to the torque created by the flux wave, B_{a-} and current wave, A_+. The total 120-Hz vibratory torque component resulting from the interactions between B_{a+} and A_- and B_{a-} and A_+ will always be present in a single-phase motor.[17] We must live with it as best we can. The resulting noise and vibration, if they become a problem, can be effectively minimized by means of elastic motor supports.

8.8.5 An Explanation of Dual-Drive Torque Behavior

The ability of the system shown in Figure 8.29, to develop a driving torque in either direction has its identical counterpart in the single-phase motor. We need to explain this important phenomenon from a physical point of view. We need to examine the relationship between the magnitudes of the rotating fluxes of the two motors. When we spin the dual-drive rotor in the direction of one of the two counterrevolving fluxes the magnitude of that flux will increase, while the magnitude of the other flux will decrease. The reason can be seen from the equivalent circuit shown in Figure 8.31.

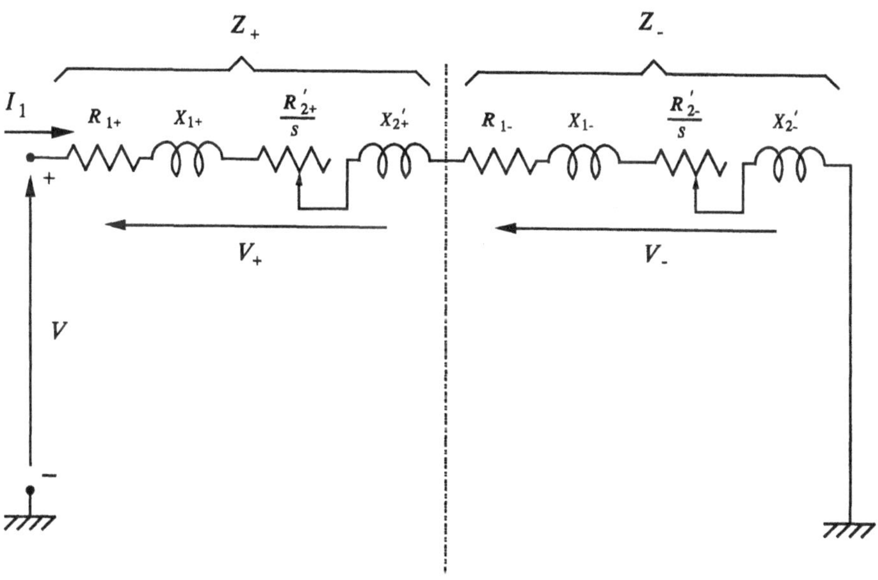

Figure 8.31

[17] The analog of two mechanically coupled three-phase motors for the single-phase motor is accurate with regard to its constant running torque component. Because the three-phase induction motors lack the pulsing torque component, they cannot simulate the 120 Hz torque in the single-phase motor.

The total source voltage V will divide between the two motors in the following proportions:

$$V_+ = \frac{Z_+}{Z_+ + Z_-} V \qquad [\mathrm{V}],$$
$$V_- = \frac{Z_-}{Z_+ + Z_-} V \qquad [\mathrm{V}], \tag{8.55}$$

where Z_+ and Z_- represent the "per phase" impedances of the stator windings. For simplicity, we use the IM model of each motor, which gives[18]:

$$Z_+ = R_1 + \frac{R_2'}{s_+} + j(X_1 + X_2') \qquad [\Omega/\text{phase}];$$
$$Z_- = R_1 + \frac{R_2'}{s_-} + j(X_1 + X_2') \qquad [\Omega/\text{phase}]. \tag{8.56}$$

The magnitude of the flux in a motor is directly proportional to the rms value of its stator voltage. This means that the flux in M_+ is proportional to $|V_+|$, and the flux in M_- to $|V_-|$.

At standstill, $s_+ = s_- = 1$ and consequently $Z_+ = Z_-$. The voltage V divides equally between the two motors. Their fluxes are equal and so are their torques.

If we give the motor a spin in the (+)-direction the slip s_+ will decrease and s_- will increase. Consequently, Z_+ will increase and Z_- will decrease. The motor, M_+ will receive a greater share of the source voltage than M_- and the magnitude of its flux will grow as that of M_- decreases. The torque, T_+ will grow as T_- is diminished.

We summarize the above in an abbreviated form as follows: When the rotation is in (+) direction,

$$\begin{array}{ll} s_+ \downarrow; & s_- \uparrow \\ Z_+ \uparrow; & Z_- \downarrow \\ V_+ \uparrow; & V_- \downarrow \\ T_{m+} \uparrow; & T_{m-} \downarrow \end{array}$$

Since T_m is proportional to the flux, Φ_a,

$$\Phi_{a+} \uparrow; \qquad \Phi_{a-} \downarrow$$

[18] In applying these equations to the single-phase motor, the impedances R_1, R_2', X_1, and X_2' should be replaced by *half* the corresponding values of the three-phase machine. A strict proof of this makes use of Kron's generalized machine transformation theory (Kron, 1930; White-Woodson, 1959) and this would go beyond the scope of this book. However, the reader should not have any difficulties in accepting the *heuristic* explanation that the single-phase motor, in essence consists of two identical series-connected three-phase motors each associated with one of the two counterrotating fluxes.

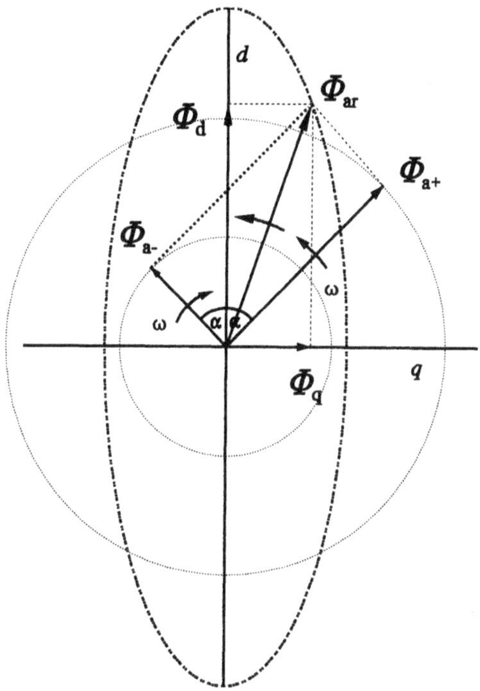

Figure 8.32

We now have two unequal rotating fluxes, Φ_{a+} and Φ_{a-}. These are shown in Figure 8.32.

Example 8.5

Assume that the speed of the dual-drive rotor shown in Figure 8.29 is 97% of $+n_s$

1. Find the ratio of $|V_+|$ to $|V_-|$.
2. Find the ratio of T_+ to T_-.
3. Find the net torque in newton-meters.
4. Compare the net torque with the torque obtainable if only M_+ were connected to the source.

Use the IM model for each machine and the data given in Example 8.1.

Solution: We first find the impedance per phase of each stator. From equation (8.56) we have

$$Z_+ = 0.295 + \frac{0.150}{0.03} + j(0.510 + 0.210) = 5.295 + j0.720 \quad [\Omega \text{ per phase}]; \tag{8.5.1}$$

$$Z_- = 0.295 + \frac{0.150}{1.97} + j(0.510 + 0.210) = 0.371 + j0.720 \quad [\Omega \text{ per phase}]. \tag{8.5.2}$$

1. From equation (8.55) we obtain

$$\frac{V_+}{V_-} = \frac{Z_+}{Z_-}. \tag{8.5.3}$$

Thus,

$$\frac{|V_+|}{|V_-|} = \frac{|5.295 + j0.720|}{|0.371 + j0.720|} = 6.598. \tag{8.5.4}$$

2. From equation (8.53) the ratio of the torques is found to be

$$\frac{T_+}{T_-} = \frac{2 - 0.03}{0.03} = 65.7. \tag{8.5.5}$$

The stator current is

$$|I_1| = \frac{|V|}{|Z_+ + Z_-|} = \frac{127.0}{5.846} = 21.72 \quad [\text{A per phase}]. \tag{8.5.6}$$

3. From equations (8.48) and (8.49) we get

$$T_+ = \frac{90}{\pi \cdot 1200} \cdot (21.72)^2 \cdot \frac{0.150}{0.03} = 56.31 \quad [\text{N} \cdot \text{m}]. \tag{8.5.7}$$

$$T_- = \frac{1}{65.7} \cdot 56.31 = 0.86 \quad [\text{N} \cdot \text{m}]. \tag{8.5.8}$$

$$T_{tot} = T_+ - T_- = 55.45 \quad [\text{N} \cdot \text{m}]. \tag{8.5.9}$$

4. If only M_+ were connected to the source and running at a slip 0.03, the stator current would be

$$|I_1| = \frac{|V|}{|Z_+|} = \frac{127.0}{|5.295 + j0.720|} = 23.77 \quad [\text{A per phase}]. \tag{8.5.10}$$

According to equation (8.48) its torque would be

$$T_{m+} = \frac{90}{\pi \cdot 1200} \cdot (23.77)^2 \cdot \frac{0.150}{0.03} = 67.42 \quad [\text{N} \cdot \text{m}]. \tag{8.5.11}$$

From this example we can make the following observations:

1. If one of the motors operates close to synchronous speed, its stator winding will have almost all of the available source voltage across it.

2. Consequently the flux and the torque of the other machine will be almost negligible.
3. The value of the net torque approaches the torque available from a single machine operating at the same speed.

If we were to perform the computations in the above example for several values of slip, then we would obtain a torque-speed curve as shown in Figure 8.30. We note several points of interest:

1. Whereas the torque of a single three-phase motor passes through zero at exactly the synchronous speed, the zero torque for the dual-drive system occurs below n_s.
2. For speeds close to n_s the torque of the dual-drive system has nearly the same magnitude as that of a single three-phase motor.
3. The torque for positive speeds is equal to the torque for negative speeds but with the opposite sign.

Returning to Figure 8.32, the two unequal rotating fluxes give the resultant flux Φ_{ar} is a rotating flux whose magnitude changes with time. At any instant, Φ_{ar} can be resolved into Φ_d and Φ_q.[19] Note that Φ_d and Φ_q are two sinusoidally pulsating fluxes at right angles to each other with a phase shift of 90°. This means that if we can obtain two pulsating fluxes spatially displaced at right angles, we have a rotating flux whose magnitude varies with time. Such a rotating flux can be used to start the single-phase induction motor.

8.8.6 Sinusoidally Pulsating and Rotating Fluxes

Figure 8.33 shows the loci of the combined effect of two sinusoids acting spatially at right angles (along the d and q axes) to each other.

In Figure 8.33a, the two sinusoids of equal amplitude with a phase difference of 90° are shown, one with its time axis along the x axis and the other along the y axis. The combined locus of the two sinusoids is obtained by plotting the corresponding points on the x and y sinusoids and drawing projection lines parallel to the x and y axes. The locus in the x–y plane is the point of intersection of the projection lines. It can be seen in Figure 8.33a that the locus is a circle. This means that if we have two sinusoidally pulsating fluxes oriented in space at right angles to each other *and* they have a phase difference of 90°, we obtain a rotating flux of constant magnitude rotating at the angular frequency of the sinusoidal fluxes. This would provide an ideal starting torque.

[19] We can actually demonstrate the existence of this flux by measuring an emf induced in a test coil placed on the stator having a magnetic axis coincident with the q axis. As Φ_q increases with the speed of the rotor, the magnitude of the induced emf will increase with speed. (AC tachometers are based on this phenomenon.)

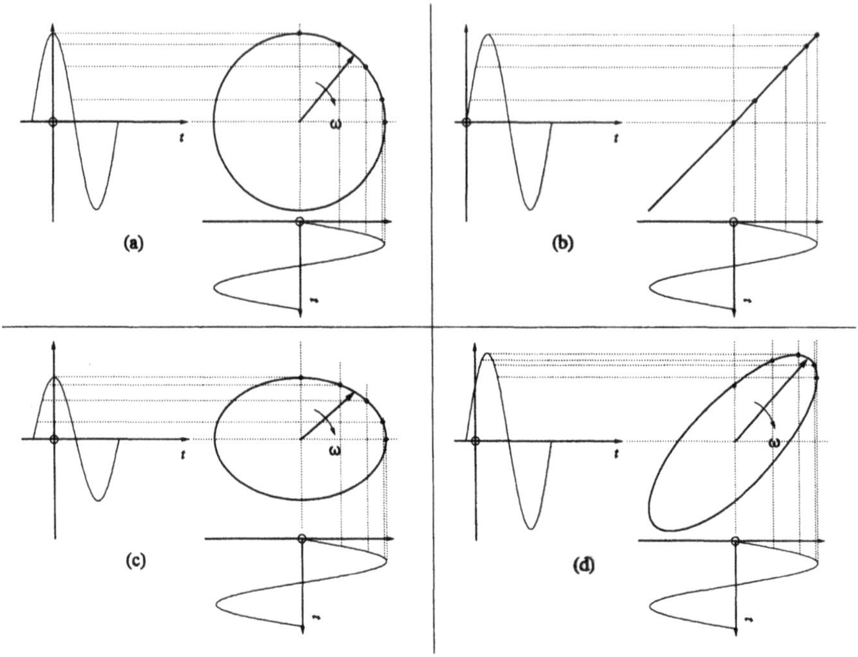

Figure 8.33

Figure 8.33b shows two sinusoids of the same amplitude with a phase difference of 0°. The locus of the resultant flux is a straight line at 45° to the axes. In terms of generating a rotating flux, this is no help!

Figure 8.33c shows two sinusoids of unequal amplitudes but with a phase difference of 90°. The locus of the resultant is an ellipse. We have a rotating flux with a variable magnitude. This could be useful.

Figure 8.33d shows two sinusoids of equal amplitudes with a phase difference $0° < \theta < 90°$. The locus is also an ellipse. Again we have a rotating flux with a variable magnitude. This could be useful for starting the motor.

We conclude that we can generate a rotating flux, with a variable magnitude, if we have two coils spatially displaced by 90° (along the d and q axes) and fed with sinusoidal currents that are *not* in phase. In terms of generating the ideal starting torque, two sinusoids of equal amplitudes with a phase difference of 90° is highly desirable but quite unnecessary. We can generate a starting torque as long as we feed the two coils with currents that are not in phase. However, the closer the phase difference is to 90°, the better the starting torque.

8.8.7 Motor Starting Techniques

Figure 8.30 shows that the single-phase motor (except for the vibratory torque component) produces a good torque, particularly at high speed. But no electric motor is acceptable that is not self-starting. Just as we are forced to provide an automobile engine with a starting system, so we must devise some starting scheme for the single-phase motor. In fact, single-phase induction motors can be grouped on the basis of their starting methods. The names given to the different types of motors are often descriptive of the starting techniques employed.

All starting methods in use are based on the following principle:

> If it is true that two revolving fluxes, in opposite directions, of unequal magnitude give rise to two pulsating fluxes of unequal phase in the d and q directions (Figures 8.32 and 8.33), then the opposite is also true—two pulsating fluxes in the d and q directions of unequal phase will give rise to two revolving fluxes of unequal magnitude in opposite directions.

The application of this principle is shown in Figure 8.34. In addition to the main winding whose magnetic axis coincides with the d axis, there is the *auxiliary* or *starter* winding with its magnetic axis in the q direction. Assume that somehow, we have arranged the magnetization currents in the windings to be of different phase. In Figure 8.34, the current in the q coil is leading the current in the d coil by $\alpha°$. The current I_d gives rise to a sinusoidally pulsating flux in the d direction. Similarly, current I_q causes a sinusoidally pulsating flux in the q direction. If

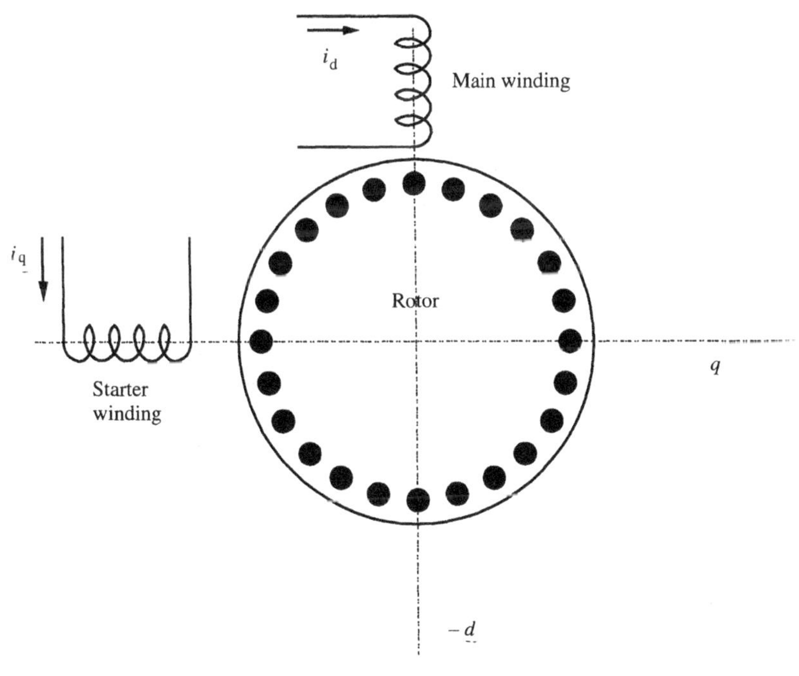

Figure 8.34

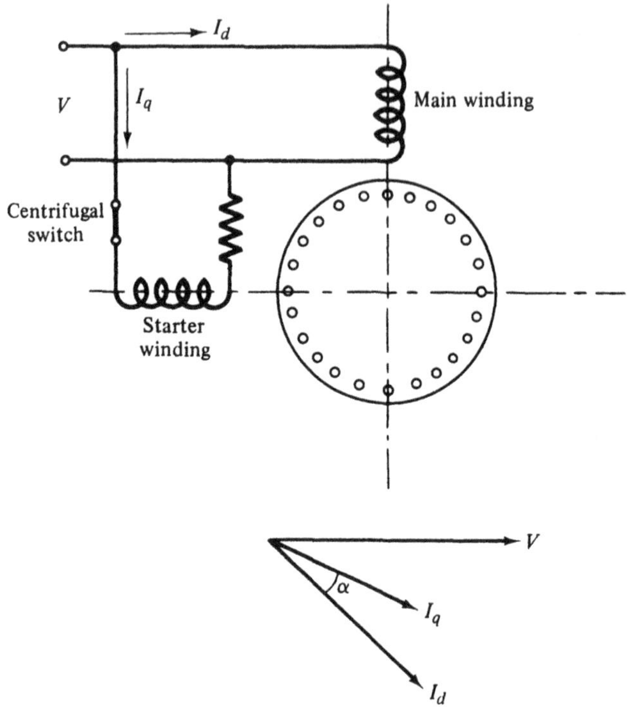

Figure 8.35

a phase difference exists between I_d and I_q, then according to Figure 8.33, a starting torque will be generated. A general rule is that the motor will start in the direction of the winding carrying the lagging current.

8.8.7.1 Resistance Split-Phase Motor

The simplest way to obtain the required phase differential, α between the two winding currents is to insert a resistance[20] in the starter winding (Figure 8.35). The added resistance makes the q winding less inductive than the d winding. Consequently its current I_q will lead I_d. This method results in a fairly small phase angle, α, and thus a relatively weak starting torque. The method is used for motors of less than 0.33 hp rating. Washing machines and dishwashers are typical applications.

Once the motor has started, the main winding produces the necessary torque. The starter winding is then unnecessary and may actually reduce the overall torque of the motor. For this reason, it is usually disconnected by means of a centrifugal switch, which operates at, for example, 70% of full speed.

[20] The winding can also be designed with a high resistance: reactance ratio).

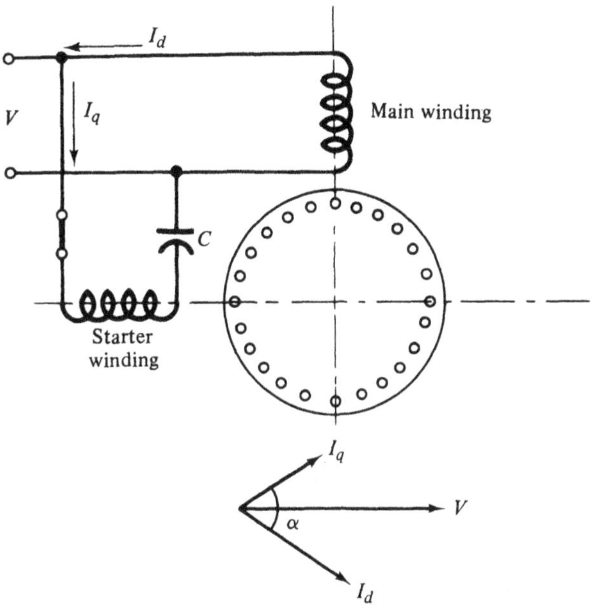

Figure 8.36

8.8.7.2 Capacitor-Start Motor

The most effective way to make the phase angle, $\alpha = 90°$ (and even larger), is to inserting a capacitor in series with the starter winding. The capacitor-start motor (see Figure 8.36) therefore has the best possible starting torque and it is one of the most common type of single-phase motor. It comes in sizes as large as 5 hp. As the capacitor must have a fairly large value, it is usually of the electrolytic type.

Example 8.6

The two windings of a single-phase motor have the following impedances, measured at 60 Hz:

$$\text{main winding: } Z_m = 3.1 + j2.9\ \Omega;$$

$$\text{starter winding: } Z_s = 7.0 + j3.1.\ \Omega.$$

Find the size of the capacitor that will produce a phase angle, $\alpha = 90°$.

Solution: From Figure 8.36 we have

$$\angle I_d = \angle V - \angle Z_m = \angle V - \angle 3.1 + j2.9; \tag{8.6.1}$$

$$\angle I_q = \angle V - \angle Z_s + 1/j\omega C = \angle V - \angle 7.0 + j3.1 + 1/j\omega C. \tag{8.6.2}$$

For the angle, α we get

$$\alpha = 90° = \underline{/I_q} - I_d = \underline{/3.1 + j2.9} - \underline{/7.0 + j(3.1 - 1/\omega C)}. \quad (8.6.3)$$

This can be written as

$$90° - \tan^{-1}\left(\frac{2.9}{3.1}\right) = \tan^{-1}\left(\frac{1/\omega C - 3.1}{7.0}\right). \quad (8.6.4)$$

Solving for the unknown, C, gives

$$C = 250.7 \quad [\mu F]. \quad (8.6.5)$$

8.8.7.3 Shaded-Pole Motor

Very small single-phase induction motors obtain their starting torque by means of *magnetic shading*, the principle of which is shown in Figure 8.37. The flux created by the main winding splits into two parts, Φ_d and Φ_q. The component of the flux, Φ_q, passes through the parallel magnetic path encircled by the *shading coil*, which consist of one short-circuited turn of copper wire.

A current is induced in the shading coil and, according to Lenz's law, it will have such a direction as to oppose the change in the flux from taking place. The net effect is that Φ_q will lag Φ_d by an angle, α. The conditions are therefore met for the creation of a starting torque. The motor will run in the direction of the shading or q coil, that is, in the CCW direction.

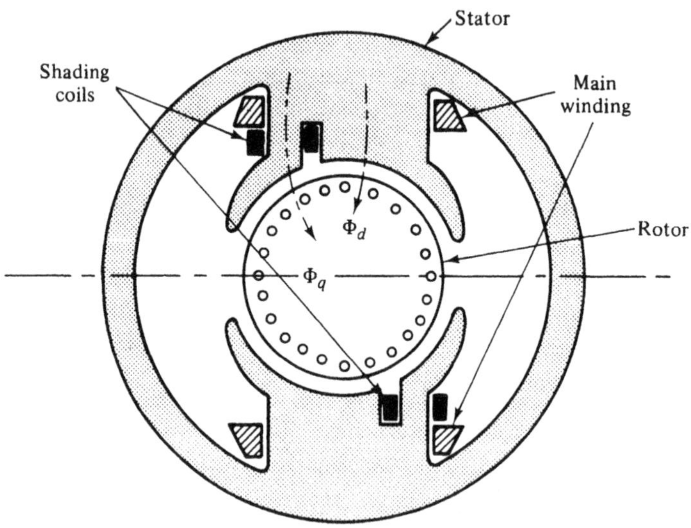

Figure 8.37

8.8.8 Induction-Start Synchronously Running Motors

In many applications it is very important that the speed of the motor be constant. For example, the requirement for accuracy of an electric clock requires a motor that runs at synchronous speed. The need to reproduce frequency accurately in a high-quality audio recording system excludes the use of an asynchronous induction motor. In such cases, one may choose a motor that starts as an induction motor but runs as a synchronous motor. Figure 8.38 shows the details of a rotor[21] with these particular features.

The basic difference between a normal induction motor and an induction-start synchronously run motor is the *salient* design of the rotor. The number of salient rotor poles must match the number of poles of the stator winding. The induced currents in the squirrel-cage winding will provide the torque necessary for starting the machine. When the rotor reaches speeds close to the synchronous speed and if the slip is below a critical value the rotor "snaps" and locks into synchronism with the stator flux. The torque necessary for this "lock-in" is of the *reluctance type* discussed in Section 3.26.5.3. This type of motor lacks the brushes and slip-rings of a normal synchronous machine. Its rotor winding has the simplicity and ruggedness of a squirrel-cage induction motor. The motor is widely used in applications where constant speed is required and the demand for high torque is not too severe.

Synchronous machines (see Chapter 4) lack starting torque and require a prime mover to bring them up to synchronous speed before being synchronized to

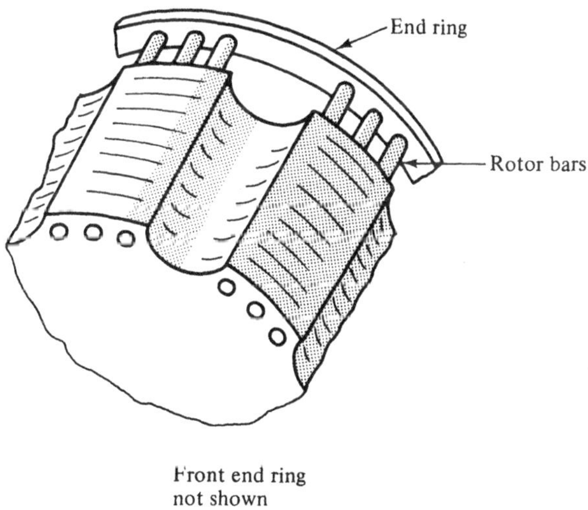

Figure 8.38

[21] The stator may be either three-phase or single-phase (with a starter winding). Three-phase units are built in sizes up to about 150 hp.

the grid. For some applications the synchronous machine must be self-starting. To achieve this the rotor is provided with a squirrel-cage as well as an appropriate rotor winding with suitable slip rings through which dc can be supplied. The rotor starts as an induction motor propelled by the squirrel cage when its speed is close to the synchronous value, the dc supply to the rotor winding is switched on.[22] It snaps into synchronism due to the synchronous torque emanating from the dc in the rotor winding. The damper winding can sometimes be used as a *starter winding*.

8.9 Summary

Alternating current induction motors are the most common electrical motors in use. Larger sizes (in excess of about 5 hp) are invariably of three-phase design. Smaller units, usually "fractional hp," are typically made single-phase.

Three-phase induction motors deliver a constant torque and load the ac power supply network symmetrically. Single-phase motors deliver a torque containing a 120-Hz pulsating torque component. In addition, of course, they load the supply network unsymmetrically. Single-phase motors must be equipped with special starter windings.

Compared to dc motors, three-phase induction motors have low starting torque. It is possible, however, by secondary resistor control to make the motor deliver its maximum torque at startup. The attraction of ac induction motors is the simplicity of their design, ruggedness, low price, and ability to run directly off the ac power network.

EXERCISES

8.1 It was explained in the text that the induction motor cannot reach synchronous speed because the rotor currents are zero for $s = 0$. If a dc source (via slip-rings) were used to inject a dc current into the rotor winding, explain how the motor would behave. Compare this to the synchronous motor. What changes would you have to make to the normal rotor windings of the induction motor?

8.2 A 6-pole, 3-phase, 60-Hz induction motor is running at a speed of 1162 rpm. Two stator phases are suddenly reversed.

a) What is the motor slip, s before the phase reversal?

b) What is the slip immediately after the phase reversal?

c) Use the motor data given in Example 8.1 to compute the current immediately following the phase reversal. (Neglect any transients.) Use the IM model.

8.3 Consider the 10-hp induction motor in Example 8.1. Compute the motor torque, motor power, power drawn from the network, and ohmic power loss when running at the following speeds:

[22] Large machines in the megawatt range can rarely stand this treatment.

(a) 1150 rpm
(b) 1250 rpm
(c) −1200 rpm
(d) ∞ rpm.

Use the IM model and identify the cases in which the machine acts as a generator.

8.4 Consider the circle diagram in Figure 8.17. At what slip, s will the magnitude of the stator current reach its maximum value?

a) At this slip, show that the real power supplied to or from the supply network is equal to zero.

b) What will be the ohmic power loss?

c) From which source will the ohmic power loss be supplied? Use the IM model in your analysis.

8.5 A 3-phase, 8-pole, squirrel-cage induction motor is designed to run off a 440-V, 60-Hz network. The motor has the following equivalent circuit parameters, expressed in ohms per phase, referred to the stator:

$$R_1 = 0.087, R_2' = 0.070, X_1 = 0.200, X_2' = 0.168.$$

a) Construct a circle diagram for the machine. Use the IM model.

b) Compute the horsepower output, stator current, power factor, efficiency, and torque for a rotor slip of 3.1%. Identify the stator current phasor on the circle diagram.

8.6 What will be the starting current drawn by the motor in Exercise 8.5? Identify this current on the circle diagram.

8.7 In this exercise we account for the motor losses. Tests show that the motor in Exercise 8.5 has a stray loss of 520 W. The motor draws 2.72 kW at a power factor of 0.31 in a no-load test.

Correct the model of the motor and the circle diagram for the above nonideal features and then redo Exercises 8.5 and 8.6. Assume that half of the no-load losses are made up of core losses.

8.8 Find the maximum torque of the motor in Exercise 8.5.

8.9 The induction machine in Exercise 8.5 is run as an "induction generator" on a 440-V network at a speed of 930 rpm, driven by a gas turbine. What power does it deliver to the network? What is the efficiency of the generator? Use the IM model in your analysis.

8.10 The 3-phase induction motor in Exercise 8.5 is connected to an 11-kV bus via a 100-kVA, 3-phase transformer of voltage rating 11/0.44kV.

a) Compute the voltage on the LV side at direct start. The transformer impedance is $0.75 + j5.1\%$ based on its ratings.

b) What is the LV-side voltage when the motor is running at a slip of 3.1%? *Assumption:* The 11-kV bus experiences no voltage drop.

8.11 Equation (8.35) for maximum torque was derived on the assumption that the motor terminal voltage was fixed.

Compute the maximum torque for the motor in Exercise 8.10 if the transformer impedance is taken into account.

8.12 It is often found useful to express the induction motor torque, T_m in relation to its maximum torque, T_{max}. Use equations (8.32) and (8.35) to show that the torque-ratio T_m/T_{max} is

$$\frac{T_m}{T_{max}} = \frac{1 + \sqrt{R^2 + 1}}{1 + \frac{1}{2}\sqrt{R^2 + 1}\,[(s/s_{max}) + (s_{max}/s)]} \tag{8.57}$$

where

$$R \equiv \frac{X_1 + X_2'}{R_1} \tag{8.58}$$

and s_{max} is determined by equation (8.34).

8.13 Plot T_m/T_{max} for the range of slip,

$$0.1 > \frac{s}{s_{max}} > 10 \tag{8.59}$$

In computing the torque ratio, use the following widely different R values:

(a) $R = 0$
(b) $R = 3$
(c) $R = \infty$

Your three plots will not show a significant difference which would indicate that the parameter R does not have a great influence on the ratio of the torques. How would you interpret this fact?

8.14 A 3-phase, 100-hp induction motor develops its rated power at a rotor slip of 1.8%. The maximum torque is 250% of rated torque (that is, the torque developed at rated power). The motor has an R ratio of 8.

Using the IM model, find

a) slip, s_{max}, at maximum torque,
b) stator current at maximum torque,
c) starting torque,
d) starting current.

Express your answers in parts b, c, and d in terms of the current and torque at rated speed.

References

Alger, Philip L. *Induction Machines,* New York: Gordon and Breach, Science Publishers, Inc., 1970.

Bergseth, F.R., Venkata, S.S. *Introduction to Electric Energy Devices.* Englewood Cliffs, NJ: Prentice-Hall, 1987.

Kabisama, H.W. *Electric Power Engineering.* New York: McGraw-Hill, 1993.

Kron, G. Generalized theory of electrical machinery. *AIEE Trans.* 1930; 49: 666–683.

Lindsay, J.F., Rashid, M.H. *Electromechanics and Electrical Machinery.* Englewood Cliffs, NJ: Prentice-Hall, 1986.

White, D.C. and Woodson, H.H. *Electromechanical Energy Conversion*, New York: Wiley, 1959.

9

Electric Motors for Special Applications

In this chapter we discuss various types of electric motors that have applications in other branches of engineering, especially in control systems and robotics. The major objective of this chapter is to introduce the student to the basic structure, the principles of operation, and the capabilities and limitations of these machines. Wherever possible the simplest possible model of the machine is used to explain its operation. Discussion of details such as departure from ideal behavior and static and dynamic errors have been excluded in favor of brevity, simplicity, and clarity.

9.1 Linear Induction Motor

9.1.1 Introduction

In Section 4.7.1, it was demonstrated that when three coils placed in a stator and spatially displaced by 120° are fed from a three-phase balanced source, a rotating flux, Φ_T of constant amplitude is obtained. In Section 8.4 it was shown that when a short-circuited coil supported on two bearings, is placed in the path of the rotating flux, a voltage is induced, and a current flows in the coil. The coil current will generate a flux, Φ_I, of its own, the magnitude and direction of which will be determined by the magnitude of the induced voltage, the impedance of the short-circuited coil, and the orientation of the coil to the flux, Φ_T. The two magnetic fluxes will interact to produce a torque, which tends to align them with each other. The rotating flux, Φ_T, represented by a permanent horseshoe magnet driven at a constant angular velocity by a prime mover and the short-circuited coil, which produces the flux Φ_I, are shown in Figure 9.1.

Since Φ_T is rotating at a speed determined by the frequency of the three-phase balanced power source, it follows that the torque generated by the interaction of the coil flux, Φ_I and Φ_T, will attempt to rotate the coil at the same speed as Φ_T. It

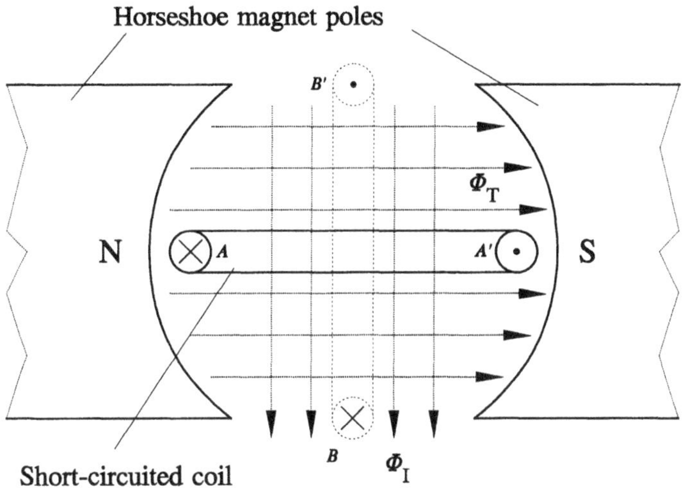

Figure 9.1

follows that the coil cannot actually rotate at the same speed as Φ_T as $d\Phi/dt$ will approach zero and hence the induced voltage will also approach zero. The coil (rotor) will rotate at a speed slightly below that of Φ_T. In a practical motor, the single short-circuited coil is replaced by several short-circuited coils, which then constitute a "squirrel-cage" rotor. The advantage of the squirrel cage is that at any instant in time, the net torque is the sum of the contributions from the "individual" coils, and hence the net torque is not a function of the relative position of Φ_T with a single-coil flux, Φ_I. Note that in Figure 9.1, the coil marked A–A' produces maximum torque, while the coil B–B' generates minimum (zero) torque.

Since the magnitude of Φ_I is a constant and it rotates about the axis of the machine, it follows that at any instant in time, Φ_T has a sinusoidal distribution around the periphery of the stator. Figure 9.2a shows the profile of the flux Φ_T at any instant for a 2-pole machine, and Figure 9.2b shows the profile for a 4-pole machine. For comparison, Figure 4.6 shows the profile of the magnetic flux in an 8-pole synchronous machine. It should be noted that all three profiles of Φ_T are sinusoids with their x axes "wrapped" along the circumference of the dotted line circle in Figure 9.2a, b. As Φ_T rotates about the center of the circle, the rotor will experience a sinusoidal traveling flux wave.

9.1.2 Conversion From Rotating to Linear Motor

In a linear induction motor, the stator is split longitudinally along the axis X–X' as shown in Figure 9.3a, and the structure is straightened as shown in Figure 9.3b.

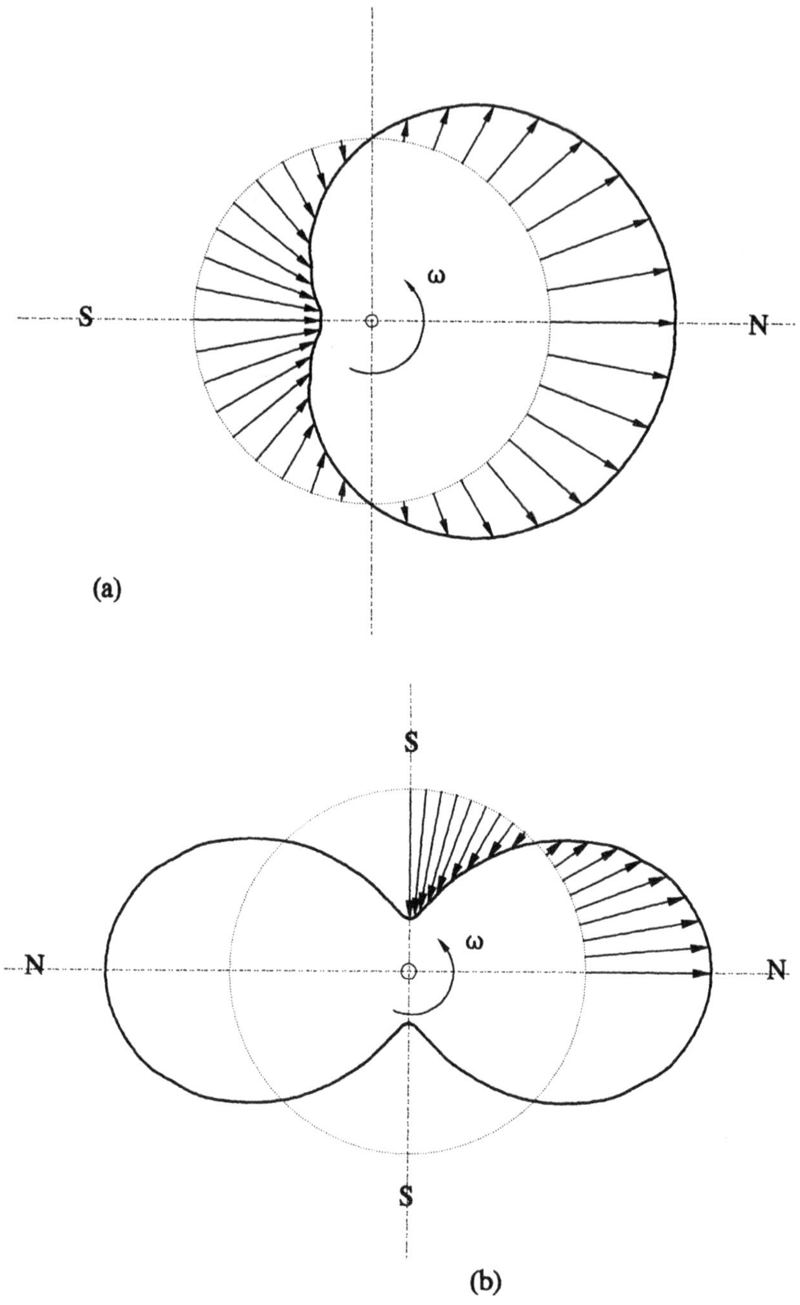

Figure 9.2

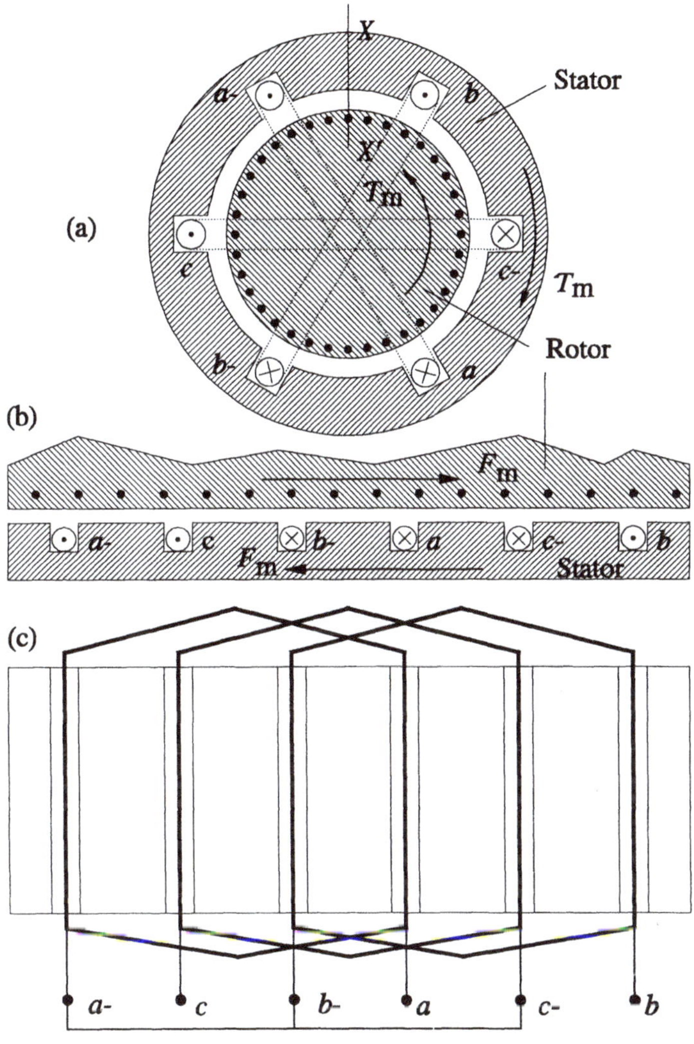

Figure 9.3

Figure 9.3c shows the plan view of Figure 9.3b and the connection of the three coils in a Y configuration. The associated travelling flux wave is shown in Figure 9.3d, at two instants one-quarter of a cycle apart. When the rotor is straightened, it becomes a ladder structure. When a balanced three-phase current is supplied to the stator, the rotor will experience a sinusoidal traveling wave along its length, and, instead of the rotational motion of the rotor in an counterclockwise direction, the "straightened" rotor will move in a straight line from left to right, relative to

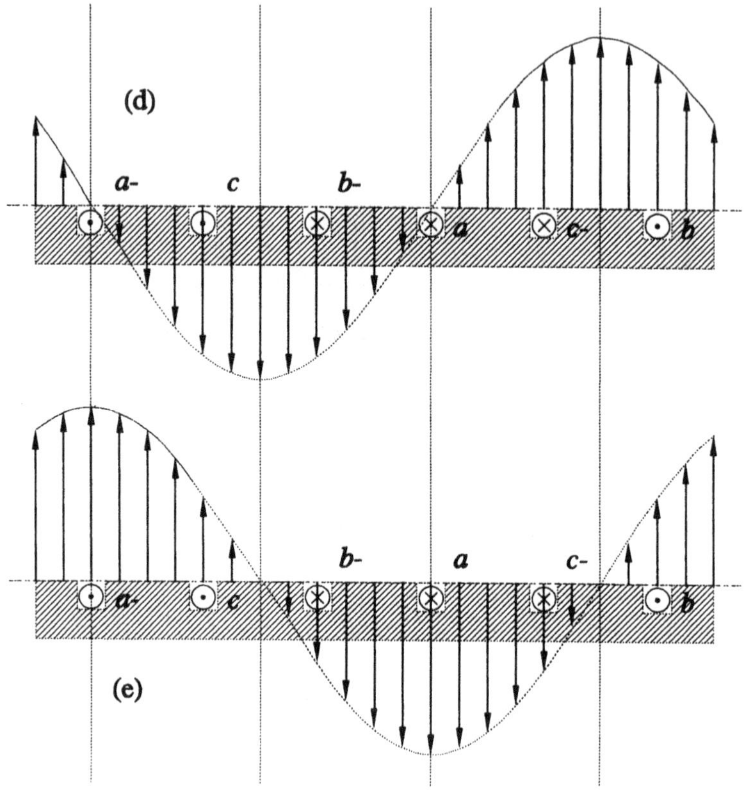

Figure 9.3 (cont.)

the stator. In most applications of the linear induction motor, it is an advantage to fix the rotor and allow the stator to move. The stator will therefore move from right to left relative to the rotor.

9.1.3 Applications

The most common application of the linear induction motor is in electric traction. Direct current is supplied to the moving locomotive from an overhead conductor by means of a *pantograph*. An *inverter* on the locomotive converts the dc into a three-phase ac of variable frequency. The three-phase ac is fed to the *stator*. The *rotor* is usually made up of a solid piece of conducting material instead of the ladder structure described earlier. This change does not alter significantly the nature of the currents flowing in and hence the force on the *rotor*. The *rotor* can be made of copper, aluminum, or steel and it can form part of the *rail* system of the loco-

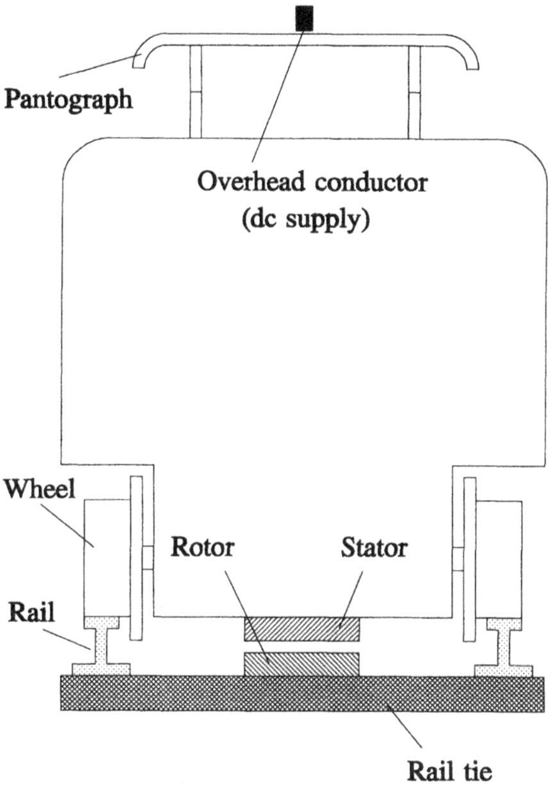

Figure 9.4

motive. Figure 9.4 shows a cross section of a locomotive driven by a linear induction motor.

Linear induction motors have been used to drive experimental electric trains weighing over 20 tons and traveling at speeds in excess of 200 km/h. The major advantage of the linear induction motor, when used for traction purposes, is the elimination of the coefficient of friction between the driven wheels and the rail as a limiting factor on the load that can be pulled.

9.2 Stepper Motor

9.2.1 Introduction

The stepper motor belongs to a class of motors that can be described as *digital*. They operate on the basis of a pulse supplied from a controller; advancing one step when the digit 1 is applied and not advancing when the digit 0 is applied.

Stepper motors have become very important mainly because of advances made in digital computers and their use in control systems such as robotics, computer disc drives, and so on. They are particularly useful in open-loop control systems such as that commonly employed in satellite dish antenna positioning systems for geostationary communication satellites.

9.2.2 Principle of Operation

Figure 9.5 shows the cross section of a stepper motor with a 5-pole permanent magnet rotor and a 4-pole stator. There are many types of stepper motors, some with a soft iron rotors (reluctance type) instead of permanent magnets and others with different designs of the stator. This particular example is chosen because its operation is easy to understand.

From Figure 9.5 it can be seen that the 4-pole stepper motor has a rotor that has five equally spaced "teeth" with "north" and "south" designations and four poles on the stator as shown. The coils on the stator poles A–A' are connected in series and the direction of the current can be reversed. The coils on stator poles B–B' are also in series, and the direction of the current can be reversed.

In the initial position, the permanent magnet ensures that the rotor teeth such as $S5$–$N5$ line up with the stator coils such as B–B'. This is because the reluctance of the magnetic flux path is a minimum in this position. This is shown in Figure 9.6a.

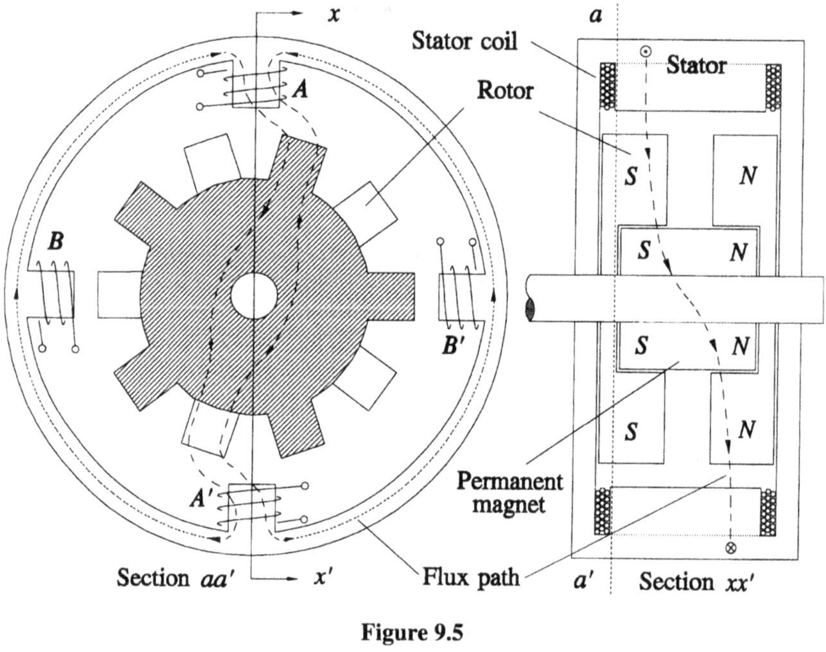

Figure 9.5

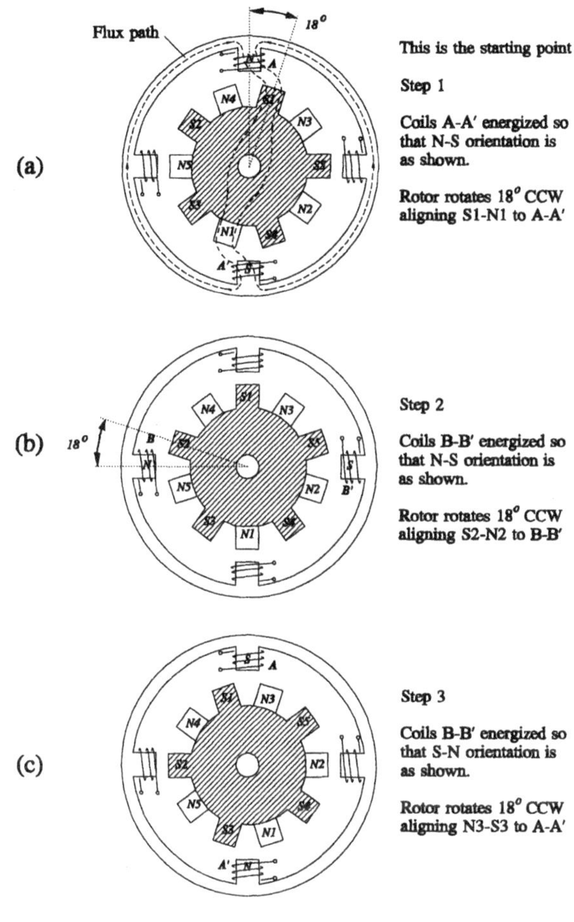

Figure 9.6

When the stator coils A and A' are energized to produce a north and a south pole, respectively, the pattern of the major flux paths, just before the rotor moves, are as shown in Figure 9.5.

Step 1: Coils A–A' are energized such that the pole shoe A produces a north pole, while A' is the south pole. The north pole A exerts a force of attraction on $S1$ and an equal force of repulsion on $N4$, causing a CCW torque. It also attracts $S2$ and repulses $N3$, causing a CW torque. At the same time it attracts $S5$ and repulses $N5$, causing a CCW torque. Similarly, the south pole A' repulses $S4$ and attracts $N1$, causing a CCW torque. It also attracts $N2$ and repulses $S3$, causing a CW torque. Simultaneously, it attracts $N5$ and repulses $S5$, causing a CCW torque. The resulting torque, the sum of all the torques (CCW is positive; CW is negative) moves the rotor to align $S1$ and $N1$ with the poles A–A'—an angle of 18° counterclockwise. This is shown in Figure 9.6b.

Step 2: Coils B–B' are energized with pole shoe B, producing a north pole and B' a south pole. The north pole B exerts a force of attraction on $S2$ and an equal force of repulsion on $N5$, causing a CCW torque. It also attracts $S3$ and repulses $N4$, causing a CW torque. At the same time it attracts $S1$ and repulses $N1$, causing a CCW torque. Similarly, the south pole B' repulses $S5$ and attracts $N2$, causing a CCW torque. It also attracts $N3$ and repulses $S4$, causing a CW torque. Simultaneously, it attracts $N1$ and repulses $S1$, causing a CCW torque. The net torque causes a further 18° counterclockwise rotation of the rotor. This is shown in Figure 9.6c.

Step 3: Current is passed through coils A—A' in the opposite direction to that in Step 1. $N3$ and $S3$ are aligned with A–A'; a further counterclockwise rotation of 18° occurs. This is shown in Figure 9.6d.

Step 4: Current is passed through coils B–B' in the opposite direction to that in Step 2. $N4$ and $S4$ are aligned with B–B'; a further counterclockwise rotation of 18° occurs. This is shown in Figure 9.6e.

Step 5: This is the same as the starting point, except that $N5$ and $S5$ occupy the positions of $N1$ and $S1$, respectively. The rotor has turned through a total of 72°. The process can now be repeated.

(d)

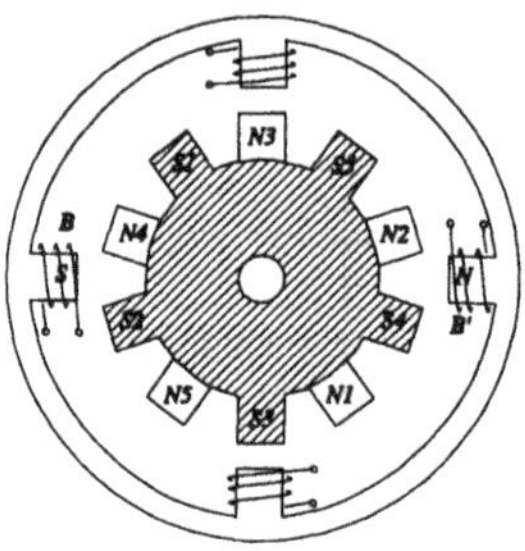

Step 4

Coils B-B′ energized so that S-N orientation is as shown.

Rotor rotates 18° CCW aligning N4-S4 to B-B′

(e)

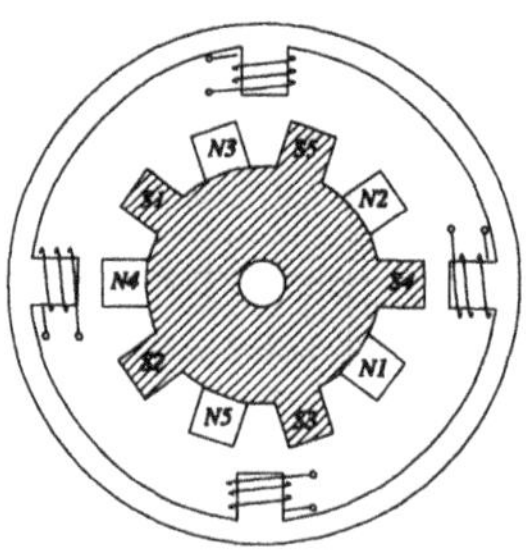

Step 5

This position is the same as the starting point except S5-N5 occupies the position of S1-N1.

Figure 9.6 (cont.)

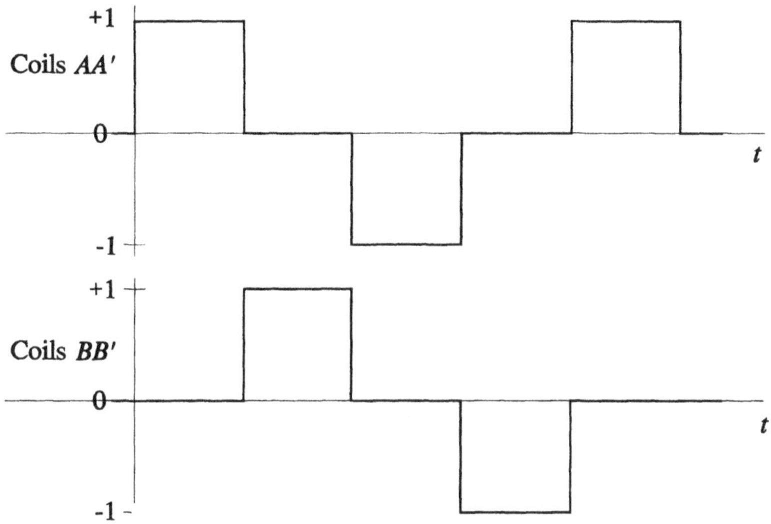

Figure 9.7

The current required to implement the steps above are shown in Figure 9.7 in the form of a current timing diagram.

The description of the steps required to drive the motor in a clockwise direction is left as an exercise to the reader. It should be noted that the motor can be designed to rotate through an angle much less than the 18° shown in Figure 9.5. Naturally, the smaller the angle of rotation per pulse, the larger the number of pulses required for a complete revolution. The electrical time constant of the coils and the mechanical time-constant of the rotor set a limit to how fast the pulses can be applied while avoiding errors.

9.2.3 Advantages and Disadvantages

The stepper motor is a variable speed machine; its speed is controlled by the pulse rate of the driving circuit. Its direction of rotation depends on the sequence in which the pulses are applied. It is very highly accurate in position location operations since every pulse moves the rotor a precise number of degrees and it is not subject to accumulated errors. It has found general application in servo systems, numerical control machinery, and computer peripherals. Its disadvantages are the complexity of the control circuit and the fact that care should be exercised not to operate it close to its resonant frequency, where serious errors can occur. It is usually found in sizes from 0.01 to 0.5 hp.

9.3 Brushless Direct Current Motors

9.3.1 Introduction

The torque–speed characteristics and the flexibility of the dc motor have made it a very attractive choice for applications in control systems, robotics, and traction. However, it comes with a number of disadvantages, such as commutators and brushes, which tend to wear out and must be serviced periodically. Then there is arcing at the trailing edges of the commutator segments, which is a definite fire hazard in the presence of volatile materials. The development of the brushless dc motor is driven by the wish to eliminate the commutator and brushes while keeping the highly desirable characteristics of the dc motor.

Before we examine the design of the brushless dc machine, it is advantageous to return to Chapter 7 and examine Figure 7.13. A careful look reveals that all the armature currents under the influence of the north pole flow into the plane of the page, while those under the influence of the south pole flow out of the plane of the page. If we apply the right-hand rule to all the conductors in the armature together, we find that the flux produced by the armature currents will be at right angles to the flux produced by the field winding. The two fluxes interact to produce a torque, which tends to align them to each other. However, as soon as one armature coil side is about to leave the vicinity of the north pole and come under the influence of the south pole, the current in it is reversed by the commutator/brush mechanism. From Figure 7.13, it can be seen that as the armature turns and commutator segments *d* and *a* come into contact with the brushes, the currents in the conductors in slots 3 and 9 will change direction. They then produce a torque, which adds to the counterclockwise torque produced by the other coils. Thus the rotor keeps turning. The commutator/brush mechanism is simply a mechanical switch that ensures that the currents in the armature coils are reversed at the appropriate time to maintain the torque in the chosen direction.

What is required to convert the normal dc motor into a brushless motor is a system to replace the commutator/brush mechanism with a sensor that senses the position of the coils and a switch to change the direction of the current in the coils at the appropriate time. In the brushless dc motor, the sensing and switching are done electronically.

In general, large dc machines require a field winding and field current to produce the field flux. The power required to maintain the flux is lost in the form of heat. At first sight, it might appear to be advantageous to use a permanent magnet in place of the field circuit. However, the heat and vibration associated with the normal operation of an electric motor represent a hostile environment for the long-term survival of a permanent magnet. Besides, the field current is a very important control parameter for the dc motor. In small dc motors such as those used in control systems, permanent magnet machines are quite common.

9.3.2 Principles of Operation

Figure 9.8 shows the cross section of a simple two-segment permanent magnet brushless dc motor and a schematic diagram of the electronic control system. The rotor is a permanent magnet; in this case, a 2-pole model with north and south poles as shown. The stator has 12 slots carrying the stator windings and are divided into four segments.

When current is supplied to the coils in segments A–A' as shown, the application of the right-hand rule shows that the stator flux will be vertical. The flux from the permanent magnet rotor will interact with the stator flux to give a clockwise torque. The rotor will take up a new position at right angles to the starting position with the north pole vertically above the south pole. A pulse of current supplied to the coils in segments B–B' with the current flowing into the coil sides in segment B and out of the coil sides in segment B' will cause a further clockwise rotation of 90°. It is unusual to have only two segments in a machine, as it leads to *cogging*. However, it should not be difficult to visualize a machine with four, six, or higher segments, which would make the machine run more smoothly. The price for this is a more complicated electronic circuit.

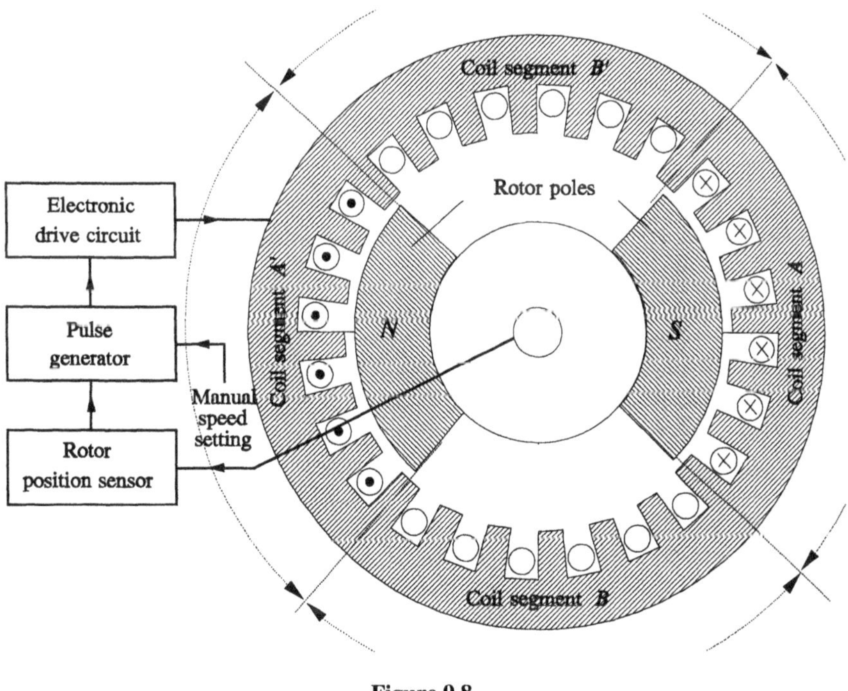

Figure 9.8

The angular position sensor uses a Hall effect transducer or an optical sensor to determine which coils should be energized. This information, together with the speed setting, are fed to the pulse generator that drives the microcircuit source, which in turn supplies the current to the stator.

It is very important to keep the inductance of the stator windings as low as possible so as to minimize the energy stored when the coil is energized. This energy has to be dissipated when the current is switched off, and commutation is easier when the stored energy is low.

9.3.3 Advantages and Disadvantages

The brushless dc motor comes with all the advantages of the common dc motor but without the drawback of the mechanical switching system made up of the commutator and brushes. The rotor of the brushless dc machine has a lower inertia compared to an equivalent dc machine, since it has no windings and commutator. It is therefore a prime choice for applications in servo systems and for driving computer peripherals. The major disadvantage is the complexity of the electronic control circuit especially in applications where smooth rotation is required and hence a large number of coils have to be switched in sequence. Electronic switching limits the range of the size of the machine from 0.1 to a maximum of about 1 hp.

9.4 Synchros

9.4.1 Introduction

Synchros, unlike the other machines discussed above do not normally perform rotary or linear motion and therefore do not strictly fit into the category of electric motors. However, the principles on which they are based is the same as all the other types of machines discussed in this and earlier chapters. Synchros are generally used in servomechanisms as error detectors and in the transmission of torque over distances in which a mechanical shaft would be impractical.

9.4.2 Principles of Operation

A synchro is best thought of as a synchronous machine. However, instead of supplying dc current to the rotor and driving it with a prime mover, the rotor is fed from a sinusoidal source through a set of slip-rings.

The basic features of the synchro can be seen in Figure 9.9. The three-phase windings are represented by the three identical coils spatially displaced from each other by 120° and embedded in the stator slots. In practice, the stator coils are distributed. They are assumed to be Y-connected for simplicity.

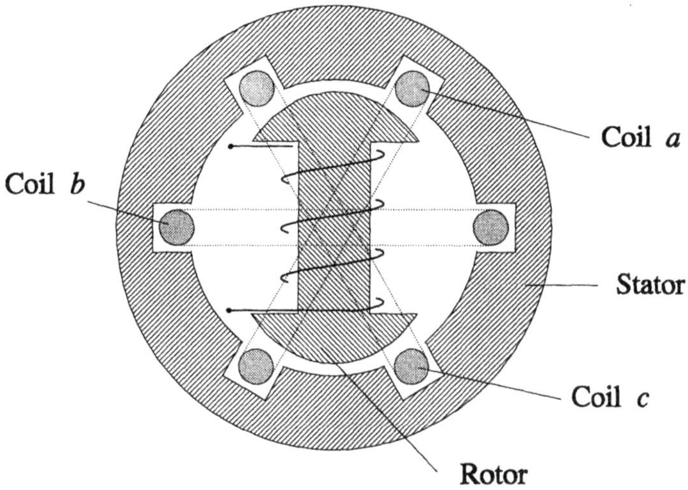

Figure 9.9

The sinusoidal current supplied to the rotor sets up a flux that links with the stator. The system behaves like a transformer with one primary and three secondary windings. Assuming that the rotor is at rest in any given position, it follows that the voltages induced in the three stator coils will be different in both magnitude and phase since the flux linkage to each coil is different.

We can define the turns ratio of the "transformer" as

$$a = \frac{\text{number of turns in stator}}{\text{number of turns in rotor}} = \frac{N_{\text{sta}}}{N_{\text{rot}}}. \tag{9.1}$$

Let the applied voltage

$$v_{\text{rot}} = V_{\text{rot}} \sin \omega t \qquad [\text{V}]. \tag{9.2}$$

If the rotor is placed in the position in which it induces the maximum voltage in the stator coil a, it follows that the voltage induced will be

$$v_{\text{sta}} = aV_{\text{rot}} \sin \omega t \qquad [\text{V}]. \tag{9.3}$$

If the rotor is now displaced by angle α from coil a, the induced voltage will be

$$v_{\text{sta(a)}} = aV_{\text{rot}} \sin \omega t \cos \alpha \qquad [\text{V}]. \tag{9.4}$$

It is evident that the voltage induced in coil b will lag that in coil a by 120°, hence

$$v_{\text{sta(b)}} = aV_{\text{rot}} \sin \omega t \cos(\alpha - 120°) \qquad [\text{V}], \tag{9.5}$$

and for coil c we have

$$v_{\text{sta(c)}} = aV_{\text{rot}} \sin \omega t \cos(\alpha - 240°) \qquad [\text{V}]. \tag{9.6}$$

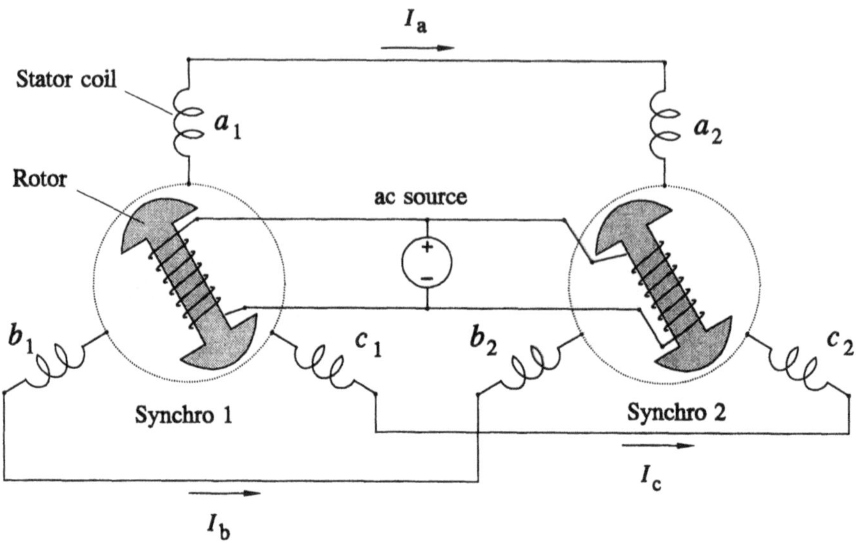

Figure 9.10

9.4.3 Applications

9.4.3.1 Transmission of Torque

Figure 9.10 shows two synchros connected in parallel. Assuming that the rotor of synchro 1 is at rest with its axis at an angle, α relative to phase coil a_1 and the rotor of synchro 2 is also at rest with its axis at an angle, β relative to its phase coil a_2. When their rotors are fed from the same sinusoidal source, it follows that the induced voltages in the stator of synchro 1 will be as described by equations (9.4), (9.5), and (9.6). The corresponding induced voltages for synchro 2 will be

$$v_{\mathrm{sta(a)}} = aV_{\mathrm{rot}} \sin \omega t \cos \beta \qquad [\mathrm{V}]. \tag{9.7}$$

$$v_{\mathrm{sta(b)}} = aV_{\mathrm{rot}} \sin \omega t \cos (\beta - 120°) \qquad [\mathrm{V}]. \tag{9.8}$$

$$v_{\mathrm{sta(c)}} = aV_{\mathrm{rot}} \sin \omega t \cos (\beta - 240°) \qquad [\mathrm{V}]. \tag{9.9}$$

As the corresponding phase voltages are different, the line currents I_a, I_b, and I_c will flow. If the shaft of synchro 1 is fixed and that of synchro 2 is free to rotate it follows that synchro 2 will rotate into a new position where its induced voltages are identical in magnitude and phase to those of synchro 1. In this position, all the line currents will be zero. The only possibility for all three pairs of induced voltages to be equal simultaneously, exists when the rotor of synchro 2 takes a position identical to that of synchro 1, that is, the rotor will move and come to rest when its axis is at an angle, α relative to its phase coil a_2, that is, when $\alpha = \beta$. The torque required to move the rotor of synchro 2 is generated by the interaction of the flux due to the line currents and the flux generated by the rotor current. When the line currents are all equal to zero, the torque is zero.

The system shown in Figure 9.10 is a good example of an "open loop" control system. Such a control system is commonly used in earth satellite dish antenna.

9.4.3.2 Error Detector

The function of the error detector is shown in Figure 9.11. It converts a mechanical command in the form of the displacement of the shaft of the transmitter (synchro 1) into an electrical (error) signal appearing across the terminals of the rotor of the receiver (synchro 2). The error signal is obtained by subtracting the signal applied to synchro 1 from the output of synchro 2. The difference signal is amplified and fed to the servomotor as shown in Figure 9.11. The servomotor moves the load, which is mechanically connected to the the rotor of synchro 2 (usually through a set of gears), in an attempt to minimize the error signal. The system comes to rest when the error signal is equal to zero, that is, synchro 2 has taken up a position identical to that of synchro 2.

In Figure 9.11 the rotor of synchro 1 is fed from a suitable sinusoidal source. The induced voltages described by equations (9.4), (9.5), and (9.6) are applied to the terminals of synchro 2 as shown, and they supply magnetization currents I_a,

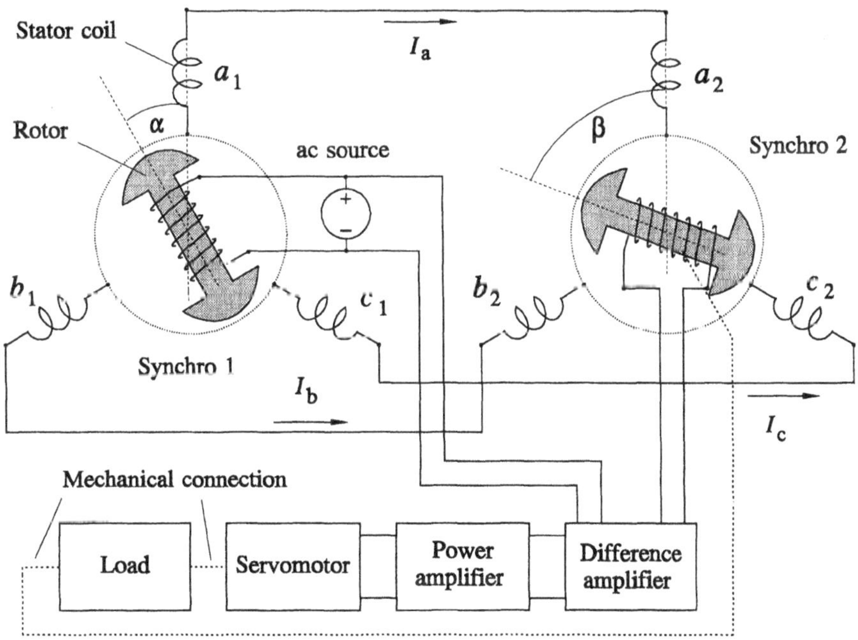

Figure 9.11

I_b, and I_c to synchro 2. Using the position of synchro 1 as a reference and assuming that the rotor of synchro 2 is in the position where it is at an angle β to its own phase coil a_2, then the combined effect of the currents will be the induction of a voltage in the rotor of synchro 2 equal to

$$v_{rot} = V_{rot} \sin(\omega t + \phi) \qquad [\text{V}], \tag{9.10}$$

where

$$\phi = \alpha - \beta. \tag{9.11}$$

Note that equation (9.10) is similar to (9.2) except for the phase shift, ϕ, which is equal to zero when $\alpha = \beta$. It is assumed that the synchros behave like ideal transformers.

When the signal applied to synchro 1 is represented by a phasor $V_{rot}\underline{/0^\circ}$, the output of the rotor of synchro 2 is given by $V_{rot}\underline{/\phi}$. The output of the difference amplifier is given by

$$e = V_{rot}[1 - \cos\phi - j\sin\phi] \qquad [\text{V}]. \tag{9.12}$$

This is the error signal. The relationship between the two rotor signals and the error signal is shown in Figure 9.12.

The amplified error signal from the difference amplifier is fed to the power amplifier, which drives the servomotor coupled to the load. The load is mechanically coupled to the rotor of synchro 2. As long as the error signal is nonzero, the servomotor will continue to drive the load and change the orientation of the rotor of synchro 2 in the process. When the error message is zero, the servomotor will stop and the rotor of synchro 2 will remain in this position until the position of the rotor of synchro 1 is changed.

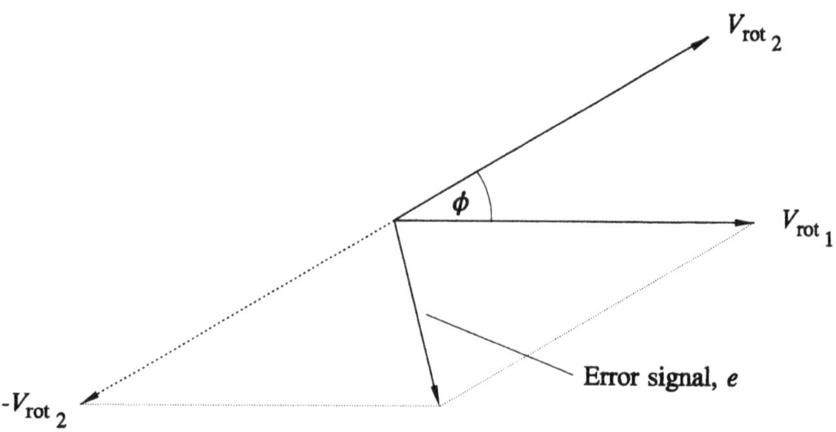

Figure 9.12

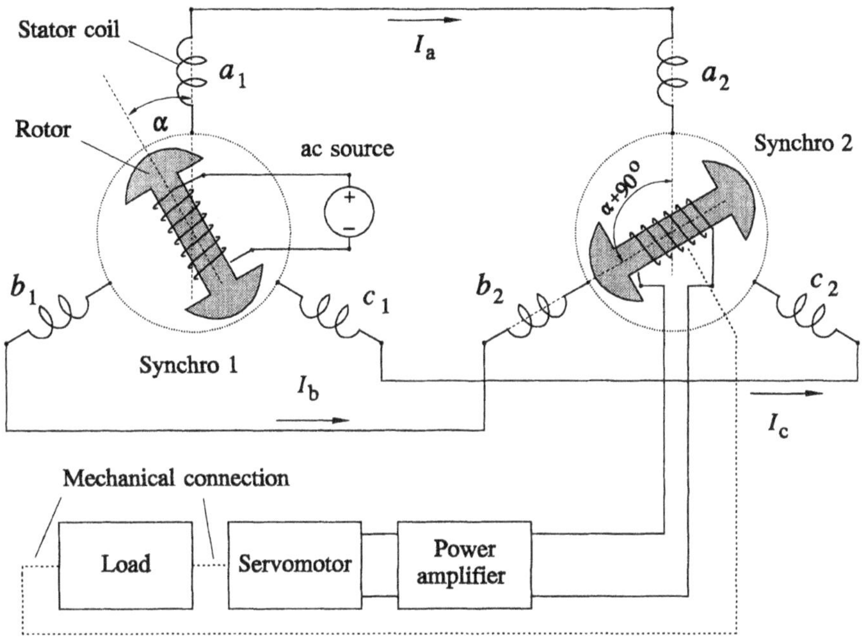

Figure 9.13

An alternate system diagram is shown in Figure 9.13. Note that the difference amplifier has been eliminated. In this case the error signal becomes zero when the rotor of synchro 2 takes up a position $(\alpha + 90°)$ or $(\alpha - 90°)$.

This type of control system is an example of a "closed loop" system, and it can be used to adjust the orientation of dish antennas for radar and earth satellite communication systems.

The synchro has applications in measurements and instrumentation.

9.5 Summary

In this chapter a number of electric machines used in special applications have been discussed. Some of these are variations on the machines discussed in earlier chapters and others are new but they all conform to the basic principles of electromagnetism. We trust that electrical engineering students and others from subdisciplines such as control systems and robotics will gain extra insight into their subdiscipline from an appreciation of how these special machines are designed and their operational characteristics.

EXERCISES

9.1 Determine the displacement angle for the following permanent magnet stepper motors:
a) 10-pole rotor with 8-pole stator
b) 3-pole rotor with 4-pole stator
c) 9-pole rotor with 4-pole stator
Draw a timing diagram for each case.

9.2 Determine the displacement angle for the following reluctance type stepper motors:
a) 8-pole rotor with 6-pole stator
b) 6-pole rotor with 8-pole stator
c) 2-pole rotor with 6-pole stator.
Draw a timing diagram for each case.

9.3 How could you change the displacement angle of the reluctance type stepper motor of Exercise 9.2(c) to half its original value without changing the physical structure of the motor? [HINT: Drive more than one pair of stator poles at a time.]

9.4 A permanent magnet stepper motor has an angular displacement of 1.8° per step. Calculate the number of poles in the rotor and stator. This motor is used to drive a lead screw on a lathe with a pitch of 1 cm. How many pulses are required to move the cutting tool a distance of 0.35 cm?

9.5 Design the rotor and stator of a stepper motor with a displacement angle of 30° per step. Explain its operation for one cycle.

References

Del Toro, V. *Electric Machines and Power Systems.* Englewood Cliffs, NJ: Prentice-Hall, 1985.

Del Toro, V. *Basic Electric Machines.* Englewood Cliffs, NJ: Prentice-Hall, 1990.

de Silva, C.W. *Control Sensors and Actuators.* Englewood Cliffs, NJ: Prentice-Hall, 1989.

Sokira, T.J., Jaffe, W. *Brushless DC Motors.* Blue Ridge Summit, PA: Tab Books, 1989.

Wildi, T. *Electric Machines, Drives and Power Systems*, 2nd ed. Englewood Cliffs, NJ: Prentice-Hall, 1991.

Appendix A

Phasor Analysis

A.1 Vector Representation of Sinusoids: The Concept of Phasors

Consider the harmonic, or sinusoidal, time functions

$$\begin{aligned} u(t) &= u_{max} \sin \omega t; \\ v(t) &= v_{max} \sin(\omega t - \alpha). \end{aligned} \tag{A.1}$$

These functions may represent steady-state voltages, currents, velocities, or any other physical variables. Graphically, the functions may be depicted (Figure A.1) as the vertical projections of the rotating vectors U and V. These vectors have the lengths u_{max} and v_{max}, respectively, and rotate with the angular velocity, ω, [rad/s].

The algebraic sum

$$w(t) = u(t) + v(t) \tag{A.2}$$

is also sinusoidal and can be represented by the projection of a third rotating vector, W. This vector is obtained as the vectorial sum of the vectors U and V as shown in Figure A.1.

It is clear that analysis of sinusoidally varying quantities can be carried out by a geometric study of vectors. The studies can be further simplified by "freezing" the rotating vectors in a given position. Such "time-frozen" vectors are often referred to as *phasors*. One may think of a phasor as a snapshot of a rotating vector.

Example A-1

Find the sum w of the two sinusoids

$$\begin{aligned} u(t) &= 6 \sin \omega t, \\ v(t) &= 5 \sin(\omega t - 37°). \end{aligned} \tag{A.1.1}$$

Solution: The two vectors U and V are shown in an x–y coordinate system (Figure A.2). For simplicity they have been "frozen" at the instant when the U vector

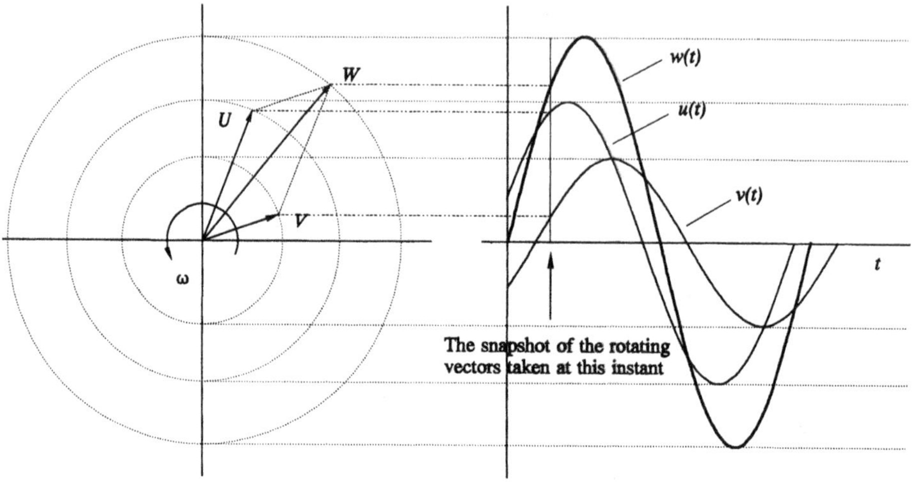

Figure A.1

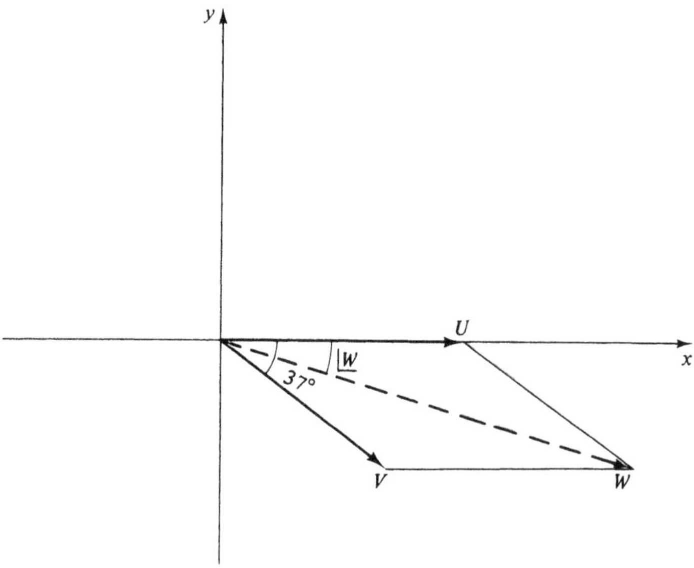

Figure A.2

coincides with the x axis. According to our definition, the rotating vectors have been turned into phasors. The U phasor, which coincides with the x axis, is referred to as the *reference* phasor.

If the sum of the two phasors U and V is the phasor, W we first determine its components, W_x and W_y along the x and y axes, respectively, and obtain

$$W_x = 6 + 5\cos 37° = 9.993,$$
$$W_y = 0 - 5\sin 37° = -3.009. \tag{A.1.2}$$

$|W|$ is given by

$$|W| = \sqrt{9.993^2 + 3.009^2} = 10.436. \tag{A.1.3}$$

The angle $\angle W$ of phasor, W relative to the reference phasor is

$$\angle W = \tan^{-1}\left(\frac{W_y}{W_x}\right) = \tan^{-1}\left(\frac{-3.009}{9.993}\right) = -16.76°. \tag{A.1.4}$$

For the time function, $w(t)$, we can write

$$w(t) = 10.436\sin(\omega t - 16.76°). \tag{A.1.5}$$

Example A.2

A sinusoidal current,

$$i = i_{max}\sin\omega t \qquad [\text{A}], \tag{A.2.1}$$

is injected into a circuit consisting of a resistor, R, in series with an inductor, L, and a capacitor, C. Describe the steady-state voltages v_R, v_L, and v_C defined in Figure A.3.

Solution: The voltage across the resistor, according to Ohm's law, equation (3.23) is

$$v_R = Ri = Ri_{max}\sin\omega t \qquad [\text{V}]. \tag{A.2.2}$$

The voltage across the inductor, from equation (3.70), is

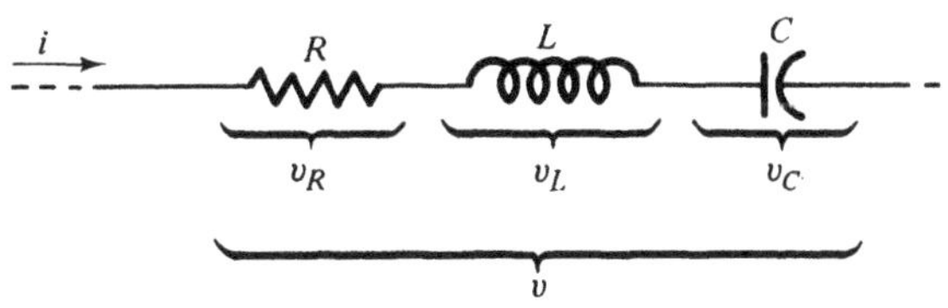

Figure A.3

$$v_L = L\frac{di}{dt} \quad [\text{V}],$$
$$= \omega L i_{\max} \cos \omega t \quad [\text{V}], \tag{A.2.3}$$
$$= \omega L i_{\max} \sin(\omega t + 90°) \quad [\text{V}].$$

The voltage across the capacitor, from equation (3.11) is

$$v_C = \frac{1}{C}Q = \frac{1}{C}\int i\, dt$$
$$= \frac{1}{C}\int i_{\max} \sin \omega t\, dt = -\frac{i_{\max}}{\omega C}\cos \omega t \tag{A.2.4}$$
$$= \frac{i_{\max}}{\omega C}\sin(\omega t - 90°) \quad [\text{V}].$$

Summary: The resistor voltage v_R is a sinusoid and in phase with the current i. The inductor voltage v_L and capacitor voltage v_C are likewise sinusoidal but, respectively, leading and lagging the current by 90°. If we represent the above time variables by the phasors V_R, V_L, V_C, and I, respectively, we obtain the *phasor diagram* shown in Figure A.4.

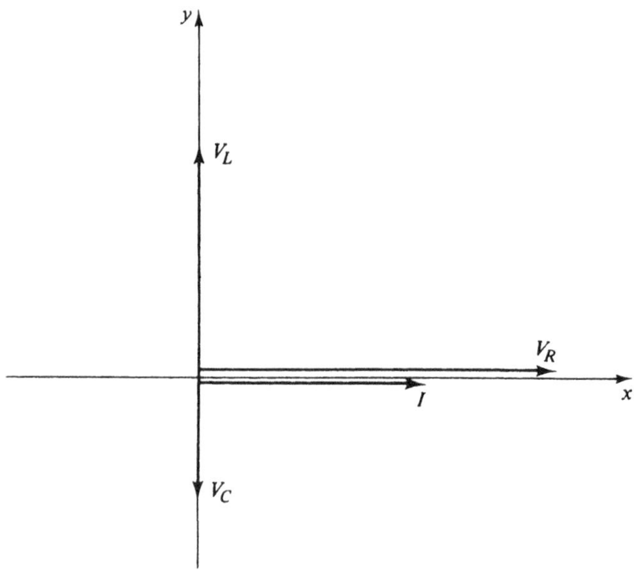

Figure A.4

Example A.3

For the circuit in Example A.2, we have the following numerical values:

$$\begin{aligned} i_{max} &= 10 \quad [A], \\ R &= 20 \quad [\Omega], \\ \omega &= 377 \quad [rad/s] \quad (60\ Hz), \\ L &= 0.040 \quad [H], \\ C &= 300 \quad [\mu F]. \end{aligned} \tag{A.3.1}$$

Find the amplitudes of the voltages across *R*, *L*, and *C*, and also the total voltage across the circuit.

Solution: From equation (A.2.2) we get,

$$v_{R\,max} = Ri_{max} = 20 \cdot 10 = 200 \qquad [V]. \tag{A.3.2}$$

From equation (A.2.3),

$$v_{L\,max} = \omega L i_{max} = 377 \cdot 0.040 \cdot 10 = 150.8 \qquad [V]. \tag{A.3.3}$$

From equation (A.2.4),

$$v_{C\,max} = \frac{i_{max}}{\omega C} = \frac{10}{377 \cdot 300 \cdot 10^{-6}} = 88.4 \qquad [V]. \tag{A.3.4}$$

The total voltage v across the circuit in Figure A.3 is the sum of the individual voltages. By representing v by its phasor V, we obtain the latter simply by vectorial addition (Figure A.5) of the phasors V_R, V_L, and V_C. We get

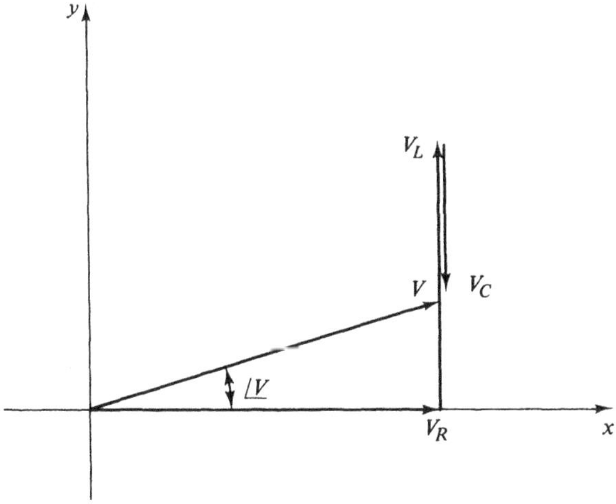

Figure A.5

$$|V| = \sqrt{200^2 + (150.8 - 88.4)^2} = 209.5 \qquad [\text{V}]. \tag{A.3.5}$$

$$\underline{/V} = \tan^{-1}\left(\frac{150.8 - 88.4}{200}\right) = 17.33^\circ. \tag{A.3.6}$$

For the total voltage we can write,

$$v = 209.5 \sin(\omega t + 17.33^\circ) \qquad [\text{V}]. \tag{A.3.7}$$

A.2 Phasor Representation Using Complex Numbers

Complex algebra is a most valuable analytical tool for the study of phasors. We first give a very brief exposé of complex algebra followed by examples.

A.2.1 Complex Numbers: Definition

Figure A.6 shows a *complex number plane*. The complex number z is defined by its coordinates, x and jy, along the real and imaginary axes, respectively. We write the complex number z in the cartesian form:

$$z = x + jy. \tag{A.3}$$

The factor j is defined by

$$j^2 = -1. \tag{A.4}$$

We say that x is the "real part of z," and y the "imaginary part." The distance $|z|$ from the origin to the coordinate point z is referred to as the modulus or magnitude of z.

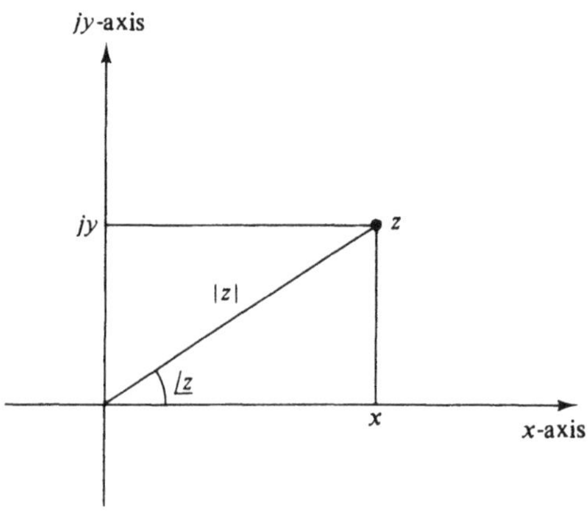

Figure A.6

We have

$$|z| = \sqrt{x^2 + y^2}. \tag{A.5}$$

The angular orientation $\angle z$ of z in reference to the real axis is obtained from

$$\angle z = \tan^{-1}\left(\frac{y}{x}\right) \tag{A.6}$$

From Figure A.6 we have

$$\begin{aligned} x &= |z| \cos \angle z, \\ y &= |z| \sin \angle z. \end{aligned} \tag{A.7}$$

Euler's theorem[1] states that

$$e^{j\angle z} = \cos \angle z + j \sin \angle z \tag{A.8}$$

The complex number z can be written in the alternate *polar* form:

$$\begin{aligned} z &= x + jy, \\ &= |z| \cos \angle z + j|z| \sin \angle z, \\ &= |z|(\cos \angle z + j \sin \angle z) = |z|e^{j/z}. \end{aligned} \tag{A.9}$$

A.2.2 Complex Algebra

Addition, subtraction, multiplication, and division of complex numbers are defined in terms of the same operations valid for real numbers. We give one example of each operation.

Example A.4

Given the two complex numbers

$$z_1 = 3 + j1 \quad \text{and} \quad z_2 = 4 - j2, \tag{A.4.1}$$

[1] This theorem can be proved as follows: From the series

$$e^x = 1 + x + \frac{x^2}{2!} + \frac{x^3}{3!} + \frac{x^4}{4!} + \cdots,$$

we obtain

$$\begin{aligned} e^{j\phi} &= 1 + j\phi - \frac{\phi^2}{2!} - j\frac{\phi^3}{3!} + \frac{\phi^4}{4!} + \cdots \\ &= 1 - \frac{\phi^2}{2!} + \frac{\phi^4}{4!} - \cdots + j\left(\phi - \frac{\phi^3}{3!} + \cdots\right) \\ &= \cos\phi + j \sin\phi \end{aligned}$$

find:

(i) $z_1 + z_2$,
(ii) $z_1 - z_2$,
(iii) $z_1 z_2$,
(iv) z_1 / z_2.

Solution:

(i) The sum of the complex numbers is obtained as follows:

$$\begin{aligned} z_1 + z_2 &= (3 + j1) + (4 - j2) \\ &= (3 + 4) + j(1 - 2) = 7 - j1. \end{aligned} \tag{A.4.2}$$

Note that the real and imaginary parts are added separately.

Note also (Figure A.7) that if each complex number is associated with a phasor, then the complex sum of the numbers corresponds to a vectorial addition of the phasors.

(ii) The difference:

$$\begin{aligned} z_1 - z_2 &= (3 + j1) - (4 - j2) \\ &= (3 - 4) + j(1 + 2) = -1 + j3. \end{aligned} \tag{A.4.3}$$

(iii) The product:

$$\begin{aligned} z_1 z_2 &= (3 + j1)(4 - j2) \\ &= 3 \cdot 4 + (j1)(-j2) + j(1 \cdot 4 - 3 \cdot 2) \\ &= 14 - j2. \end{aligned} \tag{A.4.4}$$

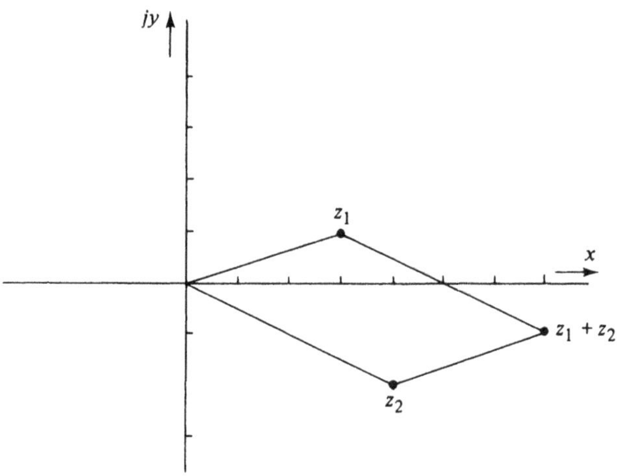

Figure A.7

(iv) The quotient:

$$
\begin{aligned}
\frac{z_1}{z_2} &= \frac{3 + j1}{4 - j2} = \frac{(3 + j1)(4 + j2)}{(4 - j2)(4 + j2)} \\
&= \frac{3 \cdot 4 + (j1)(j2) + j(1 \cdot 4 + 2 \cdot 3)}{16 + 4} \\
&= \frac{10}{20} + j\frac{10}{20} = 0.5 + j0.5.
\end{aligned} \tag{A.4.5}
$$

By using the polar form of complex numbers, we obtain

(v) The product:

$$
\begin{aligned}
z_1 z_2 &= (3 + j1)(4 - j2) = 3.162e^{j18.435^\circ} \cdot 4.472e^{-j26.565^\circ} \\
&= 14.140e^{-j8.130^\circ} = 14.140(0.990 - j0.141) \\
&= 14.00 - j2.00 \quad \text{(as before).}
\end{aligned} \tag{A.4.6}
$$

(vi) The quotient:

$$
\begin{aligned}
\frac{z_1}{z_2} &= \frac{3.162e^{j18.435^\circ}}{4.472e^{-j26.565^\circ}} = 0.707e^{j45^\circ}, \\
&= 0.5 + j0.5
\end{aligned} \tag{A.4.7}
$$

(as before).

Example A.5

Use complex algebra to solve Example A.3.

Solution: If we express the voltage phasors, V_R, V_L, and V_C as complex numbers we get

$$
\begin{aligned}
V_R &- 200 \quad [\text{V}]. \\
V_L &= j150.8 \quad [\text{V}]. \\
V_C &= -j88.4 \quad [\text{V}].
\end{aligned} \tag{A.5.1}
$$

The applied voltage represented by the phasor V is then obtained by the application of KVL:

$$
\begin{aligned}
V &= V_R + V_L + V_C = 200 + j150.8 - j88.4, \\
&= 200 + j62.4 = 209.5e^{j17.33^\circ} \quad [\text{V}].
\end{aligned} \tag{A.5.2}
$$

When the total voltage is expressed as a function of time, we get

$$
v(t) = 209.5 \sin(\omega t + 17.33^\circ) \quad [\text{V}]. \tag{A.5.3}
$$

Compare this to equation (A.3.7).

A.3 Impedances

Consider the phasor P and the complex number z. The product zP can be written as

$$zP = |z|e^{j\angle z} \cdot |P|e^{j\angle P} = |z||P|e^{j(\angle z + \angle P)}. \tag{A.10}$$

We make the following observation:

> Multiplication by z results in a magnitude amplification of $|z|$ and a phase advancement by $\angle z$ degrees.

In particular, when a phasor is multiplied by the factor j, it is rotated counter-clockwise through the angle 90° with an unchanged magnitude. Multiplication by the factor $-j$ results in a clockwise rotation through 90°. Consider the four phasors shown in Figure A.4. In view of equations (A.2.2), (A.2.3), and (A.2.4), and in consideration of the multiplication rules above, we can write:

$$\begin{aligned} V_R &= RI \quad [\mathrm{V}]; \\ V_L &= j\omega LI \quad [\mathrm{V}]; \\ V_C &= -j\frac{1}{\omega C}I = \frac{I}{j\omega C} \quad [\mathrm{V}]. \end{aligned} \tag{A.11}$$

The voltage phasor V, representing the applied voltage across the circuit, is the sum (Figure A.5) of the individual voltage phasors, that is,

$$\begin{aligned} V &= RI + j\omega LI + \frac{1}{j\omega C}I \quad [\mathrm{V}] \\ &= \left(R + j\omega L + \frac{1}{j\omega C}\right)I \quad [\mathrm{V}]. \end{aligned} \tag{A.12}$$

The multiplication factors $R, j\omega L$, and $1/j\omega C$, which when multiplied by the current phasor, I give us the voltage phasors, are referred to as the *impedances* of the individual elements (Ohm's law applied to impedances). We use the symbol, Z for impedances and for the resistor, inductor, and capacitor, respectively, we get

$$\begin{aligned} Z_R &\equiv R \quad [\Omega] \\ Z_L &\equiv j\omega L \quad [\Omega] \\ Z_C &\equiv \frac{1}{j\omega C} \quad [\Omega]. \end{aligned} \tag{A.13}$$

The total impedance, Z, for the series circuit is, according to equation (A.12),

$$Z = R + j\omega L + \frac{1}{j\omega C} \quad [\Omega]. \tag{A.14}$$

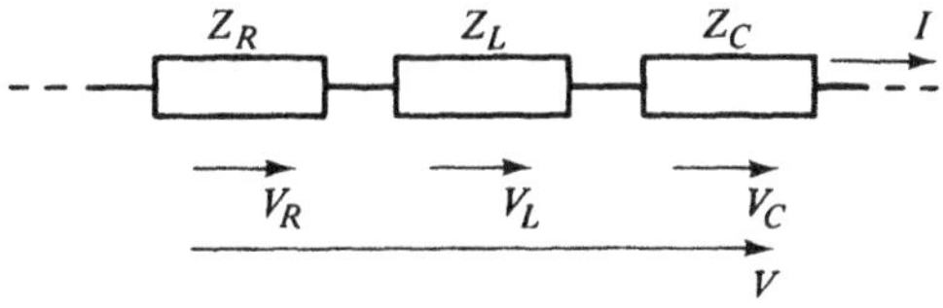

Figure A.8

In symbolic form, the relationships between impedances and current and voltage phasors are shown in Figure A.8.

Example A.6

Find the impedance of the *R–L–C* circuit assuming the numerical values given in Example A.3.

Solution: We obtain

$$Z_R = R = 20 \qquad [\Omega], \tag{A.6.1}$$

$$Z_L = j\omega L = j377 \cdot 0.040 = j15.08 \qquad [\Omega], \tag{A.6.2}$$

$$\begin{aligned} Z_C &= \frac{1}{j\omega C} = \frac{1}{j377 \cdot 300 \cdot 10^{-6}} \qquad [\Omega]; \\ &= \frac{8.84}{j} = -j8.84 \qquad [\Omega]. \end{aligned} \tag{A.6.3}$$

The total impedance is equal to

$$Z = 20 + j15.8 - j8.84 = 20 + j6.24 \qquad [\Omega]. \tag{A.6.4}$$

Example A.7

If a voltage of 100 V rms is applied across the circuit of Example A.6, find the rms values of the current and voltages across each of the three circuit elements and write all voltages and current as functions of time.

Solution: If we chose the applied voltage V as our reference phasor, that is,

$$V = 100e^{j0^\circ} = 100 \qquad [\mathrm{V}]. \tag{A.7.1}$$

(Note that the "length" of the phasor has been assumed to be proportional to its rms value, that is, $v_{max}/\sqrt{2}$. This is practical because we are usually interested in the rms values of ac variables rather than their peak values.)

From equation (A.12) we then get

$$
\begin{aligned}
I = \frac{V}{Z} &= \frac{100e^{j0^\circ}}{20 + j6.24} \qquad \text{[A]}; \\
&= \frac{100e^{j0^\circ}}{20.95e^{j17.33^\circ}} = 4.773e^{-j17.33^\circ} \qquad \text{[A]}.
\end{aligned} \tag{A.7.2}
$$

(Note that the value of the current is in rms because the voltage was expressed in rms.)

For the voltages across the components, we have

$$
\begin{aligned}
V_\text{R} &= Z_\text{R} I = 20 \cdot 4.773e^{-j17.33^\circ} = 95.47e^{-j17.33^\circ} \qquad \text{[V]}; \\
V_\text{L} &= Z_\text{L} I = j15.08 \cdot 4.773e^{-j17.33^\circ} = 71.98e^{j72.67^\circ} \qquad \text{[V]}; \\
V_\text{C} &= Z_\text{C} I = -j8.84 \cdot 4.773e^{-j17.33^\circ} = 42.19e^{j107.33^\circ} \qquad \text{[V]}.
\end{aligned} \tag{A.7.3}
$$

Note that if we show the above phasors in a phasor diagram, we obtain the same diagram as in Figure A.5 but reduced in scale and rotated clockwise by 17.33°.

If we express the voltages and current as functions of time we obtain

$$
\begin{aligned}
v(t) &= 100\sqrt{2} \sin \omega t = 141.4 \sin \omega t \qquad \text{[V]}, \\
i(t) &= 6.750 \sin(\omega t - 17.33^\circ) \qquad \text{[A]}, \\
v_\text{R}(t) &= 135.0 \sin(\omega t - 17.33^\circ) \qquad \text{[V]}, \\
v_\text{L}(t) &= 101.8 \sin(\omega t + 72.67^\circ) \qquad \text{[V]}, \\
v_\text{C}(t) &= 59.67 \sin(\omega t - 107.33^\circ) \qquad \text{[V]}.
\end{aligned} \tag{A.7.4}
$$

Example A.8

The spring–mass–dashpot system shown in Figure A.9 is subjected to a sinusoidally varying force, $f(t)$. Find the displacement, x, and the velocity, s, of the mass when it is in the steady state.

Solution: The restraining forces of the spring and dashpot are

$$
f_\text{sp} = kx \qquad \text{[N]} \tag{A.8.1}
$$

and

$$
f_\text{dp} = k_\text{d} \frac{dx}{dt} = k_\text{d} s \qquad \text{[N]}, \tag{A.8.2}
$$

respectively, where k and k_d are constants.

According to Newton's laws of motion, the acceleration, d^2x/dt^2, of the mass will be proportional to the force acting on it, hence

$$
f(t) - kx - k_\text{d} \frac{dx}{dt} = m \frac{d^2x}{dt^2} \qquad \text{[N]}, \tag{A.8.3}
$$

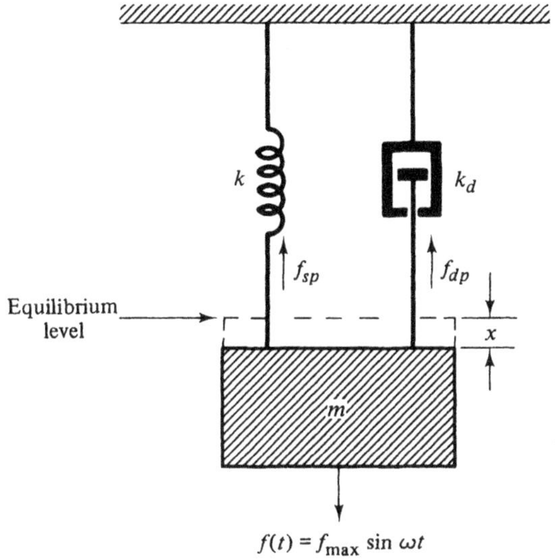

Figure A.9

or, in terms of the velocity,

$$f(t) = f_{\max} \sin \omega t = k \int s\, dt + k_d s + m \frac{ds}{dt} \qquad [\text{N}]. \qquad (\text{A.8.4})$$

In steady state, the velocity will be sinusoidal and of the form:

$$s = s_{\max} \sin(\omega t + \underline{/s}). \qquad [\text{m/s}]. \qquad (\text{A.8.5})$$

By substitution of equations (A.8.5) into (A.8.4) we obtain

$$\begin{aligned} f_{\max} \sin \omega t &= \frac{k}{\omega} s_{\max} \sin(\omega t + \underline{/s} - 90^\circ) \qquad [\text{m/s}]; \\ &+ k_d s_{\max} \sin(\omega t + \underline{/s}) \qquad [\text{m/s}]; \qquad (\text{A.8.6}) \\ &+ m\omega s_{\max} \sin(\omega t + \underline{/s} + 90^\circ) \qquad [\text{m/s}]. \end{aligned}$$

We now represent the force, $f(t)$, by the phasor, F, and the velocity, $s(t)$, by the phasor, S, in accordance with

$$\begin{aligned} F &= f_{\max} e^{j0^\circ} \qquad [\text{N}]; \\ S &= s_{\max} e^{j\underline{/s}} \qquad [\text{m/s}] \end{aligned} \qquad (\text{A.8.7})$$

These phasors are shown in Figure A.10, where we have also shown the displacement phasor, X. (X lags S by 90°. Why?)

In terms of these phasors we can write equation (A.8.6) as follows:

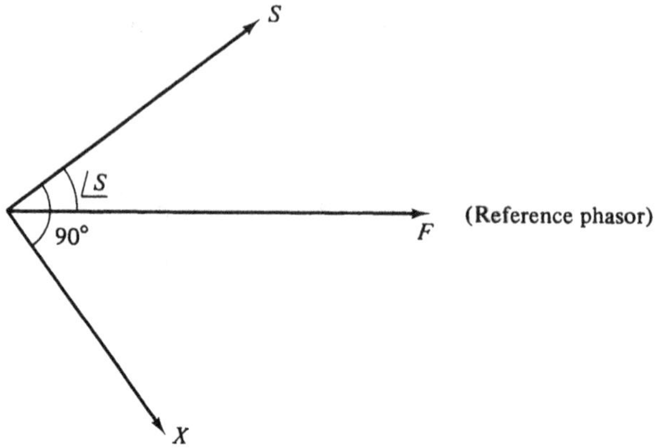

Figure A.10

$$F = \frac{k}{j\omega} S + k_d S + j\omega m S \qquad [\text{N}]. \tag{A.8.8}$$

This equation is similar to equation (A.12) and we can therefore define the *mechanical impedance*, Z_m as

$$Z_m \equiv k_d + j\omega m + \frac{k}{j\omega} \qquad [\text{N} \cdot \text{s/m}]. \tag{A.8.9}$$

We then can write equation (A.8.8) in the shorter form:

$$S = \frac{F}{Z_m} \qquad [\text{m/s}]. \tag{A.8.10}$$

Example A.9

For Example A.8 we have the following numerical values:

$$\begin{aligned} f_{max} &= 10 \qquad [\text{N}], \\ m &= 1 \qquad [\text{kg}], \\ k &= 100 \qquad [\text{N/m}], \\ k_d &= 1 \qquad [\text{N} \cdot \text{s/m}]. \end{aligned} \tag{A.9.1}$$

Determine the amplitude of the velocity, s_{max}, as a function of ω.

Solution: We have

$$Z_m = 1 + j\omega + \frac{100}{j\omega} \qquad [\text{N} \cdot \text{s/m}]. \tag{A.9.2}$$

Combining equations (A.8.7) and (A.8.10) we obtain

$$S = s_{max} e^{j\angle s} = \frac{10e^{j0°}}{1 + j(\omega - 100/\omega)} \qquad [\text{m/s}] \qquad (A.9.3)$$

that is

$$s_{max} = \frac{10}{\sqrt{1 + (\omega - 100/\omega)^2}} \qquad [\text{m/s}]. \qquad (A.9.4)$$

$$\angle s = -\tan^{-1}\left(\omega - \frac{100}{\omega}\right) \qquad (A.9.5)$$

We have plotted s_{max} and $\angle s$ against ω in Figure A.11. Note that the peak of the velocity occurs at $\omega = 10$ rad/s. This is the *resonance frequency*. It is interesting to note that $\angle s = 0$ at resonance; that is, at resonance the velocity and the force are in phase:

$$\begin{aligned} f_{res} &= f_{max} \sin \omega t \qquad [\text{N}]; \\ s_{res} &= s_{max} \sin \omega t \qquad [\text{m/s}]. \end{aligned} \qquad (A.9.6)$$

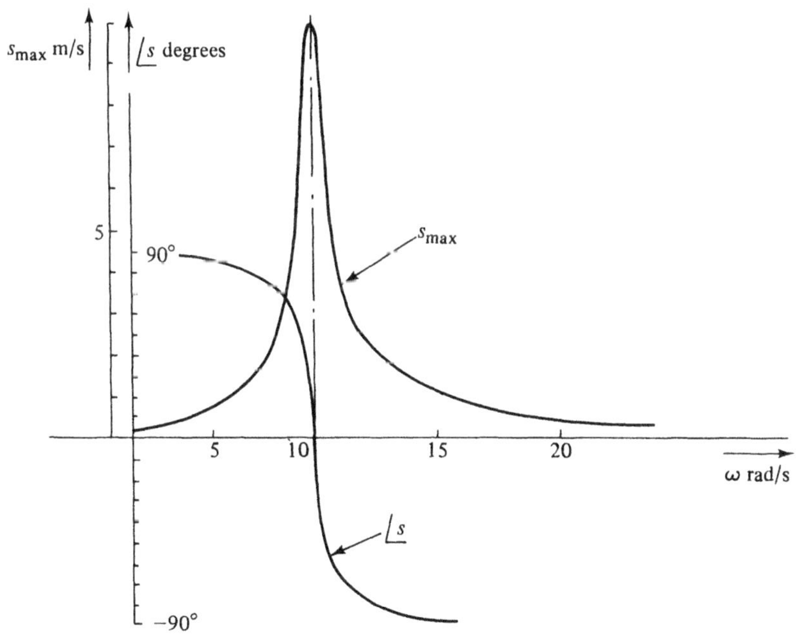

Figure A.11

A.4 Admittances

Consider the parallel circuit shown in Figure A.12. For the currents, I_1 and I_2 flowing in the impedances Z_1 and Z_2 we have

$$I_1 = \frac{V}{Z_1}, \quad I_2 = \frac{V}{Z_2} \qquad [A]. \tag{A.15}$$

The total current, I is

$$I = I_1 + I_2 = \frac{V}{Z_1} + \frac{V}{Z_2} = V\left(\frac{1}{Z_1} + \frac{1}{Z_2}\right) \qquad [A]. \tag{A.16}$$

We introduce the concept of *admittance* as follows:

$$Y_1 \equiv \frac{1}{Z_1}, \quad Y_2 \equiv \frac{1}{Z_2} \qquad [S]. \tag{A.17}$$

and the total admittance,

$$Y = Y_1 + Y_2 \qquad [S]. \tag{A.18}$$

Equation (A.16) can then be written in the form:

$$I = V(Y_1 + Y_2) = VY \qquad [A]. \tag{A.19}$$

Example A.10

Connect the three impedances in Example A.6 in parallel and feed this circuit from an ac generator delivering 100 V rms, at 60 Hz. Find the current drawn from the generator.

Solution: The three parallel elements represent a total admittance of

$$\begin{aligned} Y &= \frac{1}{20} + \frac{1}{j15.08} + \frac{1}{-j8.84} \qquad [S] \\ &= 0.0500 - j0.06631 + j0.1131 \qquad [S] \qquad (A.10.1) \\ &= 0.0500 + j0.0468 \qquad [S] \end{aligned}$$

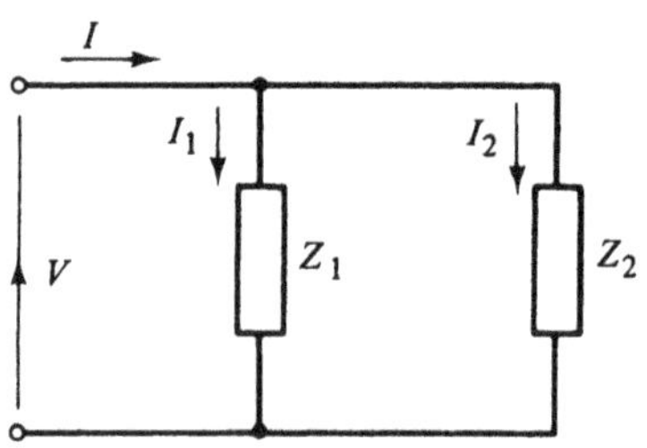

Figure A.12

Equation (A.19) then gives

$$I = 100(0.0500 + \mathrm{j}0.0468) = 5.00 + j4.68 = 6.85e^{j\,43.11^\circ} \qquad [\mathrm{A}]. \qquad (\mathrm{A}.10.2)$$

Note: The current I leads the voltage V by 43°, that is, the parallel circuit draws a leading current—it is capacitive. When the three circuit elements were connected in series across a 100-V source (Example A.7), the current was 4.7 A, and it lagged the voltage by 17°—the series circuit was inductive. The explanation of this phenomenon is left as an exercise for the reader.

Appendix B

Spectral Analysis

B.1 Periodic Waveforms

A function or "wave," $f(t)$, is said to be periodic if, for all values of t, it satisfies the equation

$$f(t) = f(T + t), \tag{B.1}$$

where T is the *period* (see Figure B.1).

According to Fourier's theorem, it is possible to express a periodic function as an infinite sum of *harmonic components* in accordance with

$$f(t) = A_0 + \sum_{\nu=1}^{\infty} A_\nu \sin(\nu\omega t + \phi), \tag{B.2}$$

where

$$\omega = \frac{2\pi}{T} \qquad \text{[rad/s]}.$$

is the *base* or *fundamental radian frequency*.

If we make use of the trigonometric identity

$$\sin(\alpha + \beta) = \sin\alpha\cos\beta + \cos\alpha\sin\beta, \tag{B.4}$$

we can write the series (B.2) in the alternate form:

$$f(t) = A_0 + \sum_{\nu=1}^{\infty} (B_\nu \sin\nu\omega t + C_\nu \cos\nu\omega t). \tag{B.5}$$

The amplitudes A_ν and phase angles ϕ_ν in the series (B.2) are related to the coefficients B_ν and C_ν in the series (B.5) by

$$\begin{aligned} A_\nu &= \sqrt{B_\nu^2 + C_\nu^2} \\ \tan\phi_\nu &= \frac{C_\nu}{B_\nu}. \end{aligned} \tag{B.6}$$

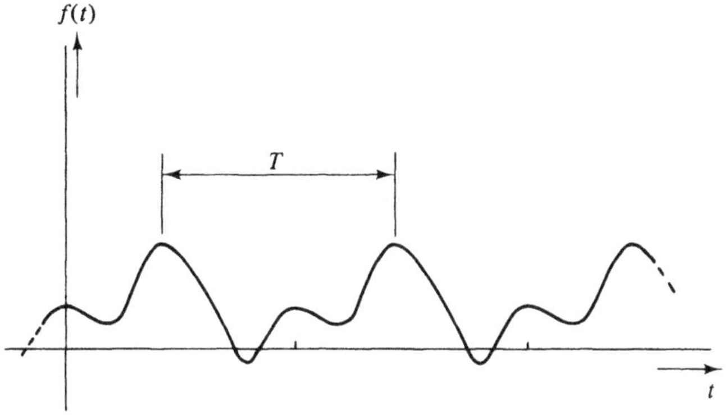

Figure B.1

The constant A_0 represents the "dc" component or average value of the periodic wave. The coefficients $A_1, A_2, \ldots$ are the amplitudes of the first, second, ... harmonics. The "first" harmonic (of frequency ω) is also referred to as the "fundamental" or "base" component.

B.2 Finding the Amplitudes of the Harmonics

It can be shown[1] that the coefficients in equation (B.5) may be derived from the following:

$$B_\nu = \frac{2}{T}\int_0^T f(t)\sin\nu\omega t\,dt;$$
$$C_\nu = \frac{2}{T}\int_0^T f(t)\cos\nu\omega t\,dt. \tag{B.7}$$

The dc component is computed from

$$A_0 = \frac{1}{T}\int_0^T f(t)\,dt. \tag{B.8}$$

[1] The proof is carried out as follows:

1. Multiply both sides of equation (B.5) by the factor sin $(\nu\omega t)$.
2. Integrate both sides over one full time period. In so doing, one finds that all integrals on the right side are equal to zero except the integral:

$$\int_0^T \sin^2\nu\omega t\,dt$$

which is equal to $B_\nu T/2$. The first line of equation (B.7) is obtained.

The second line of equation (B.7) can likewise be confirmed by following the above steps, but using cos $(\nu\omega t)$ as a multiplication factor.

Finally, equation (B.8) is obtained by integrating both sides of equation (B.5) over one full cycle.

If the average value of the periodic wave is zero, it clearly has no dc component. Certain features of the wave (symmetry) may simplify the task of finding the harmonics, as demonstrated in the following example.

Example B.1

Consider the periodic "triangular" wave shown in Figure B.2. Find the fundamental component and all higher harmonics of this wave.

Solution: We first explore the symmetry features of this wave. We note in particular that the wave is characterized by

$$f(t) = -f(T-t), \tag{B.1.1}$$

and

$$f(t) = f\left(\frac{T}{2} - t\right) \tag{B.1.2}$$

On the basis of equations (B.1.1) and (B.1.2) we can make the following observations:

Observation 1. Write the integral (B.8) in two parts:

$$A_0 = \frac{1}{T}\int_0^T f(t)\,dt = \frac{1}{T}\left[\int_0^{T/2} f(t)\,dt + \int_{T/2}^T f(t)\,dt\right] \tag{B.1.3}$$

Because of equation (B.1.1), the two parts of the integrals are equal but of opposite sign. Therefore, $A_o = 0$; the wave contains no dc component; its average value is zero.

Observation 2. For a cosine wave we have,

$$\cos \nu\omega t = \cos \nu\omega(T - t). \tag{B.1.4}$$

In view of equation (B.1.1), it is clear that the second set of integrals (B.7) vanishes for all ν. Therefore,

$$C_\nu = 0 \quad \text{for} \quad \nu = 1, 2, 3, \ldots, \infty, \tag{B.1.5}$$

that is, the wave contains no cosine terms.

Observation 3. For a sinewave we have

$$\sin \nu\omega t = \sin \nu\omega\left(\frac{T}{2} - t\right) \text{ for odd } \nu\text{'s}, \tag{B.1.6}$$

$$\sin \nu\omega t = -\sin \nu\omega\left(\frac{T}{2} - t\right) \text{ for even } \nu\text{'s}. \tag{B.1.7}$$

Because of equation (B.1.2) it is clear that the first set of integrals of (B.7) vanishes for all even values of ν;

$$B_\nu = 0 \quad \text{for} \quad \nu = 2, 4, 6, \ldots, \infty. \tag{B.1.8}$$

It follows that the wave contains only odd harmonics.

Observation 4. In view of equation (B.1.2), (B.1.6), and (B.1.7) the product

$$f(t) \sin \nu\omega t \tag{B.1.9}$$

attains identical values for t, $\left(\frac{T}{2} - t\right)$, $\left(\frac{T}{2} + t\right)$ and $(T - t)$ (assuming ν is odd). Thus we need to perform the integration of (B.7) over only one-quarter cycle; that is,

$$B_\nu = \frac{8}{T}\int_0^{T/4} f(t) \sin \nu\omega t \, dt \quad \text{for} \quad \nu = 1, 3, 5, \ldots, \infty \tag{B.1.10}$$

In the interval $0 < t < T/4$ the function $f(t)$ (see Figure B.2) is a straight line of the form:

$$f(t) = 4\frac{\hat{A}}{T}t. \tag{B.1.11}$$

Substituting into equation (B.1.10) we get

$$B_\nu = 32\frac{\hat{A}}{T^2}\int_0^{T/4} t \sin \nu\omega t \, dt \quad \text{for} \quad \nu = 1, 3, 5, \ldots, \infty. \tag{B.1.12}$$

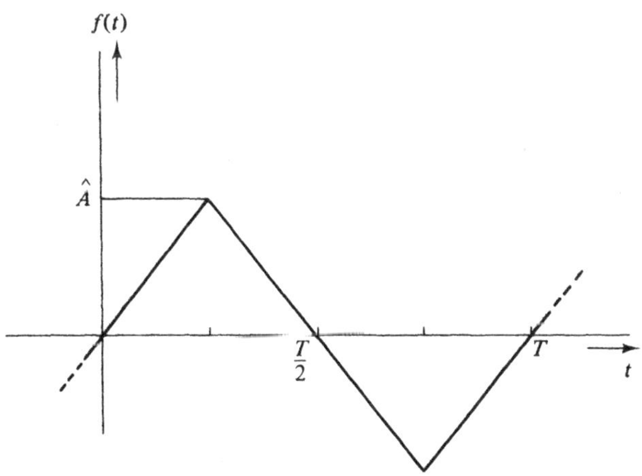

Figure B.2

The integral gives the following for B_ν :

$$B_\nu = \frac{8}{\pi^2}\frac{\hat{A}}{\nu^2}(-1)^{(\nu+3)/2} \quad \text{for} \quad \nu = 1, 3, 5, \ldots, \infty. \tag{B.1.13}$$

The triangular waveform is described by the infinite series

$$f(t) = \frac{8}{\pi^2}\hat{A}\left[\sin\omega t - \frac{1}{9}\sin 3\omega t + \frac{1}{25}\sin 5\omega t - \cdots\right] \tag{B.1.14}$$

In practice, one obtains very good accuracy by including only the first few terms in the series. Figure B.3 shows the original triangle compared to the waveform obtained by including only the first three terms in the series.

B.3 Spectral Analysis by Numerical Integration

In many practical cases one cannot perform the analysis as neatly as the previous example would have us believe. Often the periodic wave is obtained experimentally and the function $f(t)$ is available not in analytic form but as a graph.

In such situations, the spectral analysis must be carried out numerically using approximations for the integrals (B.7) and (B.8).

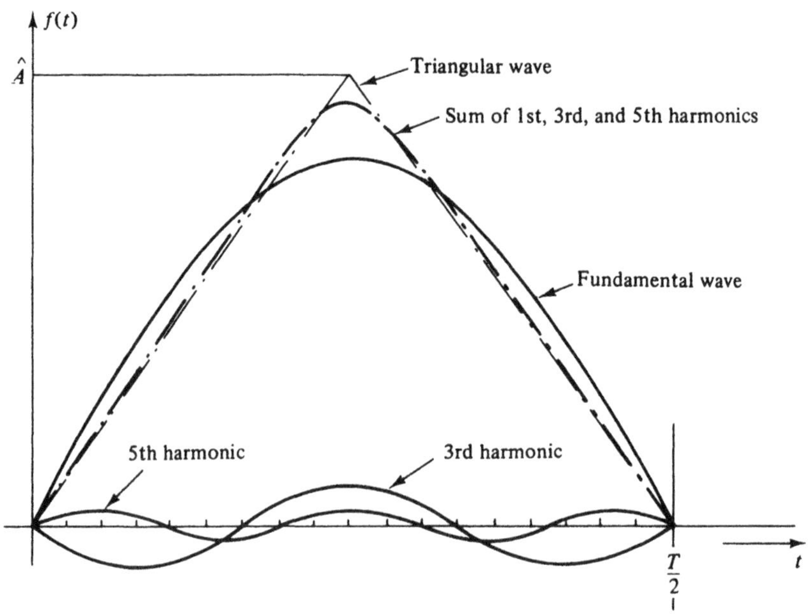

Figure B.3

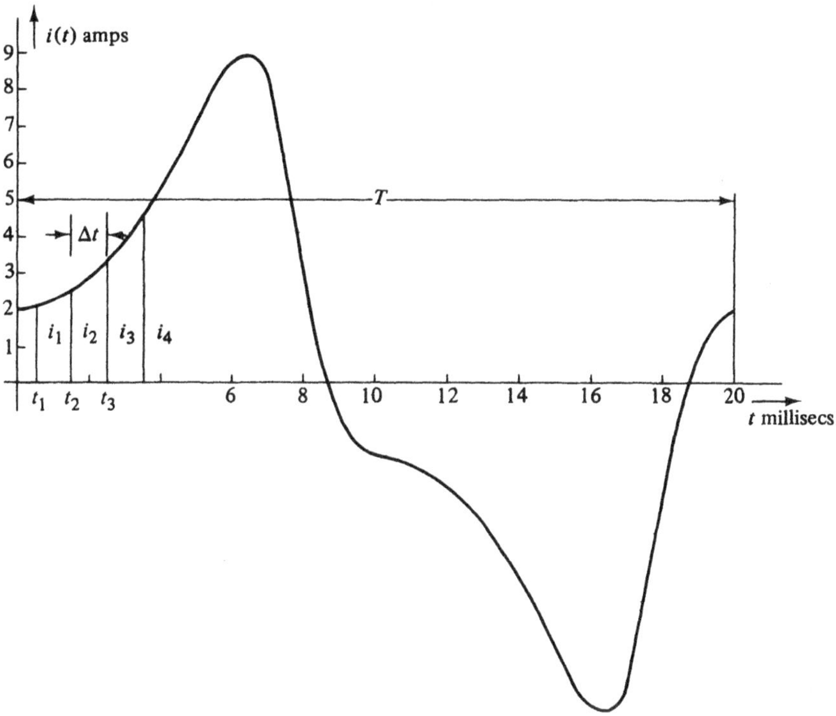

Figure B.4

Example B.2

Figure B.4 shows the magnetization current in a power transformer as recorded in an experiment (see Figure 5.9). The base frequency is 50 Hz, which means that the period, T is equal to 20 ms. Carry out a spectral analysis of the waveform. Specifically, find the amplitude of the 50-Hz component.

Solution: Because the function $f(t)$ cannot be expressed in an analytical form, we must calculate B_1 and C_1 by using the $f(t)$ values obtained from the graph.

We can write the integrals (B.7) as the approximations:

$$B_1 \approx \frac{2\,\Delta t}{T} \sum_{r=1}^{n} i_r \sin \omega t_r;$$
$$C_1 \approx \frac{2\,\Delta t}{T} \sum_{r=1}^{n} i_r \cos \omega t_r. \tag{B.2.1}$$

We have divided the time interval $0 < t < T$ into n time segments of width Δt. In Figure B.4, we have chosen $n = 20$, which corresponds to $\Delta t = 1$ ms. The value

Table B.1

r number	i_r amperes	ωt_r degrees	$\sin \omega t_r$	$i_r \sin \omega t_r$	$\cos \omega t$	$i_r \cos \omega t$
1	2.1	9	0.156	0.329	0.988	2.074
2	2.9	27	0.454	1.317	0.891	2.584
3	3.3	45	0.707	2.333	0.707	2.333
4	4.5	63	0.891	4.010	0.454	2.043
5	6.2	81	0.988	6.124	0.156	0.970
6	8.0	99	0.988	7.902	−0.156	−1.251
7	9.0	117	0.891	8.019	−0.454	−4.086
8	6.0	135	0.707	4.243	−0.707	−4.243
9	0.6	153	0.454	0.272	−0.891	−0.535
10	−1.8	171	0.156	−0.282	−0.988	1.778
Sum				34.267		1.667

of the current, i_r, in the center of each time segment is recorded. We then compute and tabulate ωt_r, $\sin \omega t_r$, and $\cos \omega t_r$ for each value of t_r in the interval. Finally the products $i_r \sin \omega t_r$ and $i_r \cos \omega t_r$ are computed. The results are shown in Table B.1. Because of the symmetrical nature of the waveform, we can use

$$f(t) = -f(T/2 + t) \tag{B.2.2}$$

and perform the summation over half the period. This means that we compute B_1 and C_1 from the following expressions:

$$\begin{aligned} B_1 &\approx \frac{4\,\Delta t}{T} \sum_{r=1}^{n/2} i_r \sin \omega t_r, \\ C_1 &\approx \frac{4\,\Delta t}{T} \sum_{r=1}^{n/2} i_r \cos \omega t_r. \end{aligned} \tag{B.2.3}$$

By using the values obtained from Table B.1 we get

$$\begin{aligned} B_1 &= \frac{4 \cdot 0.001}{0.020} \cdot 34.267 = 6.853, \\ C_1 &= \frac{4 \cdot 0.001}{0.020} \cdot 1.667 = 0.333. \end{aligned} \tag{B.2.4}$$

By using equation (B.6) we compute the amplitude and phase angle of the base harmonic or the fundamental as

$$A_1 = \sqrt{6.853^2 + 0.333^2} = 6.853;$$
$$\phi_1 = \tan^{-1}\left(\frac{0.333}{6.853}\right) = 2.78°. \qquad (B.2.5)$$

The fundamental component of the current in the waveform is

$$i(t) = 6.861 \sin(\omega t + 2.78°) \qquad [A], \qquad (B.2.6)$$

where

$$\omega = \frac{2\pi}{0.020} = 314 \qquad [rad/s]. \qquad (B.2.7)$$

B.4 Periodic Waveforms in the Space Domain

So far, we have assumed that the independent variable is time, t. This is normally the case, as harmonic analysis is most often used in communication theory where time is the variable.

However, there are areas of science and technology where the periodic phenomena to be analyzed involves space variables. For example, the periodic current, emf, and magnetic flux found in the air gaps of electric machines belong to this category.

Example B.3

Figure B.5 shows the "sheet of current" in one phase of the distributed stator winding of a three-phase induction machine. We remember from Section 8.4.2

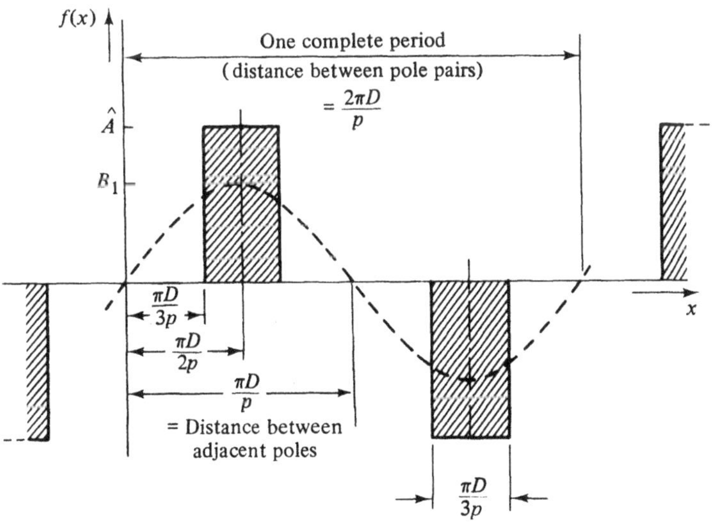

Figure B.5

that this sheet current was created by representing a macroscopic (or "smeared") view of the current distribution in the stator slots.

We also remember that, because the current is ac, the sheet current will pulsate in time. Figure B.5 is therefore a snapshot taken at a given moment in time (e.g, when the current has reached its peak).

Find the amplitude of the fundamental waveform of the sheet current.

Solution: By placing the origin as shown in Figure B.5, the waveform will have the same symmetrical features as the triangular wave in Example B.1. The observations made concerning the harmonics of the triangular wave apply here. The fundamental waveform must therefore be a sinusoid, as shown by the dashed line in Figure B.5. We want to find its amplitude, B_1.

The equations derived earlier were in terms of the independent variable, t, and the period, T. Now the independent variable is x and the period is $2\pi D/p$.[2] We can therefore use the previous formula after making the following changes to the variable:

$$t \to x$$

$$T \to \frac{2\pi D}{p} \tag{B.3.1}$$

$$\omega = \frac{2\pi}{T} \to \frac{p}{D}.$$

Substituting equation (B.3.1) into (B.1.10), we get

$$B_\nu = \frac{4p}{\pi D} \int_0^{\pi D/2p} f(x) \sin\left(\nu \frac{p}{D} x\right) dx \tag{B.3.2}$$

In this case the function $f(x)$ has the following values (see Figure B.5):

$$f(x) = 0 \quad \text{for} \quad 0 < x < \pi D/3p; \tag{B.3.3}$$

$$f(x) = \hat{A} \quad \text{for} \quad \pi D/3p < x < \pi D/2p. \tag{B.3.4}$$

The integral (B.3.2) gives

$$\begin{aligned} B_1 &= \frac{4p}{\pi D}\left[\int_0^{\pi D/3p} 0 \cdot \sin\left(\frac{px}{D}\right) dx + \int_{\pi D/3p}^{\pi D/2p} \hat{A} \cdot \sin\left(\frac{px}{D}\right) dx\right], \\ &= \frac{4p}{\pi D}\left[0 + \frac{\hat{A}D}{2p}\right] = \frac{2}{\pi}\hat{A} \qquad [\text{A/m}]. \end{aligned} \tag{B.3.5}$$

[2] The expression $2\pi D/p$ is the peripheral width of one pole pair.

Example B.4

Use the result obtained in Example B.3 to show analytically that the flux in all three stator phases of an induction motor as well as a synchronous machine, jointly create a rotating "flux wave." (In Sections 4.7.1 and and 8.3, this was shown graphically.)

Solution: The result in Example B.3 can be stated as follows:

$$\Phi_{a1}(x) = \frac{2}{\pi}\hat{\Phi}\sin\left(\frac{p}{D}x\right) \qquad \text{[Wb]}. \tag{B.4.1}$$

This represents the fundamental of the stator flux due to the current in phase a as viewed at a particular instant in time. If we multiply equation (B.4.1) by $\sin\omega t$, we express the pulsating nature of the wave as

$$\Phi_{a1}(x,t) = \frac{2}{\pi}\hat{\Phi}\sin\left(\frac{p}{D}x\right)\sin\omega t \qquad \text{[Wb]}. \tag{B.4.2}$$

The currents in phases b and c give rise to similar fluxes, but they are shifted both in space and time by $2\pi/3$ and $4\pi/3$ radians, respectively, that is,

$$\begin{aligned}\Phi_{b1}(x,t) &= \frac{2}{\pi}\hat{\Phi}\sin\left(\frac{p}{D}x - \frac{2\pi}{3}\right)\sin\left(\omega t - \frac{2\pi}{3}\right) \qquad \text{[Wb]};\\ \Phi_{c1}(x,t) &= \frac{2}{\pi}\hat{\Phi}\sin\left(\frac{p}{D}x - \frac{4\pi}{3}\right)\sin\left(\omega t - \frac{4\pi}{3}\right) \qquad \text{[Wb]}.\end{aligned} \tag{B.4.3}$$

The total flux is obtained by the sum

$$\Phi_{\text{tot }1}(x,t) = \Phi_{a1} + \Phi_{b1} + \Phi_{c1} \qquad \text{[Wb]}. \tag{B.4.4}$$

Using the same trigonometric manipulations shown in Section 4.4, we determine that the total flux of the fundamental waveform takes the form

$$\Phi_{\text{tot }1}(x,t) = \frac{3}{\pi}\hat{\Phi}\sin\left(\frac{p}{D}x - \omega t\right) \qquad \text{[Wb]}. \tag{B.4.5}$$

This is the equation for a wave revolving with constant speed and constant amplitude in the positive x direction.

Appendix C

The SI Unit System

C.1 General

The British System of Units, at present the most popular in the United States, has roots that go back to Roman times. Internationally, it is losing ground very fast to the Metric System introduced by the French Academy of Sciences, in the eighteenth century.

C.2 Basic Units

We have summarized in Table C.1 the most important features of the SI system as it is related to electric energy engineering.

C.3 Derived Units

Other units derived from the above are called "derived units" of the SI. For example, the unit of force in the SI system is the *newton* [N]. It is derived from Newton's Second Law (force = mass × acceleration) and is defined as the force

Table C.1 Basic SI Units

Physical Quantity	Name of Unit	Symbol
Length	meter	m
Mass	kilogram	kg
Time	second	s
Electric Current	ampere	A
Temperature	degree Kelvin	K

(Other basic units for radioactivity, luminous intensity, and "amount of substance" exist.)

Table C.2 Derived SI Units

Physical Quantity	Name of Unit	Symbol
Force	newton	$N = kg \cdot m/s^2$
Work (or energy)	joule	$[J] = [N \cdot m]$
Power	watt	$[W] = [J/s]$
Pressure	pascal	$[Pa] = [N/m^2]$
Electric charge	coulomb	$[C] = [A \cdot s]$
Electric potential	volt	$[V] = [J/C] = [W/A]$
Electric capacitance	farad	$[F] = [C/V]$
Electric resistance	ohm	$[\Omega] = [V/A]$
Magnetic flux	weber	$[Wb] = [V \cdot s]$
Magnetic flux density	tesla	$[T] = [Wb/m^2]$
Magnetic inductance	henry	$[H] = [Wb/A]$

needed to impart an acceleration of 1 $[m/s^2]$ to a mass of 1 [kg]. Force therefore has the dimension "kilogram times acceleration," which is written as $[kg \cdot m/s^2]$.

Consider another example. The SI unit for energy is the *joule* [J]. It is derived (see Section 2.3) from the basic equation for energy (energy = force × distance) and is defined as the work done or energy generated by a force of 1 [N] acting over a distance of 1 [m]. Energy has the dimension "force times distance," which is written as $[N \cdot m]$.

Table C.2 shows some of the more important derived SI units.

C.4 Multiplication Factors and Prefixes

The SI system is based on decimal units, formed by multiplying or dividing a single base unit by powers of 10. In this manner, widely different unit ranges can be accommodated. For example, a communications engineer is interested in microwatts [μW] of power, and his power colleague works with megawatts [MW]. These two units differ by a factor of 10^{12}.

Table C.3 shows the prefixes and letter symbols for the unit multipliers.

C.5 Conversion Between Unit Systems

Conversion between other unit systems and SI units is possible from a knowledge of the basic conversion constants. The most useful ones are given in Table C.4.

The conversion factors for the most commonly used energy and power units are given in Appendix D.

Table C.3 Multiplication Factors and Prefixes for Forming Decimal Multiples and Submultiples of the SI Units

Multiplier	Exponent	Prefix	Symbol
1 000 000 000 000 000 000	10^{18}	exa	E
1 000 000 000 000 000	10^{15}	peta	P
1 000 000 000 000	10^{12}	tera	T
1 000 000 000	10^{9}	giga	G
1 000 000	10^{6}	mega	M
1 000	10^{3}	kilo	k
100	10^{2}	hecto	h
10	10^{1}	deca	da
0.1	10^{-1}	deci	d
0.01	10^{-2}	centi	c
0.001	10^{-3}	milli	m
0.000 001	10^{-6}	micro	μ
0.000 000 001	10^{-9}	nano	n
0.000 000 000 001	10^{-12}	pico	p
0.000 000 000 000 001	10^{-15}	femto	f
0.000 000 000 000 000 001	10^{-18}	atto	a

Table C.4 Some Useful Conversion Constants

1 m = 3.28083990 ft = 39.37007874 inches
1 ft = 0.3048000 m
1 kg = 2.20462262 lb = 35.27396195 oz
1 lb = 0.45359237 kg
1 m^3 = 264.1720 gallons (U.S.)
1 gallon (U.S.) = 0.003785412 m^3
degrees Kelvin = 5/9 (degrees Fahrenheit) + 255.37
degrees Fahrenheit = 9/5 (degrees Kelvin) − 459.67

References

ASTM/IEEE Standard Metric Practice, IEEE Standard STD 268-1986. This Booklet may be obtained from IEEE Service Center, 445 Hoes Lane, Piscataway, NJ 08854.

Appendix D

Table D.1 Conversion of Units of Energy

Unit	Symbol	eV	J	Btu	Wh	kcal	toe
electron-volt	eV	1	1.602×10^{-19}	1.519×10^{-22}	4.448×10^{-23}	3.826×10^{-23}	3.81×10^{-30}
joule	J	6.24×10^{18}	1	9.479×10^{-4}	0.278×10^{-3}	0.239×10^{-3}	2.38×10^{-11}
British thermal unit	Btu	6.586×10^{21}	1.055×10^{3}	1	0.2931	0.252	2.51×10^{-8}
watt-hour	Wh	2.247×10^{22}	3.60×10^{3}	3.412	1	0.860	8.57×10^{-8}
kilocalorie	kcal	2.614×10^{22}	4.185×10^{3}	3.968	1.163	1	9.97×10^{-8}
tonnes oil equiv.	toe	0.262×10^{30}	4.2×10^{10}	3.985×10^{7}	1.166×10^{7}	1.003×10^{7}	1

Table D.2 Conversion of Units of Power

Unit	Symbol	W	hp	Btu/sec	kcal/sec
watt	W	1	1.34×10^{-3}	0.948×10^{-3}	0.239×10^{-3}
horsepower	hp	746	1	0.707	0.178
British thermal units per second	Btu/sec	1.055×10^{3}	1.414	1	0.252
kilocalories per second	kcal/sec	4.184×10^{3}	5.607	3.966	1

To convert British thermal units to joules, one must go to the left-hand side of Table D.1 and find the row for Btu (row 3). One then follows the row to the column corresponding to joules (column 4). The number given there (1.055×10^3) is the multiplicant for the conversion.

To convert kilocalories per second to horsepower, one must go to the left-hand side of Table D.2 and find the row for kilocalories per second (row 4). One then follows the row to the column corresponding to horsepower (column 4). The number given there (5.607) is the multiplicant for the conversion.

Answers To Selected Exercises

Chapter 1

1.1 $m = 445.1$ kg

1.2 About 18.31 billion dollars

Chapter 2

2.1 **a)** 16.5% (or 1.62 [m/s^2])
b) 1.62 [J/kg]

2.2 **a)** 0
b) 17,900 [MW] (or 24.0 million hp)

2.3 678.6 [kW]

2.4 The kinetic energy of the elevator must also be included. This energy component amounts to

$$w_{\text{kin}} = \tfrac{1}{2} \cdot 5000 \cdot 5^2 = 62{,}500 \qquad \text{[J]}.$$

2.5 9.26 miles per gallon

2.7 $\ddot{x} = \dfrac{(M - m)g_0}{m + M + I/R^2}$ [m/s]

2.10 $\Delta T = 1.55$°C

2.12 **a)** 30.5 [MJ/kg]
b) 4.8% and 95.2%, respectively

Chapter 3

3.1 **a)** $3v_0$ [V].
b) $w_e = (3/2)\,Cv^2$ [J].
c) A force is required to pull the plates apart (the opposite charges attract). The work done by this force adds to the energy of system.

3.3 **a)** $i(t) = \frac{V}{R} e^{-2t/RC}$ [A]

b) The two capacitor voltages can be written as

$$\frac{V}{2}(1 + e^{-2t/RC}) \quad \text{and} \quad \frac{V}{2}(1 - e^{-2t/RC}),$$

respectively.

c) $w_\Omega = \frac{1}{4}CV^2$ [J]

d) Note that w_Ω is independent of the R value. Thus, charge redistribution via a very small R (short circuit) results in the same heat loss as a large R. (How is this possible?)

3.7 **a)** $15.6 \cdot 10^{-3}$ [N]. (The force on each charge is directed outward in a direction perpendicular to the line between the other two charges.)

b) 11.98 [kV/m] (directed perpendicular to the base)

c) Each charge contributes a field component vector directed away from the charge. As the three-component vectors have equal magnitudes they cancel.

3.9 **a)** $f = 50$ [Hz]

b) $d = 0.00398$ [mm]

3.10 **a)** $i = 0.686$ [A]

b) $p = 8.23$ [W]

c) $p = 0.00823$ [W/m]

3.12 **a)** 1.333 [kA]

b) 44.8 [MW]

c) 633.6 [kV]

d) 844.8 [MW]

e) 94.7%

3.15 $T = 1000$ [N · m]

3.18 $L = 62.5$ [mH] (without air gap)

$L = 2.42$ [mH] (with 2-mm air gap)

Chapter 4

4.1 $Q = \pm 13.85$ [kVAr]

4.3 Induced voltage = 0. (Note that flux linked with stator coil is zero for all rotor positions.)

4.4 **a)** 459.3 [kW]

b) 1377.8 [kW]

c) 11022 [kWh]

4.5 **a)** 15.29 [kV]

b) 7.647 [kV per phase] (or 13.24 [kV line to line])

4.8 **a)** 84.8 [A]
b) 146.9 [A]
c) $S = 215.9 + j\,134.7$ (3-phase, kilo values)

4.10 **a)** $\delta = 38.68°$
b) $|I| = 918$ [A] (leading the voltage by 19.34°)
c) $Q = -6.319$ MVAr. (Generator absorbs reactive power, acting like a shunt reactor.)

4.12 **a)** $\sin\delta = \dfrac{PX_s}{|V|\,|E|}$

If P, $|V|$, and X_s are constants, it follows that sin δ (and thus δ) will decrease if $|E|$ is increased.

b) $Q = \dfrac{|V|\,|E|\cos\delta - |V|^2}{X_s}$

The generator absorbs reactive power if $Q < 0$; that is, if $|E|\cos\delta < |V|$. If $|E|$ is increased, δ will decrease (see part **a**) and $\cos\delta$ will increase. Thus the product $|E|\cos\delta$ will increase, meaning that $||E|\cos\delta - |V||$ will decrease, thus decreasing $|Q|$

c) By 17.9%

4.13 **a)** $\delta = 16.48°$
b) $\phi = 26.57°$
c) $|E| = 14.69$ [kV]

4.14 $|E| = 10.76$ [kV]

Chapter 5

5.1 **a)** 60 [A] and 150 [A], respectively
b) $|Z| = 1.33$ [Ω]

5.2 **a)** 120% of normal flux value
b) 100% (200 [V]). The high flux may possibly result in core losses (and temperatures) that may damage the transformer.

5.3 **a)** $Z_s = 0.0806 + j0.418$ [Ω] (on HV side) $Z_s = 0.0129 + j0.0669$ [Ω] (on LV side)
b) 5.11% of normal value
c) Because the core flux is only 5% of normal value, the core losses likewise are minute (actually less than 5% of normal values because core losses increase almost by the square of the flux).

5.5 **a)** 63.66 [A] and 159.2 [A], respectively; Secondary voltage = 200 [V]
b) 61.23 [A] and 153.1 [A], respectively; Secondary voltage = 192.4 [V]
c) Yes, slightly.

5.7 105 [kVA]

5.8 **a)** 1167 [V]
b) Due to the high flux densities (233% of normal. Why?) the core losses will be very high with overheating as a result.

5.11 150 [kVA] tertiary resistive load in combination with 150 [kVA] secondary resistive load will result in 300 [kVA] primary [kVA]. This is the limit of the primary.

5.12 **a)** $R = 538.2$ [Ω]
b) 74.7 [A]
c) 545.4 [A] (primary); 129.3 [A] (secondary)

5.14 By 0.953%

Chapter 6

6.1 The primary winding carries 0.358 [A]. The secondary winding carries 22.6 [A] in one section and 13.5 [A] in the other.

6.3 **a)** 12.0 [MW] (3-phase)
b) 10.83 [MW] (3-phase)

6.4 Increase by 0.604%

6.7 **a)** 149.48 [kV]
b) Sending end powers: 102.48 [MW], 17.85 [MVAr]. Receiving end powers: 98.01 [MW], 0 [MVAr].

6.8 **a)** 27.1 [A per phase]
b) Line consumes 5.04 [kW] and generates 6.58 [MVAr]
c) $|V_2| = 141.19$ [kV]

6.10 **a)** $C = 106.1$ [μF per phase]
b) 31.18 [kV]

6.13 The load flow will appear as follows:

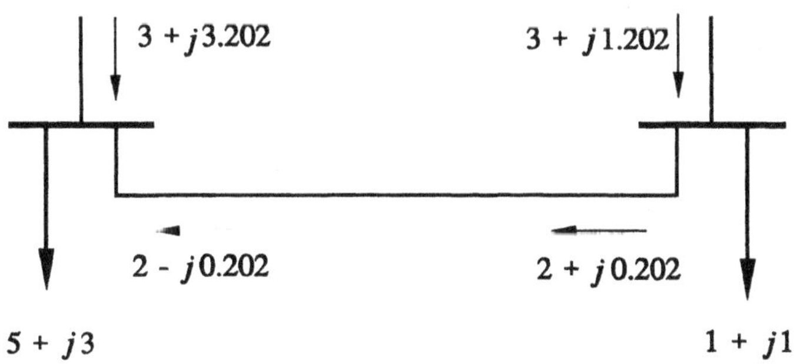

Figure Ans.1

6.15 The load flow will appear as follows:

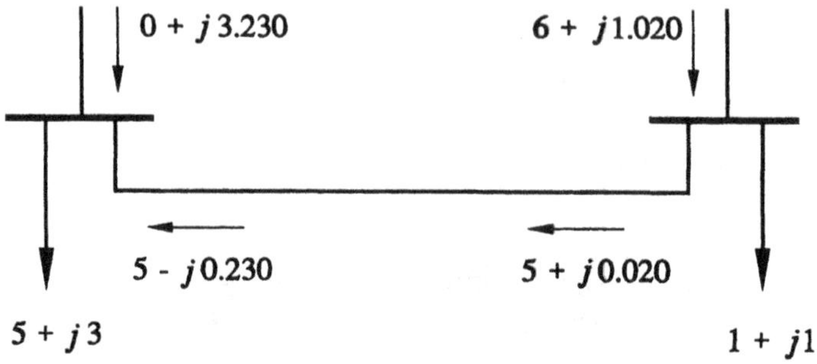

Figure Ans.2

Chapter 7

7.1 a) Maximum force occurs for $s = 0$

b) $f_{\text{m max}} = \dfrac{BLV}{R}$ [N]

7.2 a) Maximum power occurs for $s = \frac{1}{2}s_0$

b) $p_{\text{m max}} = \dfrac{1}{4}\dfrac{V^2}{R}$

c) $p_\Omega = \dfrac{1}{4}\dfrac{V^2}{R}$

d) $\eta = 50\%$

7.5 $p_{\text{m max}} = \dfrac{1}{4}\dfrac{V_a^2}{R_a}$

No. It would be overheated. For example, the motor in Example 7.8 would deliver 13.7 [kW] and dissipate an equal amount in ohmic heat. (Its normal heat losses are only 256 [W] according to Example 7.8.)

7.6 $x = 2.193$

7.7 a) $e = 498.0$ [V]; $i_a = 9.30$ [A]

b) $e = 523.0$ [V]; $i_a = 9.86$ [A]

7.9 The motor will deliver 86.79 [kW] (116 [hp]) to the load. Operating efficiency = 84.67%. Shaft torque = 361 [N · m].

7.12 $R_{\text{min}} = 2.29$ [Ω]. Generated power = 91.56 [kW]. Diesel output = 105.55 [kW].

Chapter 8

8.2 a) 3.17%

b) 196.83%

c) 156.8 [A per phase]

8.3 **a)** $P_1 = 12.01$ [kW]
$P_m = 10.64$ [kW]
$P_\Omega = 1.37$ [kW]
$T_m = 88.39$ [N · m]
b) $P_1 = -13.98$ [kW] (delivers power to netvork)
$P_m = -15.86$ [kW] (draws this power from prime mover)
$P_\Omega = 1.88$ [kW]
$T_m = -121.2$ [N · m] (instead of delivering torque to load, will now require torque from a prime mover to run at this speed)
c) $P_1 = 27.33$ [kW]
$P_m = -5.54$ [kW] (needs power from prime mover)
$P_\Omega = 32.87$ [kW] (With these losses the motor would burn up in a hurry. Note that 83.2% of this lost power is drawn from the network and 16.8% from prime mover.)
$T_m = -44.1$ [N · m]
d) $P_1 = 23.57$ [kW]
$P_m = -11.99$ [kW]
$P_\Omega = 35.56$ [kW] (see comment under part *c*)
$T_m = 0$

8.5 $P_m = 75.15$ [kW] = 100.7 [hp]
$|I_1| = 107$ [A per phase]
$\cos\phi = 0.988$
$\eta = 93.3\%$
$T_m = 823$ [N · m]

8.6 635 [A] (if IM model is used); 646.5 [A] (if magnetization current is included)

8.7 Stator current = 111 [A per phase]
Motor output power = 73.30 [kW] = 98 [hp]
Efficiency = 89.5%
Power factor = 0.970

8.9 93.07 [kW]

8.11 Let the transformer series impedance be

$$Z_T = R_T + jX_T \qquad [\Omega \text{ per phase}].$$

(We neglect its magnetizing impedance.) Then the formula for maximum torque will be

$$T_{max} = \frac{45}{\pi n_s} \frac{|V|^2}{\sqrt{(R_1 + R_T)^2 + (X_1 + X_2' + X_T)^2} + R_1 + R_T}$$

where $|V|$ is the primary transformer voltage (which we consider constant).

INDEX

www.ingramcontent.com/pod-product-compliance
Lightning Source LLC
Chambersburg PA
CBHW060057031025
33327CB00054BA/359
* 9 7 8 0 4 1 2 1 0 5 5 1 7 *